JET PROPULSION

A Simple Guide to the Aerodynamic and Thermodynamic Design and Performance of Jet Engines

This is the second edition of Cumpsty's excellent self-contained introduction to the aerodynamic and thermodynamic design of modern civil and military jet engines. Through two engine design projects, first for a new large passenger aircraft, and second for a new fighter aircraft, the text introduces, illustrates and explains the important facets of modern engine design. Individual sections cover aircraft requirements and aerodynamics, principles of gas turbines and jet engines, elementary compressible fluid mechanics, bypass ratio selection, scaling and dimensional analysis, turbine and compressor design and characteristics, design optimisation, and off-design performance. The book emphasises principles and ideas, with simplification and approximation used where this helps understanding. This edition has been thoroughly updated and revised, and includes a new appendix on noise control and an expanded treatment of combustion emissions. It is suitable for student courses in aircraft propulsion, but also an invaluable reference for engineers in the engine and airframe industry.

JET PROPULSION

A Simple Guide to the Aerodynamics and Thermodynamic Design and Performance of Jet Engines

NICHOLAS CUMPSTY

University of Cambridge

CAMBRIDGE
UNIVERSITY PRESS

PUBLISHED BY THE PRESS SYNDICATE OF THE UNIVERSITY OF CAMBRIDGE
THE PITT BUILDING, TRUMPINGTON STREET, CAMBRIDGE, UNITED KINGDOM

CAMBRIDGE UNIVERSITY PRESS
The Edinburgh Building, Cambridge CB2 2RU, UK
40 West 20th Street, New York, NY 10011-4211, USA
477 Williamstown Road, Port Melbourne, VIC 3207, Australia
Ruiz de Alarcón 13, 28014 Madrid, Spain
Dock House, The Waterfront, Cape Town 8001, South Africa

http://www.cambridge.org/

First published 2003
Third printing 2005

Printed in the United Kingdom at the University Press, Cambridge

Typeface Times 9/13pt   System QuarkXPress® [UPH]

A catalogue record for this book is available from the British Library

ISBN 0 521 54144 1   paperback

The publisher has used its best endeavours to ensure that the URLs for
external websites referred to in this book are correct and active at the
time of going to press. However, the publisher has no responsibility for the
websites and can make no guarantee that a site will remain live or that the
content is or will remain appropriate.

CONTENTS

PREFACE TO SECOND EDITION

The book has been well received and Cambridge University Press approached me with the invitation to bring out a second edition. This was attractive because of the big events in aerospace, most significantly the decision by Airbus Industrie at the end of 2000 to launch their new large aircraft, the A380. This meant that some changes in the first ten chapters were needed. Another major development is the decision to develop an American Joint Strike Fighter, the F-35.

Another more personal change took place when I left academia to become Chief Technologist of Rolls-Royce from the beginning of 2000. It should be noted, however, that the character and ideas of this second edition remain those of the university professor who wrote the first edition and do not reflect my change of role.

The aim and style of the book is unchanged. The primary goal of creating understanding and the emphasis remains on simplicity, so far as this is possible, with the extensive use of relevant numerical exercises. In a second edition I have taken the opportunity to update a number of sections and to include some explanatory background on noise; noise has become a far more pressing issue over the last four or five years. The book remains, however, very similar to the first edition and, in particular, numerical values have been kept the same and the exercises have not been changed. Fortunately I do not think that the changes are not large enough to mislead the reader.

In writing the first edition I was grateful for the help of many people. Mention should be made here of help from Professor Mike Owen of the University of Bath and from the students who took courses given at Rensselaer Polytechnic in Hartford Connecticut, leading to changes to the 2000 revision of the first edition. For the second edition I would like to acknowledge the additional help received in preparing the second edition from colleagues in Rolls-Royce, notably Nigel Birch, Andrew Bradley, Chris Courtney, Jason Darbyshire, Peter Hopkins, Andrew Kempton, Paul Madden, Steve Morgan, Mike Provost, Joe Walsh and Eddie Williams. From outside the Company the suggestions of George Aigret were gratefully received. Comments and corrections from readers will continue to be welcomed.

PREFACE TO FIRST EDITION

This book arose from an elementary course taught to undergraduates, which forms the first ten chapters concerned with the design of the engines for a new 600-seat long range airliner. Introductory undergraduate courses in thermodynamics and fluid mechanics would provide the reader with the required background, but the material is also presented in a way to be accessible to any graduate in engineering or physical sciences with a little background reading. The coverage is deliberately restricted almost entirely to the thermodynamic and aerodynamic aspects of jet propulsion, a large topic in itself. The still larger area associated with mechanical aspects of engines is not covered, except that empirical information for such quantities as maximum tip speed are used, based on experience. To cover the mechanical design of engines would have required a much bigger book than this and would have required a mass of knowledge which I do not possess.

In preparing the course it was necessary for me to learn new material and for this I obtained help from many friends and colleagues in industry, in particular in Rolls-Royce. This brought me to realise how specialised the knowledge has become, with relatively few people having a firm grasp outside their own speciality. Furthermore, a high proportion of those with the wide grasp are nearing retirement age and a body of knowledge and experience is being lost. The idea therefore took hold that there is scope for a book which will have wider appeal than a book for students – it is intended to appeal to people in the aircraft engine industry who would like to understand more about the overall design of engines than they might normally have had the opportunity to master. My ambition is that many people in the industry will find it useful to have this book for reference, even if not displayed on bookshelves.

The original course, Chapters 1–10, was closely focused on an elementary design of an engine for a possible (even likely) new large civil aircraft. Because the intention was to get the important ideas across with the least complication, a number of simplifications were adopted, such as taking equal and constant specific heat capacity for air and for the gas leaving the combustor as well as neglecting the effect of cooling air to the turbines.

Having decided that a book could be produced, the scope was widened to cover component performance in Chapter 11 and off-design matching of the civil engine in Chapter 12. Chapters 13 – 18 look at various aspects of military engines; this is modelled on the treatment in Chapters 1–10 of the civil engine, postulating the design requirement for a possible new fighter aircraft. In dealing with the military engine some of the simplifications deliberately adopted in the early chapters are removed; Chapter 19 therefore takes some of these improvements from Chapters 13 – 18 to look again at the civil engine.

Throughout the book the emphasis is on being as simple as possible, consistent with a realistic description of what is going on. This allows the treatment to move quickly, and the book to be brief. But more important it means that someone who has mastered the simple formulation can make reasonably accurate estimates for performance of an engine and can estimate changes in performance with alteration in operating condition or component behaviour. Earlier books become complicated because of the use of algebra; furthermore to make the algebra tractable frequently forces approximations which are unsatisfactory. The present book uses arithmetic much more – by taking advantage of the computer and the calculator the numerical operations are almost trivial. The book contains a substantial number of exercises which are directed towards the design of the civil engine in the early chapters and the military engine in the later chapters. The exercises form an integral part of the book and follow, as far as possible, logical steps in the design of first the civil engine and then later the military combat engine. Many of the insights are drawn from the exercises and a bound set of solutions to the exercises may be obtained from the author.

Because Chapters 1–10 were directed at undergraduates there are elementary treatments of some topics (most conspicuously, the thermodynamics of gas turbines, compressible fluid mechanics and turbomachinery) but only that amount needed for understanding the remainder of the treatment. I decided to leave this elementary material in, having in mind that some readers might be specialists in areas sufficiently far from aerodynamics and thermodynamics that a brief but relevant treatment would be helpful.

ACKNOWLEDGEMENTS

It is my pleasure to acknowledge the help I have had with this book from my friends and colleagues. The largest number are employed by Rolls-Royce (or were until their retirement) and include: Alec Collins, Derek Cook, Chris Freeman, Keith Garwood, Simon Gallimore, John Hawkins, Geoffery Hodges, Dave Hope, Tony Jackson, Brian Lowrie, Sandy Mitchell, James Place, Paul Simkin, Terry Thake and Darrell Williams. Amongst this group I would like to record my special gratitude to Tim Camp who worked through all the exercises and made many suggestions for improving the text. I would also like to acknowledge the late Mike Paramour of the Ministry of Defence. In the Whittle Laboratory I would like to record my particular debt to John Young and also to my students Peter Seitz and Rajesh Khan. I am also grateful for the help from other students in checking late drafts of the text. In North America I would like to mention Ed Greitzer and Jack Kerrebrock (of the Gas Turbine Laboratory of MIT), Bill Heiser (of the Air Force Academy), Phil Hill (of the University of British Columbia), Bill Steenken and Dave Wisler (of GE Aircraft Engines) and Robert Shaw. Above all I would like to express my gratitude to Ian Waitz of the Gas Turbine Laboratory of MIT who did a very thorough job of assessing and weighing the ideas and presentation – the book would have been very much the worse without him. In addition to all these people I must also acknowledge the help and stimulus from the students who took the course and the people who have added to my knowledge and interest in the field over many years.

THE EXERCISES

An important part of the book are exercises related to the engine design. To make these possible it is necessary to assume numerical values for many of the parameters, and appropriate values are therefore assumed to make the exercises realistic. These values are necessarily approximate, and in some cases so too is the model in which they are used. The answers to the exercises, however, are given to a higher level of precision than the approximations deserve. This is done to assist the reader in checking solutions to the exercises and to ensure some measure of consistency. The wise reader will keep in mind that the solutions are in reality less accurate than the number of significant figures seems to imply.

The usefulness of the book will be greatly increased if the exercises are undertaken. In some cases one exercise leads to another and a few simple calculations on a hand-held calculator suffice. In others it is desirable to carry out several calculations with altered parameters, and such cases call out of a computer and spread sheet.

SOLUTIONS TO THE EXERCISES

Solutions to all the exercises may be obtained from the publisher by e-mailing
solutions@cambridge.org

GLOSSARY

afterburner a device common in military engines where fuel is burned downstream of the turbine and upstream of the final propelling nozzle. Also known as an augmentor or as reheat.

aspect ratio the ratio of one length to another to define shape, usually the ratio of span to chord

blades the name normally given to the aerofoils in a turbomachine (compressor or turbine). Sometimes stationary blades are called stator vanes (or just vanes) and rotor blades are called buckets.

booster a name given to compressor stages on the LP shaft in two-shaft engines. The booster stages only affect the core flow.

bypass engine an engine in which some of the air (the bypass stream) passes around the core of the engine. The bypass stream is compressed by the fan and then accelerated in the bypass stream nozzle. These are sometimes called turbofan engines or fan engines.

bypass ratio the ratio of the mass flow rate in the bypass stream to the mass flow rate through the core of the engine.

chord the length of a wing or a turbomachine blade in the direction of flow.

combustor also known as a combustion chamber. The component where the fuel is mixed with the air and burned.

compressor the part of the engine which compresses the air, a turbomachine consisting of stages, each with a stator and rotor row.

core the compressor, combustion chamber and turbine at the centre of the engine. The core turbine drives only the core compressor. A given core can be put to many different applications, with only minor modifications, so it could form part of a high bypass ratio engine, a turbojet (with zero bypass ratio) or part of a land-based power generation system. The core is sometimes called the gas generator.

drag the force D created by the wings, fuselage etc. in the direction opposite to the direction of travel.

fan the compressor operating on the bypass stream; normally the pressure ratio of the fan is small, not more than about 1.8 for a modern high bypass civil engine (in one stage with no inlet guide vanes) and not more than about 4.5 in a military engine in two or more stages.

gross thrust the thrust F_G created by the exhaust stream without allowing for the drag created by the engine inlet flow; for a stationary engine the gross thrust is equal to the net thrust.

HP the **high-pressure** compressor or turbine are part of the engine core. They are mounted on either end of the **HP** shaft. In a two-shaft engine they form the core spool.

incidence sometimes called angle of attack, is the angle at which the wing is inclined to the direction of travel or the angle at which the inlet of a compressor or turbine blade is inclined to the inlet flow direction.

IP   the **intermediate-pressure** compressor or turbine, mounted on the **IP** shaft. There is only an **IP** shaft in a three-shaft engine.

jetpipe   the duct or pipe downstream of the LP turbine and upstream of the final propelling nozzle.

LCV   the **lower calorific value** of the fuel; the energy released per unit mass of fuel in complete combustion when the products are cooled down to the inlet temperature but none of the water vapour is allowed to condense.

lift   the force L created, mainly by the wings, perpendicular to the direction of travel.

LP   the **low-pressure** compressor and turbine are mounted on either end of the **LP** shaft. Combined they form the **LP** spool.

nacelle   the surfaces enclosing the engine, including the intake and the nozzle.

net thrust   the thrust F_N created by the engine available to propel the aircraft after allowing for the drag created by the inlet flow to the engine. (Net thrust is equal to gross thrust minus the ram drag.)

ngv   the **nozzle guide vane**, another name for the stator row in a turbine.

nozzle   a contracting duct used to accelerate the stream to produce a jet. In some cases for high performance military engines a convergent-divergent nozzle may be used.

payload   the part of the aircraft weight which is capable of earning revenue to the operator (can be freight or passengers).

pylon   the strut which connects the engine to the wing.

ram drag   the momentum of the relative flow entering an engine.

sfc   **specific fuel consumption** (actually the *thrust* specific fuel consumption) equal to the mass flow rate of fuel divided by net thrust. The units should be in the form (kg/s)/kN, but are often given as lb/h/lb or kg/h/kg.

specific thrust   the net thrust per unit mass flow through the engine, units m/s.

spool   used to refer to the compressor and turbine mounted on a single shaft, so a two-spool engine is synonymous with a two-shaft engine

stagnation   stagnation temperature is the temperature which a fluid would have if brought to rest adiabatically. The stagnation pressure is the pressure if the fluid were brought *isentropically* to rest. Stagnation quantities depend on the frame of reference and are discussed in Chapter 6.

static   static temperature and pressure are the actual temperature and pressure of the fluid, in contrast to the stagnation quantities defined above.

turbine   a component which extracts work from a flow. It consists of rotating and stationary blades. The rotating blades are called rotor blades and the stationary blades are called stator blades or nozzle guide vanes.

turbofan   a jet engine with a bypass stream.

turbojet   a jet engine with no bypass stream – these were the earliest type of jet engines and are still used for very high speed propulsion.

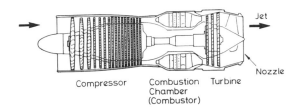

A single-shaft turbojet engine (no bypass)

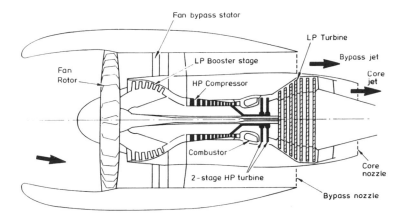

A two-shaft high bypass engine

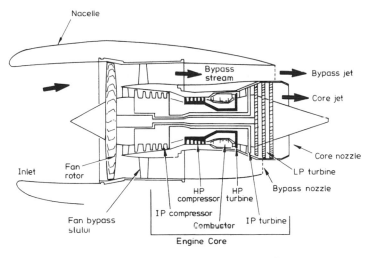

A three-shaft high bypass engine

NOMENCLATURE

a	speed of sound $\sqrt{\gamma RT}$		s	entropy
A	Area		SEP	specific excess power
bpr	bypass ratio		sfc	specific fuel consumption
c	chord of wing or blade		T	static temperature
c_p	specific heat at const. pressure		T_0	stagnation temperature
C_D	drag coefficient		U	blade speed
C_L	lift coefficient		V	velocity
D	drag (force opposing motion)		V_j	jet velocity
D	diameter		V^{rel}	velocity relative to moving blade
E	energy state $m(gh + V^2/2)$		w	weight
E_s	specific energy state $gh + V^2/2$		W	work
F_G	gross thrust		$\dot{W}$	work rate, power
F_N	net thrust			
g	acceleration due to gravity		α	flow direction (measured from axial)
h	static enthalpy		α^{rel}	flow direction relative to moving blades
h_0	stagnation enthalpy		γ	ratio of specific heats c_p/c_v
h,H	altitude		δ	flow deviation (Chapters 9 and 18)
h	blade height (i.e. span)		δ	p_0/p_{0ref}
i	incidence		θ	$\sqrt{T_0/T_{0ref}}$
L	lift (force perp. to direction of motion)		ρ	density
LCV	lower calorific value of fuel		ε	cooling effectiveness (Chapter 5)
m	mass			
$\dot{m}$	mass flow rate		**Subscripts**	
$\bar{m}$	non-dimensional mass flow rate, $\dot{m}\sqrt{c_pT_0}/Ap_0$		a	ambient
			ab	afterburner
M	Mach number		air	air
n	load factor		b	bypass
N	shaft rotational speed		c	core
p	static pressure		dry	no afterburner in use
p_0	stagnation pressure		e	combustion products (c_p and γ)
q	dynamic pressure $1/2\rho V^2$		f	fuel
Q	heat transfer		$isen$	isentropic (efficiency)
$\dot{Q}$	heat transfer rate		m	mean
r	radius (Chapters 9 and 18)		p	polytropic (efficiency)
r	pressure ratio		sl	sea level
R	gas constant		$therm$	thermal (efficiency)

A NOTE ON NOMENCLATURE

The various stations or positions throughout an engine are given numbers and different companies have different conventions for the many positions along the flow path of a multi-spool engine. An internationally recommended numbering scheme applies to some of the major stations and of these the most important station numbers to remember are:

2   engine inlet face;

3   compressor exit and combustion chamber inlet;

4   combustion chamber exit and turbine inlet.

The above brief list shows the one superficial snag, the inlet face of the engine is station 2, whereas most teaching courses call it station 1. The reason for this discrepancy is that for some engine installations, particularly in high-speed aircraft, there can be a substantial reduction in stagnation pressure along the inlet; station 2 is after this loss has taken place. In this book the international standard will be used, where appropriate, with 2 at the inlet to the engine, and a simplified guide is shown in Fig. 7.1. For more detailed treatment of the engine the schemes in Fig. 12.7 or Fig. 15.1 should be consulted.

Subscript zero is used to denote stagnation conditions, for example stagnation pressure, p_0 and stagnation temperature, T_0. (See Chapter 6 for an explanation of the terms *stagnation pressure* and *stagnation temperature*. Some people use the word *total* in place of *stagnation*.) The stagnation pressure at engine inlet is therefore written p_{02} and temperature at turbine entry as T_{04}.

TERMINOLOGY

There are differences between British and American usage, but usually these are small - aeroplane and airplane, for example. It may be noted that in Britain it is normal to use the word *civil* when referring to aviation, aircraft and air transport, where in the USA the word *commercial* would normally be used. In the book the British usage *civil* is adopted. However, while it is still quite common in Britain to refer to *reheat*, the corresponding American term *afterburner* is used throughout the book.

Part 1

Design of Engines for a

New 600-seat Aircraft

CHAPTER 1   THE NEW LARGE AIRCRAFT – REQUIREMENTS AND BACKGROUND

1.0 INTRODUCTION

This chapter looks at some of the commercial requirements and background to the proposals to build a new civil[1] airliner capable of carrying about 600 people. The costs and risks of such a project are huge, but the profits might be large too. In explaining the requirements some of the units of measurement used are discussed. Design calculations in a company are likely to assume that the aircraft flies in the International Standard Atmosphere (or something very similar) and this assumption will be adopted throughout this book. The standard atmosphere is introduced and discussed towards the end of the chapter. The chapter ends with brief reference to recent concerns about environmental issues.

1.1 SOME COMMERCIAL BACKGROUND

In December 2000 Airbus formally announced the plans to go ahead with a new large aircraft, dubbed the A380, intended in its initial version to carry a full payload (with 555 passengers) for a range of up to 8150 nautical miles. First flight is intended to be in 2004 and entry into service in 2006. There are already plans afoot for heavier versions, carrying more that 555 passengers and for all-freight versions with a larger payload. In December 2000 Airbus Industrie had received enough orders to justify the expected cost of over $10 billion, with an expected break-even point with a sale of 250 aircraft. They forecast delivery of the 250th aircraft in 2011.

The large capital expenditure and the long payback period highlight the risks, for cost over-run, project delay or slow sales could undermine all these estimates. Boeing, who have until now dominated the large end of the market with the Boeing 747, offered an updated version, the 747X to compete with the A380. The Boeing 747-400, currently the largest civil aircraft, was introduced into service in 1989 but it is a derivative of the 747-100 which entered service in 1970. The 747-400 incorporated some aerodynamic improvements, including improvements to existing engines, but more radical redesign would be needed to take full advantage of the developments in aerodynamics and materials since 1970. Adopting these technology developments for the A380, together with new engines, should result in a substantially more cost-efficient aircraft with about a 15% reduction in seat-mile costs, compared

---

[1] The word *civil* is used in Britain where *commercial* would be used in the USA.

with the Boeing 747-400. At the end of March 2001 Boeing had not received a single order for their 747X and the project was formally put on hold, Boeing stating that there was not an adequate market for a very large aircraft. While many expect that the 747X will ultimately be cancelled, Boeing firmly deny this. Boeing's intentions for new very large aircraft are not clear and it must be assumed that this is a topic of intense consideration within the company.

For several years Boeing and Airbus Industrie have separately and jointly discussed proposals for a much larger aircraft, with anywhere from about 600 to about 800 seats. All the proposals for very large aircraft have four engines hung from under the wing. However, at the same time as the announcement that the 747X was being postponed, Boeing announced a very different aircraft, unofficially dubbed the "sonic cruiser". This would cruise at a Mach number of at least 0.95 (whereas the 747X would have cruised at $M=0.85$) with a range of 9000 nautical miles but with only about 200 seats. The specification is still fluid at the time of writing. In any case this higher speed aircraft, which some in the industry believe will ultimately be unattractive on economic and environmental grounds, is *not* the subject of this book, though the topic of high- speed passenger carrying aircraft is returned to briefly in Chapter 19.

1.2 THE NEW LARGE AIRCRAFT

The first ten chapters of this book are concerned with a hypothetical New Large Aircraft (NLA) which bears a close resemblance to proposals put out by Airbus and by Boeing from around 1996. The final aircraft launched as the Airbus A380-100 in December 2000 differs in a number of ways from these and the A380-100 is compared with the hypothetical New Large Aircraft are compared in Table 1.1.

Table 1.1 Comparison of hypothetical NLA with Airbus A380-100

	New Large Aircraft NLA	Airbus A380-100*
No. of passengers	620	555
Range (nautical miles)	8000	8150
Payload at this range (tonne)	58.8	52.9
Max. take-off weight (tonne)	635.6	560.2
Empty weight (tonne)	298.7	274.9
Cruise Mach number	0.85	0.85
Initial cruise altitude (feet)	31000	35000
Cruise Lift/Drag	20	20
Wing area (metre^2)	790	845

* specifications as of 1 May 2001   1 tonne = 2205 lb mass

The main differences between the hypothetical New Large Aircraft and the A380-100 are the range, weight and smaller wing area of the hypothetical aircraft. The wing area assumed is close to one that Airbus first proposed before increasing it in a series of steps over the last five or so years. The larger wing area allows future growth in aircraft weight, but also allows take-off and landing at lower speeds, thereby reducing noise nuisance. The differences are sufficiently small that the aim of the book, which is the understanding of the aerodynamic and thermodynamic constraints and decisions for the propulsion of a new large civil aircraft, are not compromised by retaining the numerical values for the original hypothetical new large aircraft.

The price of a new aircraft is a complex issue, depending on the level of fittings inside the aircraft and on the various discounts offered. It may be assumed that the catalogue price of an A380 will be of the order of $200 million, with the engines costing around $12 million each. The market is variously estimated to be between 1100 and 1300 very large aircraft over the next 20 years, Boeing suggesting a much smaller number. Airbus want to be able to share in the profits from the market for large aircraft, hitherto dominated by Boeing (at present with the 747-400), with the potential this has given Boeing to cross-subsidise its smaller aircraft.

For new aircraft the manufacturers have to compete in terms of operating cost and potential revenue, as well as performance, most obviously range and payload. The proposal to increase the size of an aircraft is not without special additional constraints on size; currently the 'box' allowed at major airports is 80 m × 80 m and this limits both the length and the wingspan. In addition there are strong incentives to avoid making the fuselage higher from the ground because of the consequences for ground handling. The aspect ratio of a wing (the ratio of span to chord) has a large effect on its drag and Airbus have until now had a larger aspect ratio than Boeing, a feature which has contributed to the lower drag of Airbus aircraft. With the A380 the limit on wing span to fit in the airport 'box' has meant that its aspect ratio of 7.53 will be lower than that of the 747-400, which is 7.98. It is still reasonable to expect that the cruise lift/drag ratio for the A380 will be around 20, significantly higher than the much older 747-400.

It is essential to realise that both the new aircraft, and the engines which power it, will depend heavily on the experience gained in earlier products, particularly those of similar size and character. Most of the aircraft that Airbus have made to date have two engines (referred to as twins) and only their A340 has four engines. Airbus will be relying on their knowledge and experience gained with the earlier aircraft, but most significantly the A340-600, which had its first flight in April 2001. This is certainly a large aircraft, with a maximum take-off weight of 365 tonne, not far short of the 747-400 with a maximum take-off weight of 395 tonne. Airbus will also be looking to learn from the 747-400. In Table 1.2 below the proposed specifications for the hypothetical New Large Aircraft on which the first part of this book is based are set beside those achieved for the 747-400 as well as for the A340-500 and A340-600. For the new aircraft some of the quantities given are stipulations, such as the number of passengers and the range, whilst others, such as the lift/drag ratio (discussed below) are extrapolations of earlier experience. The proposed range of 8000 nautical miles makes possible non-stop flights between cities throughout North America and most of the major Pacific-Rim cities, even when strong head winds are liable to be encountered.

Table 1.2   Comparison of some salient aircraft parameters

	New Large Aircraft NLA	Boeing 747-400	Airbus A340 -500	-600
No of passengers	620	400	313	380
Range (nautical miles)	8000	7300	8550	7500
Payload at this range (a) (tonne)	58.8	38.5	29.7	36.1
Max. take-off weight (d) (tonne)	635.6	395.0	365	365
Empty weight* (b) (tonne)	298.7	185.7	170	177
Max. weight of fuel (c) (tonne)	275.4	174.4	171	157
Cruise Mach number	0.85	0.85	0.83	0.83
Initial cruise altitude (feet)	31000	31000	31000	31000
Cruise Lift/Drag	20	17.5	19.5	19.5
Wing area (m^2)	790	511	439	439

(Note that $d \approx a + b + c$)

* no fuel, no payload   1 tonne = 2205 lb mass

In calculating payload one passenger is taken to be 95 kg, a similar value in pounds is specified by Boeing.   In looking at the specifications for the new aircraft it is worth noting that the maximum payload is only 58.8 tonne, compared to the total weight of the new aircraft at take off, 635.6 tonne. More seriously, the payload (the total weight of passengers and freight) is not much more than one third of the fuel load. It follows from this that small proportional changes in the weight of the engine (which is 5 - 6% of the maximum take-off weight) or in the fuel consumption can have disproportionately large effects on the payload. Of the fuel carried not all could be used in a normal flight; typically about 15% ($\approx$38.6 tonne) would need to be kept as reserve in case landing at the selected destination airport is impossible. Based on past experience it may be assumed that about 4% (around 11 tonne) of fuel would be used in take off and climb to the initial cruising altitude, with the bulk of the fuel consumption being involved in the cruise portion of a long flight.

1.3 PROPULSION FOR THE NEW LARGE AIRCRAFT

It takes several years to design, develop, and certificate (i.e. test so that the aircraft is approved as safe to enter service) a new aircraft, though the length of time is becoming shorter. It seems to take even longer to develop the engines, but until the specifications of the aircraft are settled it is not clear what engine is needed. There are three major engine manufacturers (Rolls-Royce in

Britain, Pratt & Whitney and General Electric in the USA) and it is their aim to have an engine ready for whatever new large aircraft it is decided to build. The costs of developing a wholly new engine are so high that it is always the objective of a manufacturer to use whenever possible an existing engine, perhaps with some uprating. On a recent new large aircraft, the Boeing 777, all three major manufacturers offered an engine and the competition was fierce. Pratt & Whitney and Rolls-Royce offered developments of existing large engines; General Electric developed a wholly new engine, the GE90. The *Economist* of 18 September 1999 reported that the GE90 had cost General Electric $1 million per day for $4^1/_2$ years, in total about $1.6 billion; it is not clear how much extra was spent by risk sharing partner companies. This huge sum can be made more understandable if an average wage for an employee, with the appropriate overheads, is taken to be $150,000 per annum - the $1.6 billion cost then translates into over 10,000 man-years of work. To reduce the financial exposure Pratt & Whitney and General Electric have formed an alliance to produce a wholly new engine for the A380, the GP7200, in competition with Rolls-Royce, who have offered the Trent 900, a derivative of their earlier engines.

Whilst discussions are going on between aircraft manufacturers and airlines they are also going on between aircraft manufacturers and the engine manufacturers. As specifications for the 'paper' aircrafts alter, the 'paper' engines designed to power them will also change; many potential engines will be tried to meet a large number of proposals for the new aircraft before any company finally commits itself. The first ten chapters of the book will attempt, in a very superficial way, to take a specification for an aircraft and design the engines to propel it – this is analogous, in a simplified way, to what would happen inside an engine company.

Because engines are large and heavy there are good aerodynamic and structural reasons for mounting engines under the wing. For example, a Rolls-Royce Trent 800, which is the lightest engine to power the Boeing 777, weighs about 8.2 tonne when installed on the aircraft. Most of the lift is generated by the wings, so hanging the comparatively massive engines where they can most easily be carried makes good structural sense. This reduces the wing root bending moment and makes possible a reduction in the strength and weight of the whole aircraft. It is the trend for new engines to be bigger and heavier for the same thrust than the ones they replace, originally to reduce fuel consumption, but now mainly to reduce noise. This will be discussed later in Chapter 7 and in the Appendix.

The A380 is to have four engines, two slung under each wing and the same arrangement is adopted here for the New Large Aircraft. Not very long ago it would have been unthinkable to have a trans-oceanic aircraft with only two engines because the reliability of the engines was inadequate. Now two-engine aircraft are very common, being the dominant type now crossing the Atlantic, but four engines offer advantages for the New Large Aircraft for two reasons. First, every aircraft must be able to climb from take off with one engine totally disabled. For a two-engine aircraft this means that there must be twice as much thrust available at take off as that just necessary to get the aircraft safely into the air. The engines must therefore be oversized for take off, implying too much available thrust at cruise (and therefore excess weight) with the engines 'throttled back'. For a four-engine aircraft the same rule requires that there is only 4/3 times as much thrust available at take off, and for aircraft designed for very long flights it is desirable to carry as little surplus weight as possible. The success of the Boeing 777 as a very long range aircraft has undermined this argument in recent years; as is discussed in later chapters the apparent disadvantage with two engines can be mitigated by cruising at higher altitude and the benefits in reduced first cost and maintenance cost compensates for a small increase in fuel consumption.

The second reason for having four engines is that it is not considered practical to make the wing much higher off the ground than current aircraft like the 747-400, since to do so would raise the cabin and if the cabin were raised higher the existing passenger handling facilities at airports would be unusable; it would also make the undercarriage much bigger and heavier. If the New Large Aircraft, or the Airbus A380, were to have only two engines these would be too large to fit under the wings at their current height from the ground.

It should be added in parenthesis, however, that because engines are expensive to buy and to maintain it is likely that smaller aircraft than the New Large Aircraft we are considering here will have only two engines, even when they are to be operated over large distances. Recent examples are the Airbus 330 and the Boeing 777; both of these large twins are used for flights that are sufficiently long that until recently a four-engine aircraft would have been needed.

1.4 THE UNITS USED

In Table 1.1 a number of the quantities are in non-SI units. This is common because the industry is dominated by the United States which has been rather slow to see the advantages of SI units. It is helpful to remember that

1 lb mass	=	0.4536 kg,	1000 kg = 1 tonne
1 lb force	=	4.448 N	
1 foot	=	0.3048 m	(altitude in feet used for air traffic control)
1 nautical mile	=	1.829 km	(nautical mile abbreviated to nm)
1 knot	=	1 nm/hour = 0.508 m/s	

The nautical mile (abbreviated to nm) is *not* arbitrary in the way other units are, but is the distance around the surface of the earth corresponding to 1 minute of latitude (North–South). Treating the earth as a sphere this is equivalent to 1 minute of longitude (East–West) around the equator. (The circumference of the earth around the equator, or any other great circle, is therefore 360×60 nautical miles.)

The data in Table 1.1 also give the cruising speed as a Mach number, defined as V/a the ratio of the flight speed V to the local speed of sound a. Wherever possible aerodynamicists use non-dimensional numbers and Mach number is one of the most important in determining the performance of the aircraft. The speed of sound is given by

$$a = \sqrt{\gamma R\, T}$$

where T is the local atmospheric temperature (i.e. the **static** temperature) γ is the ratio of the specific heats c_p / c_v (which is taken here to be 1.40 for air) and R is the gas constant (0.287 kJ/kg K for air). Since $c_p = \gamma R / (\gamma - 1)$ this leads to $c_p = 1.005$ kJ/kg. These values will be used for the atmosphere and in Part 1 (Chapters 1–10) for the gas in the engine. These values would *not* be accurate enough for use in a real design, particularly for the products of combustion, but also for pure air at elevated temperatures. Although this simplification suffices for the treatment in Part 1 of the book it will be relaxed in later parts.

Exercise

1.1 The shortest distance between two places on the surface of the earth is the *Great Circle Distance*, which, for a perfectly spherical earth, would be equal to the radius R_e of the earth times the angle A subtended between vectors from the centre of the earth to the points on the surface.

Express the positions of points 1 and 2 on the surface of the earth in terms of Cartesian vectors about the centre of the earth, using θ_1 and ϕ_1 to denote the latitude and longitude respectively for point 1 and likewise θ_2 and ϕ_2 for point 2. Then take the dot product of the vectors to show that the cosine of the angle A is given by $\cos A = \cos \theta_1 \cos \theta_2 \cos (\phi_1 - \phi_2) + \sin \theta_1 \sin \theta_2$.

Find the shortest distance in nautical miles between London (latitude 51.5° N, longitude 0) and Sydney in Australia (latitude 33.9° South, longitude 151.3° East). (**Ans**: 9168 nm)

1.5 THE STANDARD ATMOSPHERE

The atmosphere through which the aircraft flies depends on the altitude, with the pressure, temperature and density falling as altitude increases. The temperature profile with height is determined primarily by the absorption of solar radiation by water vapour and subsequent radiation back into space. At high altitude the variation with season, location and time of day is much less than at ground level and it is normal to use a standard atmosphere in considering aircraft and engine performance. Temperature, density and pressure are plotted in Fig.1.1 according to the *International Standard Atmosphere* (ISA). Standard sea-level atmospheric conditions are defined as $T_{sl} = 288.15$ K, $p_{sl} = 101.3$ kPa, $\rho_{sl} = 1.225$ kg/m^3. In the standard atmosphere temperature is assumed to decrease linearly with altitude at 6.5 K per 1000 m below the *tropopause* (which in the standard atmosphere is assumed to be at 11000 m, that is 36089 feet), but to remain constant above this altitude at 216.65 K. (The discontinuity in temperature gradient must give a discontinuity in the pressure and density gradients too, but this is small and the curve fitting programme has smoothed it out.)

As noted above, non-SI units are common in aviation, and air traffic control assigns aircraft to corridors at altitudes defined in feet. Cruise very often begins at 31000 ft, and the corridors are separated by 2000 ft. Although this book will be based on SI units, altitudes for the civil aircraft will be given in feet. Table 1.3 may be helpful.

Table 1.3   Useful values of the International Standard Atmosphere*

| Altitude | | Temperature | Pressure | Density |
feet	km	K	10^5 Pa	kg/m^3
0	0	288.15	1.013	1.225
31000	9.45	226.73	0.287	0.442
33000	10.05	222.82	0.260	0.336
35000	10.67	218.80	0.238	0.380
37000	11.28	216.65	0.214	0.344
39000	11.88	216.65	0.197	0.316
41000	12.50	216.65	0.179	0.287
51000	15.54	216.65	0.110	0.179

*Also known as the ICAO Standard Atmosphere.

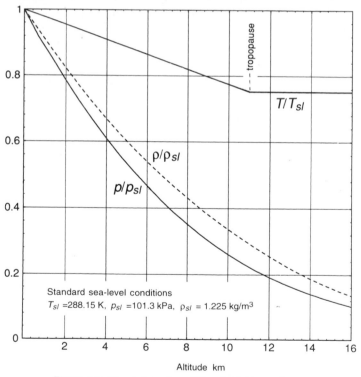

Figure 1.1 The International Standard Atmosphere

For the purpose of this book the conditions of the standard atmosphere will be assumed to apply exactly – this makes for consistency in the numbers and facilitates checking the exercises. It will be clear, however, that the standard atmosphere is at best an approximation to conditions averaged over location and season. The temperature varies more than the pressure and this variation is greatest close to the ground. It is not uncommon, for example, for the temperature at an airport in continental North America to be as low as –40 °C in winter and as high as +40 °C in summer. It is normal to refer to conditions relative to the standard atmosphere, so that if at 31000 feet altitude the temperature were 236.7 K it could, by reference to Table 1.3, be described as ISA+10°C. The corrections from standard conditions are often large for high altitude airports. Johannesburg airport, for example, is 5557 feet above sea level and the ISA temperature for this altitude is 4.0°C: suppose on a hot day that the temperature at Johannesburg airport were 35°C – in this case the conditions would be described as ISA+31°C.

Exercises

1.2 Express the maximum take off weight (mtow) for the New Large Aircraft in pounds (the units that much of the airline industry uses). Make a rough estimate of the flight time for a range of 8000 nm if cruise were at the initial altitude and Mach number for the whole flight.

(**Ans**: range = 14632 km; altitude = 9448 m; mtow = $1.4.10^6$ lb; time of flight $\approx$ 15.8 hours)

1.3[*] Find the cruising speed in m/s and km/h corresponding to the specified cruise Mach number and the initial cruise altitude. If the altitude at the end of the flight is 41000 ft, (p_a = 17.9 kPa, T_a = 216.7 K) what is the flight speed then for the same Mach number. (Note air traffic control usually allots aircraft cruising altitudes in 2000 ft steps: 31000, 35000 and 39000 going from East to West, and 33000, 37000 and 41000 going from West to East.)

(**Ans**: Initial speed at 31000 ft, 256.5 m/s, 923 km/h; at 41000 feet, speed 250.8 m/s, 903 km/h.)

1.4 The pressure change with altitude h due to hydrostatic effects is given by $dp = -\rho g \, dh$.
a) For an idealised atmosphere the temperature falls with altitude at a constant rate so that $\partial T/\partial h = -k$, where k is a constant with units K/m. Show that the pressure p at altitude H can be written

$$p = p_{sl}\{1 - kH/T_{sl}\}^{g/Rk} = p_{sl}(T/T_{sl})^{g/Rk}$$

where p_{sl} and T_{sl} are the static pressure and temperature at sea level, 101.3 kPa and 288.15 K.

For the International Standard Atmosphere the rate of change in temperature with altitude is taken to be 6.5 K per 1000 m up to the tropopause at 11 km. Show that when g =9.81 m/s^2 and R =287 J/kgK , the pressure at altitude H, in metres, is given by

$$p = p_{sl}(T/T_{sl})^{5.26} = p_{sl}\{1 - 2.26 \times 10^{-5}H\}^{5.26}$$

up to the tropopause, above which the pressure is given by

$$p = p_T \exp\{-1.58 \times 10^{-4}(H - 11.10^3)\}$$

where p_T is the pressure at the tropopause.

b) If the relationship between pressure and density were that for isentropic changes (i.e. reversible and adiabatic) p/ρ^γ = constant, show that the pressure at altitude H can then be written as

$$p = p_{sl}[1 - \frac{\gamma-1}{\gamma}\frac{gH}{RT_{sl}}]^{\gamma/(\gamma-1)}.$$

---

[*] Exercises with an asterix produce solutions which should, for convenience, be entered on the Design Sheet at the back of the book.
Note that to maintain consistency and to make checking of solutions easier, answers are given to a precision which is much greater than the accuracy of the assumptions warrants.

Plot a few values of pressure, density and temperature on Fig.1.1, the International Standard Atmosphere.

Notes: Atmospheric air is not dry. For saturated air the rate of temperature drop is given as 4.9 K per km, compared with 6.5 K per km in the International Standard Atmosphere. The isentropic calculation assumed dry air.

Different 'standard' atmospheres are sometimes used to model situations more closely: for example over Bombay in the monsoon season the atmosphere is very different from over Saudi Arabia in summer or northern Russia or America in winter.

Even below the tropopause the standard atmosphere assumes a slower reduction in temperature with altitude than that which would follow from an isentropic relation between pressure and temperature; the standard atmosphere is therefore stable. To understand this, imagine the atmosphere perturbed so that a packet of air is made to rise slowly. As the packet rises its pressure will fall to be equal to the pressure of the air that surrounds it and, as a reasonable approximation, the temperature and pressure for the packet of air will be related by the isentropic relation $p/T^{\gamma/(\gamma-1)}$ = constant. If the ascending air, which has an isentropic relation between temperature and pressure, were slightly warmer than its surroundings it would be less dense than the surrounding air and would continue to rise; such an atmosphere would be unstable. If, on the other hand, the ascending packet of air has a temperature lower than that of its immediate surroundings, as occurs in the standard atmosphere, it would be denser than the surrounding air and would fall back; such an atmosphere would be stable. In the first few hundred metres above the ground the convection frequently tends to make the atmosphere locally unstable, which is useful because it helps disperse pollutants. Stable atmospheres can occur near ground level, and frequently do at night under windless conditions when radiation leads to the ground cooling more rapidly than the air above it. Under stable conditions near the ground the natural mixing of the atmosphere is suppressed and the conditions for fog and pollution build-up are liable to occur.

1.5 ENVIRONMENTAL ISSUES

When jet propelled passenger transport was initiated, little or no thought was given to the environment, either near the airports or in the upper atmosphere. By the late 1960s the situation near airports was becoming intolerable, mainly because of the noise, but also because of pollution. The pollution involved unburned hydrocarbons, smoke (i.e. small particles of soot, which is unburned carbon) and oxides of nitrogen. Gradually steps have been taken to rein in these nuisances by international agreement with regulations both for combustion product emissions near airports and for noise during take off and landing.

The international agreements are reached so that the interests of various parts of the industry (from manufacturers of engines through to the airlines which operate rather old aircraft) are addressed. The net result is that the international agreements have lagged behind public pressure for amelioration and as a result local regulations at important airports around the world have tended to be more challenging for the makers of new engines to meet. The international limits on noise are so far above the noise produced by new aircraft with modern engines that the international limit serves merely as the benchmark from which the margin of lower noise is set. For noise the airport which tends to determine the level which new large aircraft have to achieve is London Heathrow. For products of combustion an airport which sets the level is Zurich, where charges are varied depending on the amount of pollution released in a standard landing and take-off operation. The issues and rules for emissions of pollutants are addressed briefly in Section 11.5. Noise is considered in an appendix at the end of the book.

The effect of regulations for combustion emissions has not, so far, had very much effect on the overall layout of the engine. General Electric have used a staged combustor (like two connected annular combustion chambers, one used all the time and the other only for high power) on their GE90 and it is available on the CFM-56, but so far other manufacturers have managed to avoid even this change by attention to detail in a more conventional single combustor. The 1999 report by the Intergovernmental Panel on Climate Change (IPCC)[2] may lead to greater pressure for control of oxides of nitrogen, amongst other things, during the cruise. The effect of noise regulation, however, has recently lead to very significant alterations to the engine, with consequent reduction in aircraft performance and a slightly larger fuel burn. Principally this is because at take off the largest noise source is still produced by the jet, and no method of jet-noise reduction is more certain than reducing the jet velocity. This requires bigger engines for the same thrust, engines which are bigger than those which would be chosen for optimum aircraft range, a topic taken further in Chapter 7.

SUMMARY CHAPTER 1

New engines are extremely expensive to develop and the risk of designing an engine for an aircraft which does not get built is a serious concern. Unfortunately the time to design and develop the engines has in the past been greater than for the airframe. The hypothetical New Large Aircraft which forms the basis of the first 10 chapters of this book bears some resemblance to the Airbus A380 currently under design. For such a large aircraft, intended for very long range operation, there will be four engines slung under the wings.

There is an International Standard Atmosphere used for calculating aircraft performance, which gives temperature, pressure and density as a function of altitude. Temperature is assumed to fall linearly with altitude (at 6.5 K per km) until 11 km, beyond which it is constant to 20 km.

---

[2]See bibliography

Subsonic civil air transport does not normally fly above 41000 ft (12.5 km), though business jets fly at up to 51000 ft (15.5 km). The atmospheric temperature normally falls more slowly with altitude than is implied by an isentropic variation between temperature and pressure.

Whenever possible non-dimensional variables are used, such as Mach number. When non-dimensional variables cannot be used SI units will be used throughout the book unless there is a clear reason otherwise (e.g. feet for altitude and nautical miles for range).

Environmental issues are becoming more important, with the emphasis in regulations currently being around the airport. The potentially more serious effects of emissions in the upper atmosphere will probably be the subject of future regulation. Limiting noise during take off and landing has already lead to the engine layout being modified so that it is no longer optimum for range or fuel consumption.

Chapter 1 sets out to define the needs, the operating environment and the broad specification of the aircraft. Chapter 2 moves to the next stage, which is to consider the aircraft itself.

Chapter 2

THE AERODYNAMICS OF
THE AIRCRAFT

2.0 INTRODUCTION

The engine requirements for an aircraft depend upon the size, range and speed selected, but they also depend on the aerodynamic behaviour of the aircraft and the way in which it is operated. In this chapter some very elementary aspects of civil aircraft aerodynamic performance are described; if further explanation is needed the reader is referred to Anderson(1989). These lead to a brief description of the conditions which are most critical for the engine: take off, climb and cruise. It is possible to see why cruising fast and high is desirable, and to calculate the range. Knowing the ratio of lift to drag it is possible to estimate the total thrust requirement.

2.1 SIZING THE WING

The aircraft has to be a compromise between a machine which can travel fast for cruise and relatively slowly for take off and landing. Some modification in the wing shape and area does take place for take off and landing by deploying slats and flaps, but there is a practical limit to how much can be done. As mentioned before it is normal to work with non-dimensional variables whenever possible. Lift Coefficient is defined by

$$C_L = \frac{L}{\frac{1}{2}\rho A V^2} \tag{2.1}$$

where L is the lift force, which is the force acting in the direction perpendicular to the direction of travel. In steady level flight the lift is equal to the aircraft weight. Also used in the definition of C_L are ρ the air density, A the wing area and V the flight speed.

Fig. 2.1 shows how the lift coefficient of the aircraft varies with angle of attack (i.e. incidence) at low speeds, such as at take off. It can be seen that C_L rises almost in proportion to the incidence until around the peak, beyond which it falls rapidly. This rapid fall in lift is referred to as stall and in simple terms occurs when the boundary layers separate from the upper surface of the wing. If an aircraft were to stall near the ground it would be in a desperately serious condition and it is important to make sure that this does not occur. To make sure that stalling is avoided it is essential that the flight speed is high enough for lift to be equal to aircraft weight at a value of lift coefficient which is well away from the stalling value.

The fully laden aircraft at take off is heavy and having such a heavy machine moving at high speed along the ground is potentially hazardous. It takes a considerable distance to accelerate the aircraft to its take-off speed, or to decelerate it on landing, and, apart from the cost of making the airfield very long, there is a problem in the overheating of the tyres if the speed becomes too high or is maintained for too long.

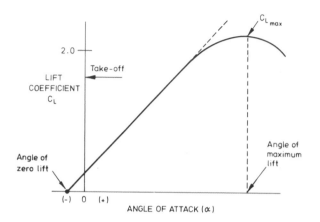

Figure 2.1. A typical curve of lift versus incidence for a large subsonic
civil transport at low speed

For these reasons, the velocity of the aircraft along the runway when the wheels leave the ground is not normally allowed to be more than about 90 m/s (just over 200 mph or 324 km/h).

Because the lift is proportional to the density of the air, problems can be encountered when the air is hot and/or the airport is high. Johannesburg is famously difficult, with an elevation of 5557 ft (1694 m). On a hot day the temperature might be as high as 35°C, (ISA+31°C) and under such conditions the density would be less than 80% that at sea level on a standard 15°C day.

Exercise

2.1a) If the lift coefficient for the NLA at take off is not to exceed 1.6 and the flight speed at lift-off (i.e. when the lift is just able to equal the weight of the aircraft) is not to exceed 90.0 m/s (177 knots), find the wing area needed, assuming a standard day at sea level. (This should agree approximately with tabulated value of wing area for the NLA.)   (**Ans:** 784 m^2)

  b) Aircraft designers often talk of wing loading defined as lift per unit area. Find this for the wing at the take-off condition. Compare this with an estimate of the weight of a car divided by the area it covers on the ground.   (**Ans:** 7.95 kN/m^2)

2.2 LIFT, DRAG, FUEL CONSUMPTION AND RANGE

It is an objective of civil aircraft to lift as much as possible with the smallest drag. Reducing the drag for the same lift allows the aircraft to use less fuel and to travel further – the impact on range is taken up in section 2.3. For steady level flight at small incidence, as for cruise, two statements[1] can be made on the basis of simple mechanics:

lift = weight   and   drag = thrust of the engines.

The lift is therefore fixed by the weight of the machine, its fuel and the payload. The drag is a quantity we wish to minimise, since this has to be matched by the engines by burning fuel. Figure 2.2 shows the ratio of lift to drag L/D versus lift coefficient C_L for a Boeing 747-400.

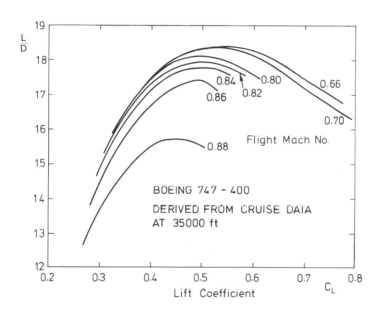

Figure 2.2. Lift–drag ratio versus lift coefficient for various flight Mach numbers

These results have been derived from flight tests at 35000 ft. The curves in Fig. 2.2 are for various Mach numbers. At each Mach number L/D rises to a maximum until further increase in lift coefficient leads to stall. At low Mach numbers stall will be caused by the separation of the

---

[1] This assumes that the engine thrust is parallel to the direction of flight (and therefore the drag). It can be shown (Exercise 14.5) that the required thrust is minimised if thrust is directed down at an angle θ, where $\tan\theta = (L/D)^{-1}$. The reduction in thrust required is approximately equal to $(L/D)^{-2}$, giving 0.25% reduction for a lift–drag ratio of 20.

boundary layers towards the rear of the wing, but at the higher Mach numbers the stall is likely to be induced by strong shock waves[2] . For this reason there is a gradual reduction in maximum L/D with Mach number up to about $M = 0.86$, beyond which there is a sharp fall in maximum L/D with Mach number. The sharp fall is because strong shock waves cause the boundary layers on the wings to separate further upstream; as well as reducing L/D the separation caused by the shocks leads to buffeting and control problems and is to be avoided Operation up to $M = 0.86$ is likely to be satisfactory, but if maximum L/D were the sole criterion one would operate at a Mach number not greater than about 0.66.

In Fig. 2.2 it is apparent that each of the curves peaks at a value of C_L around 0.5, and it is clear that the aircraft should be operating at a lift coefficient near this value. The wing area has, however, been set by conditions at take off and landing, as in Exercise 2.1a. To make the aircraft operate with a value of C_L near to the optimum one can either reduce the flight speed, which is not a very attractive option, or reduce the density by flying at high altitude. Large airliners normally begin cruising at 31000 ft or higher and then increase their altitude to maintain C_L near the value for optimum L/D as fuel is burned and the weight decreases. Ideally they would increase their altitude continuously, but air traffic control limits their movements so they normally increase altitude in 4000 ft steps up to a maximum of 41000 ft.

Exercises

2.2a) If the optimum value of C_L for cruise is 0.5, and it were decided to fly at low altitude (so that the conditions may be taken to be those of the standard atmosphere at sea level) find the velocity it would be necessary to fly at.   **(Ans:** 161 m/s)

 b) If it were intended to fly at a Mach number of 0.85 at low altitude what would be the lift coefficient?

(Ans: C_L = 0.155)

2.3* Given the wing area for the NLA derived in Exercise 2.1, show that at start of cruise at altitude 31000 ft and a Mach number of 0.85, with mass 613 tonne, the lift coefficient should be 0.527.

Find the mass at which the lift coefficient would be 0.50 at $M = 0.85$ and an altitude of 39000 ft (p_a = 19.7 kPa, T_a = 216.7 K) .   **(Ans:** 397 tonne)

The NLA is assumed to have a maximum L/D ratio of 20. Find the engine thrust needed for steady flight at start of cruise.   **(Ans:** 300.3 kN)

(Note: This effectively fixes the thrust, and ultimately the size of engines needed.)

The answers to Exercise 2.2 show that if the optimum lift coefficient is around 0.5 the cruising speed would be low near sea-level. Apart from other practical reasons, this is a good reason for cruising high. Frank Whittle seems to have been one of the first to recognise the benefits for flying fast and high in a final year thesis he wrote in 1928, at the age of 21. He recognised that at the speeds he envisaged, around 500 miles per hour, a propeller would be very inefficient.

---

[2] Even when the flight speed is subsonic, supersonic patches of flow over the wings are common since the flow is accelerated by the shape of the wing. When the lift coefficient is increased by increasing incidence, the flow is made to curve more and the acceleration is accordingly greater. Strong shocks become more of a problem when the lift coefficient is raised.

2.3 BREGUET RANGE EQUATION

Maximising the lift-drag ratio gives the maximum aircraft weight which can be kept aloft for a given engine thrust. Further consideration will reveal that this is not the primary optimum desired for a civil aircraft; instead one wants to travel the maximum possible distance. To achieve the maximum possible range the quantity to be optimised is the product of flight speed and lift-drag ratio, VL/D. Most civil airliners cruise in a band of altitudes over which the temperature does not alter very greatly, so optimising Mach number times lift-drag ratio, ML/D, is virtually equivalent and plots of this for the Boeing 747-400 are shown in Fig. 2.3. This is similar to Fig. 2.2, in that the maximum occurs for each flight Mach number at a lift coefficient of about 0.5, but it differs in having the maximum of ML/D occur for a substantially higher flight Mach number of about 0.86. At still higher Mach numbers ML/D drops precipitously. It is because of the rapid fall at higher Mach numbers that the condition for cruise has been set at $M = 0.85$.

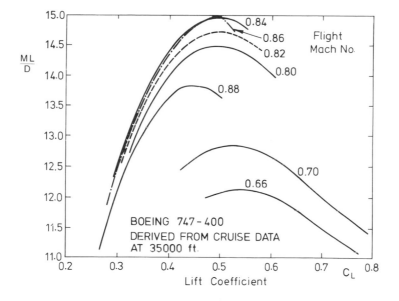

Figure 2.3. Mach number × Lift-drag versus lift coefficient for various flight Mach numbers

As a reasonably good approximation on a long flight ML/D will be maintained near the optimum, which means keeping the lift coefficient constant as fuel is burned and the aircraft weight decreases. This is achieved by increasing the altitude to reduce the air density. If the altitude were not increased the reduction in aircraft weight would cause C_L to decrease and ML/D to fall away from the optimum.

In estimating the range we consider here only the cruise part of the flight and neglect the distance travelled in the initial climb and in the final descent – though not precise, this is a reasonable approximation for a flight of perhaps 12 hours. Whereas on a long flight about 4% of the fuel will be used in take off and climb to cruising altitude, the proportion will be greater for a short flight. The proportion used in descent and landing is very much less, but reserves must be carried so that a diversion may be made to another airport if it is necessary.

To estimate the range we need to relate the fuel used to the thrust, and the conventional measure of this is the specific fuel consumption (sometimes referred to as the thrust specific fuel consumption) denoted by sfc (or tsfc). This is a quantity with dimensions and is the fuel flow rate divided by thrust. In consistent units sfc is kg/s of fuel per Newton of thrust, that is kg s^{-1}N^{-1}, or (m/s)$^{-1}$. In the English-speaking world the usual units used in the industry are (pounds of fuel per hour)÷(pounds of thrust), with units lb/hr/lb. This is numerically equal to kg of fuel per hour per kg of thrust.

During cruise the aircraft weight w changes at a rate equal to g times the mass flow rate at which fuel is burned, that is

$$\frac{dw}{dt} = -g(sfc \times \text{net thrust}) = -g(sfc \times \text{drag}) = -g\,\frac{sfc \times w}{L/D}$$

where L/D is the ratio of lift to drag and we use $w \equiv L$. Rearranging

$$\frac{dw}{w} = -\frac{g\,sfc\,dt}{L/D}.$$

This equation can then be rewritten in terms of the distance travelled s as

$$\frac{dw}{w} = -\frac{g\,sfc\,ds}{V\,L/D}.$$

An aircraft will obtain maximum range if it flies at a value of VL/D which is close to the maximum; this quantity can be kept constant during the cruise by increasing altitude as fuel is burned off. Keeping VL/D and sfc constant the above equation can then be integrated to give *Breguet's* range formula

$$s = -\frac{VL/D}{g\,sfc} \times \ln\left(\frac{w_{end}}{w_{start}}\right) \tag{2.2}$$

w_{start} and w_{end} are the total aircraft weights at the start and end of cruise respectively. As noted above, an aircraft is more likely in practice to fly at constant ML/D than constant VL/D, but the difference between the two will be small. To maintain either VL/D or ML/D constant would require a continuous very slow climb, whereas air traffic control normally requires steps of 4000 feet. Nevertheless we can see from *Breguet's* equation how the range depends on the aircraft characteristics; the structural weight is crucial, forming a large part of w_{end} , and the aerodynamic behaviour determines the product of lift-drag ratio with flight speed, VL/D. We also see how the range is inversely proportional to engine sfc.

The situation is somewhat more complicated than this because specific fuel consumption is itself a function of flight speed and it is really $(V/sfc)L/D$ which should be maximised.

Exercise

2.4 Confirm that an sfc of 1.00 lb/hr/lb is equivalent to 0.0283×10^{-3} kg $s^{-1}N^{-1}$
Estimate the range obtained with the NLA if the cruise L/D = 20, *sfc* = 0.57 lb/hr/lb, the initial mass at start of cruise is 613 tonne and the mass at end of cruise is 399.5 tonne. As a simplification assume that throughout the flight the speed is equal to that for M = 0.85 at 31000 ft.   (**Ans:** 13.9×10^{3} km)

(**Note:** This gives around 7600 nautical miles, slightly less than the specified range, but with some distance travelled during climb and descent, it should suffice.)

2.4 SELECTING THE ENGINE THRUST

Although engine size is usually referred to in terms of thrust when the engine is stationary on the ground, and is therefore available during the acceleration to take off, this is not the quantity which normally fixes the size of the high bypass ratio engine. The critical condition for sizing the engine is the top-of-climb, when the aircraft is still climbing as it approaches its cruising altitude. For a four-engine aircraft with high bypass ratio engines, such as we are considering here, engines giving adequate thrust at top-of-climb will be shown in Chapter 8 to give ample thrust at take off under normal conditions.

When the flight condition during cruise is chosen to be at maximum ML/D it follows, for a given aircraft and fixed Mach number, that the drag is proportional to the weight, and is therefore independent of air density. As we shall show later, the thrust from the engines is roughly proportional to the density and therefore falls rapidly with altitude. We therefore want to size the engine so that it is operating at an efficient condition whilst producing the necessary thrust at the altitude which will set the aircraft at its optimum ML/D .

The aircraft has to climb to its cruising altitude and there are operational advantages if it can do so quickly. It is also an advantage to have some safety margin of climbing ability. The minimum rate of climb is usually given as 300 feet per minute at cruising altitude, which corresponds to about 1.5 m/s. If the aircraft is climbing at an angle to the horizontal θ, it is easy to show that the lift L (which is perpendicular to the direction of travel through the air) is

$$L = w \cos \theta,$$

where w is the aircraft weight. The difference between the net thrust F_N and the drag D is equal to the component of weight in the direction of travel,

i.e.
$$F_N - D = w \sin \theta.$$

At altitudes for which cruise normally takes place, i.e. greater than 31000 feet, the magnitude of θ is very small. At cruise the forward speed, for a Mach number of 0.85 is about 257 m/s. For a rate of climb of 1.5 m/s the magnitude of θ is about 0.33°, so $\cos \theta \approx 1$, and a reasonable approximation is to take lift equal to weight. It then follows that

$$F_N / w = D/w + \sin\theta = 1/(L/D) + \sin\theta$$

and it is easy to find the thrust needed at the top-of-climb.

Most of the flight is spent at the cruise condition, which is the engine condition most important in determining the total fuel consumption during the flight and therefore the range. Civil engines are allowed to operate for a continuous period of no more than five minutes at the maximum take-off power, which represents only a very small proportion of the total operation time, and because comparatively little fuel is used at this condition the specific fuel consumption at take-off power is immaterial. (At the take-off power the temperatures in the engine are sufficiently high that the engine would deteriorate rapidly if allowed to operate for long periods, but at the cruise condition the temperatures are lower and the rate of deterioration is low.) The design point, in the sense of lowest fuel consumption, should therefore correspond to the cruise condition, but the engine must have the capacity to generate some additional thrust to allow the aircraft to climb and the design condition for limiting thrust output is usually the requirement at top-of-climb.

Exercise

2.5 Cruise for the NLA will begin at 31000 ft and it should arrive at this altitude with thrust available to climb at 300 ft/min (1.5 m/s) while travelling at M = 0.85. Assuming that the lift–drag ratio is 20 and that the initial cruise weight is equal to 612.9 tonne, find the thrust from each engine to achieve this rate of climb.   (**Ans**: 84.0 kN)

How much less would the required thrust be if there were no margin for climb? (See Exercise 2.3.)

2.5 ENGINE WEIGHT AND FUEL CONSUMPTION

Although it is desirable that the engine has a low specific fuel consumption, the goal is not the lowest sfc but the engine which will generate the most profit. There is a tendency for the engine weight to increase as attempts are made to increase efficiency and lower sfc, and there is a natural 'trade' between these two. The exact economic effect of changes in engine weight and sfc on direct operating cost would depend on a large number of factors, many specific to the particular airline which we cannot know, but it is possible to do a simple assessment. This starts from the recognition that the direct cost of fuel is a relatively small part of the cost of operating an airline: fuel cost varies with many factors, including the market price, but it is of the order of 20% of the direct operating cost for long-range aircraft. A 1% reduction in fuel costs does not therefore make much impression on the total costs, but a 1% reduction in the weight of fuel which must be carried at take off can be worth about a 5% increase in payload (see Table 1.1), with a much larger impact on the operating profit, when operating near the limit of range.

Suppose that during the cruise phase the average aircraft mass is m_m. If the lift–drag ratio is constant and the datum fuel consumption is $s\overline{fc}$, the mass of fuel used in a flight lasting a time T is therefore given by

$$m_f \; = \; m_m g (L/D)^{-1} s\overline{fc} \; T.$$

Now suppose that the aircraft mass is increased because a change to the engine gives a small increase in engine weight but the change also gives a small reduction in the fuel consumption so that $m \; = \; m_m(1 + \varepsilon_m)$ and $sfc \; = \; s\overline{fc}(1 - \varepsilon_f)$.

The mass of fuel consumed during cruise is therefore

$$m_f \; = \; m_m(1 + \varepsilon_m) g (L/D)^{-1} s\overline{fc}(1 - \varepsilon_f) \; T$$

and the *reduction* in fuel used is therefore given by $m_m(\varepsilon_f - \varepsilon_m) g (L/D)^{-1} s\overline{fc} \; T$. The payload will be increased if this reduction exceeds the increase in mass $\varepsilon_m m_m$. A similar argument can be used to show that the range is extended, for the same payload, under these conditions

---

Exercise

2.6 Show that the break-even point for a change in engine weight and engine sfc is given by

$$\varepsilon_f \; = \; \varepsilon_m(1 + m_m/m_{fuel}) = \varepsilon_m(1 + (L/D)/g \, s\overline{fc} \, T)$$

where m_{fuel} is the mass of fuel used in the cruise and m_m is the average mass of the aircraft during cruise, both before any perturbation in engine weight or sfc.

The aircraft mass at start of cruise is 612.9 tonne, when 264.4 tonnes of fuel are in the tanks. At the end of the flight there must be at least 38.6 tonne of fuel held in reserve in case of diversion to another airport. Find m_{fuel} and thence m_m for this limiting case. Hence find the maximum increase in the mass of each engine for no change in take-off weight, if there were a 1% reduction in sfc.

(**Ans**: 389 kg)

SUMMARY CHAPTER 2

Subsonic airliners tend to have maximum lift–drag ratio L/D and maximum VL/D when the lift coefficient is about 0.5. This requires flight at high altitude – the altitude is increased during a long flight, as the weight of fuel carried decreases, to keep near to the optimum. Maximum VL/D and maximum ML/D occur at essentially the same condition.

The range can be estimated by the Breguet range equation, which shows its dependence on VL/D and its inverse dependence on the specific fuel consumption of the engine.

If the aircraft is to cruise near the maximum of VL/D, since weight = lift, the cruise drag of the aircraft is determined for a given weight. The minimum thrust needed from the engines is therefore specified – a small additional thrust to allow climb at a rate of about 300 ft/min is normally added.

Having considered some aspects of the aircraft specification and performance we have been able to decide on the thrust needed at the altitude selected for initial cruise. We now need to look at the engine itself to decide how it works and how to specify what form it should take.

CHAPTER 3   THE CREATION OF THRUST
IN A JET ENGINE

3.0 INTRODUCTION

The creation of thrust is the obvious reason for having engines and this chapter looks at how it occurs. This is a simple consequence of Newton's laws of motion applied to a steady flow. It requires the momentum to be higher for the jet leaving the engine than the flow entering it, and this inevitably results in higher kinetic energy for the jet. The higher energy of the jet requires an energy input, which comes from burning the fuel. This gives rise to the definition of propulsive efficiency (considering only the mechanical aspects) and overall efficiency (considering the energy available from the combustion process).

3.1 MOMENTUM CHANGE

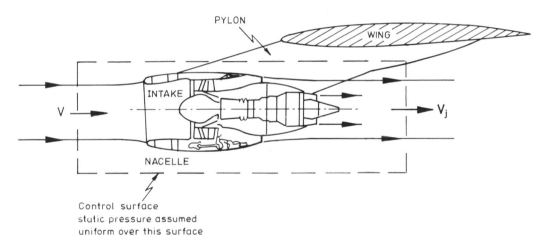

Figure 3.1. A high bypass ratio engine installed under a wing.
(A General Electric CF6 installed in the outboard position on a Boeing 747)

The creation of thrust is considered briefly here, but a more detailed treatment can be found in other texts, for example Hill and Peterson (1992). Figure 3.1 shows an engine on a pylon under a wing. Surrounding the engine a control surface has been drawn, across which passes the pylon. The only force applied to the engine is applied through the pylon. We assume that the static pressure is uniform around the control surface, which really requires that the pylon is long enough that the surface is only weakly affected by the wing. In fact we assume that the wing lift and drag are unaffected by the engine and the engine unaffected by the wing; this is not strictly true, but near enough for our purposes. A flow of fuel $\dot{m}_f$ passes down the pylon, but its velocity is low and it conveys negligible momentum. A mass flow of air $\dot{m}_{air}$ enters the engine; $\dot{m}_{air}$ is typically two orders of magnitude greater than $\dot{m}_f$.

It is assumed for simplicity here that the jet is uniform as it crosses the control surface with velocity V_j. In the case of a bypass engine this requires that any velocity difference between the core and the bypass streams has mixed out, but it would be straight forward to consider separate jets or, in principle, to integrate the momentum of a non-uniform jet.

The thrust is calculated by considering the flux of momentum across the control surface around the engine – since the pressure is assumed uniform over the control surface the pressure creates no net force. We consider momentum in the frame of reference moving with the engine, so that air enters the control surface with the flight velocity V. Most of the air entering the control surface passes around the engine and of the total only a small part, $\dot{m}_{air}$, passes through the engine. The air that passes around the engine leaves the control surface with the same velocity V as the flight speed and, since the momentum flux of this is equal on entering and leaving the surface, it does not contribute to the thrust. Considering the flow which crosses the control surface and passes through the engine, we can write down what the two fluxes are:

Flux of momentum entering the engine $= \dot{m}_{air} V.$

Flux of momentum leaving the engine $= (\dot{m}_{air} + \dot{m}_f)V_j.$

The extra mass flow in the jet is included here for completeness, but this represents a small effect for high bypass ratio engines.

The **net** thrust F_N which is available in flight is given by the difference between the two momentum fluxes, that is

$$F_N = (\dot{m}_{air} + \dot{m}_f)V_j - \dot{m}_{air} V. \qquad (3.1)$$

If the engine is operating on a stationary test bed, or with the aircraft stationary, the thrust produced is known as the **gross** thrust, which, since $V = 0$, is given by

$$F_G = (\dot{m}_{air} + \dot{m}_f)V_j . \qquad (3.2)$$

The difference between gross and net thrust is $\dot{m}_{air} V$, often referred to as the **ram drag** or the **inlet-momentum drag**, so

$$F_N = F_G - \dot{m}_{air} V.$$

Here we have ignored the drag on the outside of the engine nacelle, which leads to a ring of reduced relative velocity around the jet and a consequent reduction in useful thrust. By convention the nacelle drag is treated as part of the airframe drag and for the present we will neglect it. Because the drag from the nacelle cannot be measured independently its magnitude can give rise to serious disputes between the manufacturer of the engine and of the airframe.

Unfortunately, nacelle drag is becoming a more significant factor as the engines become larger for the same thrust, i.e. as the bypass ratios go up.

3.2 PROPULSIVE EFFICIENCY

The increase in velocity between the flow entering the engine and that leaving in the jet involves an increase in kinetic energy. This kinetic energy increase is the effect of work supplied by the engine to the air, neglecting the work which goes in loss to raise the temperature of the jet. Again assuming equal velocity in the core and bypass jets, the rate of change of kinetic energy for the flow through the engine from well ahead to downstream in the jet is given by

$$\Delta KE = \frac{1}{2} \left[(\dot{m}_{air} + \dot{m}_f) V_j{}^2 - \dot{m}_{air} V^2 \right].$$

(3.3)

The power actually associated with propelling the aircraft is given by

$$\text{Power to aircraft} = \text{flight speed} \times \text{net thrust} = V F_N$$

$$= V \left[(\dot{m}_{air} + \dot{m}_f) V_j - \dot{m}_{air} V \right].$$

(3.4)

The **propulsive** efficiency compares the power supplied to the aircraft with the rate of increase in kinetic energy of the air through the engine. Propulsive efficiency η_p is straightforward to define by

$$\eta_p = \frac{\text{Power to aircraft}}{\text{Power to jet}} = \frac{V \left[(\dot{m}_{air} + \dot{m}_f) V_j - \dot{m}_{air} V \right]}{\frac{1}{2} \left[(\dot{m}_{air} + \dot{m}_f) V_j{}^2 - \dot{m}_{air} V^2 \right]}.$$

(3.5)

Since, as already noted, the mass flow rate of fuel is very much less than that of air it is possible to write η_p with sufficient accuracy for our present purposes as

$$\eta_p = \frac{\dot{m}_{air} V [V_j - V]}{\frac{1}{2} \dot{m}_{air} [V_j{}^2 - V^2]} = \frac{2V}{V + V_j}$$

(3.6)

which is known as the *Froude* equation for propulsive efficiency. Although approximate, this equation contains the essential features associated with propulsion. If the jet velocity is nearly equal to the forward speed, the kinetic energy of the jet is used very efficiently and η_p tends to unity. Unfortunately the net thrust is given by $\dot{m}_{air} (V_j - V)$ and as V_j tends to V the **net** thrust goes to zero. For modern civil engines low fuel consumption tends to be the most important goal, and this requires a high propulsive efficiency; for military combat aircraft the principal requirement is high thrust from a compact engine, and for military applications a lower propulsive efficiency is tolerated. In terms of overall engine layout modern civil aircraft engines have bypass ratios typically greater than about 5, whereas military engines typically have bypass ratios less than unity.

3.3 OVERALL EFFICIENCY

The propulsive efficiency relates the rate at which work is done in propelling the aircraft to the rate at which kinetic energy is added to the flow through the engine, but it does not relate the work to the thermal energy made available by burning the fuel. For this we define a **thermal** efficiency by

$$\eta_{th} = \frac{\Delta KE}{\dot{m}_f \, LCV} \tag{3.7}$$

where ΔKE is the rate at which kinetic energy is added to the air, which is the work done by the engine on the air going through it. The thermal efficiency is the ratio of gas turbine work output to the energy input from burning fuel. Here LCV is the lower calorific value of the fuel, which is the chemical energy converted to thermal energy on complete combustion in air if the water in the products is not condensed but remains as vapour. (The exhaust of gas turbines is invariably hot enough that this is the case.) The **thermal** efficiency can then be written

$$\eta_{th} = \frac{\dot{m}_{air} \, [V_j{}^2 - V^2]/2}{\dot{m}_f \, LCV} \,. \tag{3.8}$$

The **overall** efficiency is given by

$$\eta_o = \frac{\text{useful work}}{\text{thermal energy from fuel}} = \eta_p \times \eta_{th} \tag{3.9}$$

which can be expanded using the above expressions into

$$\eta_o = \frac{\text{Thrust} \times \text{speed}}{\dot{m}_f LCV} = \frac{\text{Thrust}}{\dot{m}_f} \frac{\text{speed}}{LCV} = \frac{1}{sfc} \frac{V}{LCV} \,. \tag{3.10}$$

As one might expect the overall efficiency is inversely proportional to specific fuel consumption and to the heating value of the fuel; less obviously η_o is proportional to the flight speed. (Here sfc is assumed to be only a weak function of flight speed.) To take this analysis further it is necessary to understand what determines the thermal efficiency η_{th}.

Exercises

3.1 Find the propulsive efficiency for the following two engines at cruise
   a) an RB211 at 31000 ft, flight Mach number 0.85, approximate jet velocity 390 m/s.
                                                                          (**Ans:** 79.4%)
      b) an Olympus 593 (in Concorde) at 51000 ft, ($p_a = 11.0$ kPa, $T_a = 216.7$ K), flight Mach number 2.0, approximate jet velocity 1009 m/s   (**Ans:** 73.8%)
3.2 If the sfc at cruise for a version of the RB211 is about 0.60 kg/h/kg and for the Olympus 593 is about 1.19 kg/h/kg, find the overall efficiency and the thermal efficiency in each case. Take $LCV = 43$ MJ/kg.   (**Ans:** RB211: 35.1%, 44.2%;Olympus: 40.7%, 55.1%)

Note: 1) The lift–drag ratio for a Boeing 747-400 is about 16 at cruise whereas the lift–drag ratio of Concorde is between 6 and 7. Notwithstanding the high efficiency of its engines, Concorde is still an energy inefficient way to travel!

2) The thermal efficiency for the bypass engine is more accurately the product of the core thermal efficiency, LP turbine efficiency and the fan efficiency.

SUMMARY CHAPTER 3

Thrust is produced by increasing the momentum of air flow through the engine. The net thrust is that which is actually available, the gross thrust is that which would be produced under the same conditions with the engine stationary;

$$\text{Net thrust} = \text{Gross thrust} - \text{ram drag}$$

i.e. $$F_N = F_G - \dot{m}_{air} \, V.$$

For high net thrust there must either be a high jet velocity or a large mass flow of air.

Propulsive efficiency compares the rate of work done on the aircraft to the rate of kinetic energy increase of the flow through the engine. It may be approximated for the typical case, when the mass flow of fuel is much smaller than that of air, by

$$\eta_p = \frac{2V}{V + V_j}$$

and this shows that high propulsive efficiency requires the jet velocity V_j to be not much greater than the flight speed V.

Only a fraction of the energy released when the fuel is burned is converted into the kinetic energy rise of the flow, the remainder appears as internal energy in the exhaust. (In common terms, the exhaust is at a higher temperature than it would be if all of the energy had been converted into kinetic energy.) The ratio of the kinetic energy increase to the lower calorific value is denoted by the thermal efficiency η_{th}. The overall efficiency, relating the work done on the aircraft to the energy released in the fuel, is given by

$$\eta_o = \eta_p \, \eta_{th}$$

and it is easy to show that in terms of specific fuel consumption and calorific value,

$$\eta_o = \frac{1}{sfc} \, \frac{V}{LCV}$$

To make more concrete statements and to design the engine it is necessary to consider how a gas turbine works and to find a means of estimating its performance.

CHAPTER 4   THE GAS TURBINE CYCLE[1]

4.0 INTRODUCTION

The gas turbine has many important applications but it is most widely used as the jet engine. In the last few years, since the regulations changed to permit natural gas to be burned for electricity generation, gas turbines have become important prime movers for this too. Many of the gas turbines used in land-based and ship-based applications are derived directly from aircraft engines; other gas turbines are designed specifically for land or marine use but based on technology derived for aircraft propulsion.

The attraction of the gas turbine for aircraft propulsion is the large power output in relation to the engine weight and size – it was this which led the pre-Second World War pioneers to work on the gas turbine. Most of the pioneers then had in mind a gas turbine driving a propeller, but Whittle and later von Ohain realised that the exhaust from the turbine could be accelerated to form the propulsive jet.

This chapter looks at the operation of simple gas turbines and outlines the method of calculating the power output and efficiency. The treatment is simplified by treating the working fluid as a perfect gas with the properties of air, but later some examples are discussed to assess the effect of adopting more realistic assumptions. It is assumed throughout that there is a working familiarity with thermodynamics – this is not the place to give a thorough treatment of the first and second laws (something covered very fully in many excellent text books, for example Van Wylen and Sonntag, 1985). Nevertheless, in the appendix to this chapter a brief account is given to remind those whose knowledge of engineering thermodynamics is rusty or to familiarise those who have learned thermodynamics in connection with a different field.

4.1 GAS TURBINE PRINCIPLES

The essential parts of a gas turbine are shown schematically in Fig. 4.1. Air is compressed in the compressor and in the combustor (combustion chamber) fuel is burned in this high-pressure air. The hot, high-pressure gas leaving the combustor enters the turbine. In most cases there are several turbine stages, one or more to drive the compressor, the others to drive the load. The turbine driving the load may be on the same shaft as the compressor or it may be on a separate

---

[1] Purists will object to this description of a **cycle**. Strictly speaking a cycle uses a fixed parcel of fluid which in a gas turbine would be compressed, heated in a heat exchanger, expanded in a turbine and then cooled in a heat exchanger; the ideal gas turbine is sometimes called a Joule or Brayton cycle. The gas turbine 'cycle' we consider here takes in fresh air, burns fuel in it and then discharges it after the turbine: in other words it does not cycle the air. Here we are adopting the standard terminology of the industry.

shaft. The load may be an electric alternator, a ship's propeller or the fan on the front of a high bypass ratio jet engine.

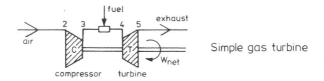

Simple gas turbine

thermodynamically equivalent to : -

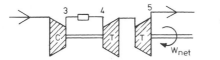

W_{net} can drive an electric generator, a propellor or a fan

or to : -

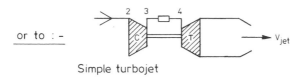

Simple turbojet

The **core** of the engine is shown by dotted lines and can be essentially identical.

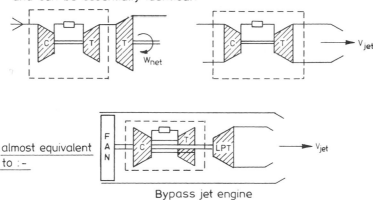

almost equivalent to : -

Bypass jet engine

Figure 4.1. Gas turbines – variations on a core theme

The central part of a gas turbine, the compressor, combustor and turbine driving the compressor, is often referred to as the **core** and the same core can be put to many different applications. In Fig. 4.1 the turbine power $\dot{W}_t$ would partly be used to supply the compressor power $\dot{W}_c$ and partly to supply the useful or net power $\dot{W}_{net}$ which is equal to $\dot{W}_t - \dot{W}_c$. At this stage we do not need to consider how $\dot{W}_{net}$ is taken from the core, but it is worth remembering that there is the special case of the pure jet engine when all the power is used to accelerate the core stream and produce a jet at exit. This was the basis of the engine envisaged and then constructed by Whittle and is still used for propulsion at supersonic speeds; Concorde, for example, is propelled by a pure jet engine. Pure jets are also used at subsonic speeds when fuel economy is unimportant but first cost and weight do matter, for example to propel missiles or target drones.

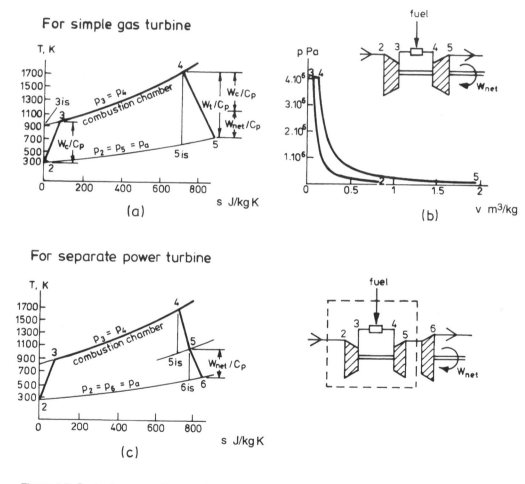

Figure 4.2. Scale diagrams of temperature–entropy and pressure–volume for gas turbine cycles. Pressure ratio 40, $T_2 = 288$, $T_4 = 1700$K, $\eta_c = \eta_t = 0.90$; $s = 0$ for $p = p_a$ and $T = 288$ K

The first law of thermodynamics can be applied to a steady process through the engine, where air enters the engine at temperature T_2 and exhaust products leave at temperature[2] T_5. (It may seem odd to make the entry condition station 2, but this is chosen here to be compatible with the normal recommended practice for aircraft engines.) If the effect of combustion is represented by an equivalent net heat transfer to the gas $\dot{Q}_{net}$, the first law may be written

$$\dot{Q}_{net} - \dot{W}_{net} = \dot{m}_{air} \Delta h \qquad (4.1)$$

where Δh is the enthalpy difference between the inlet air and the exhaust based on stagnation conditions. The mass flow of fuel is neglected in this equation as a small quantity. If the exhaust gas can be modelled as a perfect gas with the same properties as air, equation 4.1 reduces to

$$\dot{Q}_{net} - \dot{W}_{net} = \dot{m}_{air} c_p (T_5 - T_2).$$

The combustion process, which is represented as an equivalent heat transfer

$$\dot{Q}_{net} = \dot{m}_{air} c_p (T_4 - T_3)$$

can in turn be written in terms of the lower calorific value of the fuel

$$\dot{m}_f LCV = \dot{m}_{air} c_p (T_4 - T_3). \qquad (4.2)$$

For kerosene, or similar fuels used in aircraft engines, LCV = 43 MJ / kg. This magnitude is so large in relation to the specific heat of air (taken here to be c_p = 1.005 kJ/kgK) that a small flow rate of fuel is sufficient to produce a substantial temperature rise in a much greater mass flow rate of air.

Important processes in a gas turbine which burns fuel can be represented by an equivalent closed-cycle gas turbine, and by doing this it is easy to represent the processes graphically. Figure 4.2 shows the temperature–entropy (T–s) and pressure–volume (p–v) diagrams for a closed-cycle gas turbine. At entry to the compressor the temperature is T_2 and the ambient pressure p_a. The upper pressure $p_3 = p_4$ is that at which the heat transfer (equivalent to the combustion) takes place; for the present simple example it is assumed that there is no pressure drop in the combustor. In place of a heat exchanger to remove heat at the lower pressure the turbine exhausts to atmosphere in open-cycle gas turbine and the compressor draws in new air at the same pressure but at ambient temperature. After the combustion process the temperature entering the turbine is T_4. The work[3] exchanges per unit mass of air are shown on the T–s diagrams in Fig. 4.2 (though what is actually shown is work divided by specific heat, W/c_p).

---

[2] To relate this to what is discussed later in these notes it is appropriate to mention that the temperatures used in connection with the gas turbine cycle are the **stagnation** temperatures. Similarly it is **stagnation** pressures which are used here too. Stagnation and static properties are explained in Chapter 6.

[3] Work W is produced per unit mass of gas through the engine (units: Joule per kg) whereas power $\dot{W}$ is produced per unit mass flow rate (units: Watt per kg/s). The units J/kg can be shown to be identical with W/(kg/s).

In Figure 4.2 (and throughout Part 1: Design of the engines for a new 600-seat aircraft) the properties of the combustion products will be treated as pure air with the properties of a perfect gas: $c_p = 1.005$ kJ/kg K, $\gamma = 1.40$, $R = 0.287$ kJ/kg K. This is an approximation which can easily be removed (and is considered in more detail in section 4.4 below, and then in Chapter 11), but for the present purpose it is sufficiently accurate and is a substantial convenience.

4.2 ISENTROPIC EFFICIENCY AND THE EXCHANGE OF WORK

In the diagrams of Fig. 4.2 process 2–3 is the compression, and 4–5 is the expansion through the turbine. In practice the compression and expansion occur virtually without heat transfer to anything outside them, that is to say they may be taken to be adiabatic. Also shown in Fig. 4.2 are the hypothetical process 2–3is, which is an adiabatic and reversible (i.e. isentropic) compression, and the hypothetical process 4–5is, which is the isentropic expansion through the turbine. These isentropic processes are those which ideal compressors and turbines would perform. As can be seen, the actual compression process involves a greater temperature rise than that of the isentropic compressor for the same pressure rise,

$$T_3 - T_2 > T_{3is} - T_2;$$

in other words the work input to the actual compressor for each unit mass of air is greater than the work input for the ideal one. Similarly the actual turbine produces a smaller temperature drop than that in the ideal turbine, that is

$$T_4 - T_5 < T_4 - T_{5is},$$

and therefore for the same pressure ratio the actual turbine produces less work than the reversible adiabatic one.

For compressors and turbines it is normal to define efficiencies which relate actual work per unit mass flow to that of an ideal (i.e. loss-free) machine with equivalent pressure change;

$$\eta_{comp} = \frac{\text{Ideal work}}{\text{Actual work}} \quad \text{and} \quad \eta_{turb} = \frac{\text{Actual work}}{\text{Ideal work}} \cdot \tag{4.3}$$

Note that the efficiency definitions are different for a compressor or turbine so that their values are always less than unity. For an adiabatic machine the ideal equivalent process is reversible and the corresponding efficiencies are referred to as **isentropic efficiencies.**

Treating the fluid as a perfect gas, for which $h = c_p T$,

$$\eta_{comp} = \frac{T_{3is} - T_2}{T_3 - T_2} \quad \text{and} \quad \eta_{turb} = \frac{T_4 - T_5}{T_4 - T_{5is}} \cdot \tag{4.4}$$

Nowadays the isentropic efficiencies in a high quality aircraft engine for use on a civil aircraft are likely to be around 90% for compressors and turbines and this round number will normally be used in this book when a numerical value is needed. For the simple gas turbine of Fig. 4.2 the pressure rise across the compressor is equal to the pressure fall across the turbine and the corresponding pressure ratios are equal. For a jet engine, however, the pressure ratio across the turbine must be less than the pressure ratio across the compressor because some of the expansion is used to accelerate the jet. The pressure out of the turbine is p_5 and downstream of the propelling nozzle the pressure is the atmospheric static pressure, p_a.

The isentropic temperature change can very easily be found once the pressure ratio is specified. It may be recalled that for an adiabatic and reversible process

$$p / T^{\gamma/(\gamma-1)} = \text{constant}$$

which, in the present case, means

$$T_{3\text{is}} / T_2 = (p_3/p_a)^{(\gamma-1)/\gamma} \qquad \text{and} \qquad T_4/T_{5\text{is}} = (p_4/p_a)^{(\gamma-1)/\gamma}.$$

Neglecting any pressure drop in the combustor gives $p_3 = p_4$ and writing $p_3/p_a = p_4/p_a = r$ gives

$$T_{3\text{is}} / T_2 = T_4/T_{5\text{is}} = r^{(\gamma-1)/\gamma}. \tag{4.5}$$

The power which must be supplied to the compressor is given by

$$\dot{W}_c = \dot{m}_{air}\, c_p\, (T_3 - T_2) \tag{4.6}$$

and expressing this in terms of the isentropic temperature rise gives

$$\dot{W}_c = \frac{\dot{m}_{air}\, c_p\, (T_{3\text{is}} - T_2)}{\eta_{\text{comp}}}$$

$$= \frac{\dot{m}_{air}\, c_p\, T_2(T_{3\text{is}}/T_2 - 1)}{\eta_{\text{comp}}} = \frac{\dot{m}_{air}\, c_p\, T_2(r^{(\gamma-1)/\gamma} - 1)}{\eta_{\text{comp}}}. \tag{4.6}$$

Similarly, the power available from the turbine, when the mass flow of fuel in the gas stream is neglected, is given by

$$\dot{W}_t = \dot{m}_{air}\, c_p\, ((T_4 - T_5)$$

or

$$\dot{W}_t = \eta_{\text{turb}}\dot{m}_{air}\, c_p\, (T_4 - T_{5\text{is}}) = \eta_{\text{turb}}\dot{m}_{air}\, c_p T_4\, (1 - r^{-(\gamma-1)/\gamma}). \tag{4.7}$$

The turbine power must be greater than the power required to drive the compressor and the difference W_{net}, which is available to drive the load or accelerate the jet, is

$$\dot{W}_{net} = \dot{m}_{air}\, c_p\, T_2 \left(\eta_{\text{turb}}\frac{T_4}{T_2}(1 - 1/r^{(\gamma-1)/\gamma}) - \frac{(r^{(\gamma-1)/\gamma} - 1)}{\eta_{\text{comp}}} \right). \tag{4.8}$$

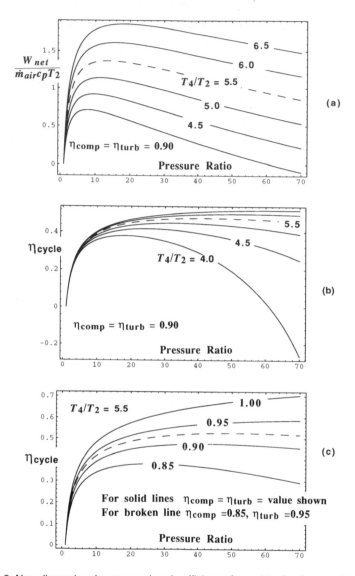

Figure 4.3. Non-dimensional power and cycle efficiency for an idealised gas turbine

---

Exercises

4.1 Air enters a compressor at a temperature of 288 K and pressure of 1 bar. If the exit pressure is 45 bar (a plausible value for take off in a new design engine) find the temperature at compressor outlet for isentropic efficiencies of 100% and 90%. What is the work input per unit mass flow for the irreversible compressor?   (**Ans:** 854.6 K; 917.5 K. W_C = 633 kJ/kg)

4.2 For the engine of Exercise 4.1 find the work per kg which could be extracted from a turbine for a pressure ratio of 45 when the turbine inlet temperature is 1700 K (a plausible value at take off) and η_{turb} = 0.90. Compare the work per kg in compressing the air in Exercise 4.1.   (**Ans:** 1019 kJ/kg)

Certain features can be determined directly from the equation for the net power per unit mass flow rate. The pressure ratio is crucial and if this tends to unity the net power goes to zero. The ratio of turbine inlet temperature to compressor inlet temperature T_4/T_2 is important and, for a given pressure ratio, increasing the temperature ratio brings a rapid rise in net power. This is effectively how the engine is controlled, because increasing the fuel flow increases T_4 and thence the power. Note, however, that it is the *ratio* T_4/T_2 which is involved, so that at high altitude, when the inlet temperature T_2 is low, a high value of temperature ratio can be obtained with a *comparatively* low value of T_4. In fact the highest values of T_4/T_2 are normally achieved at top-of-climb, the condition when the aircraft is just climbing to its cruising altitude. The dependence of power per unit mass flow rate of air, $\dot{W}_{net}$, on T_4/T_2 and pressure ratio is shown by the curves in Fig. 4.3(a). At small values of pressure ratio, increasing the pressure ratio brings a rapid increase in $\dot{W}_{net}$ but the rate of increase falls and there is a pressure ratio at which $\dot{W}_{net}$ peaks; the value of the pressure ratio for peak net power increases as T_4/T_2 rises but never exceeds about 20 for all practical values of temperature ratio.

4.3 THE GAS TURBINE THERMAL AND CYCLE EFFICIENCY

The efficiency of a closed-cycle gas turbine, such as are displayed in the T–s and p–v diagrams in Fig. 4.2, can be written as the ratio of net power out to the heat transfer rate to the air in the process replacing combustion,

$$\eta_{cycle} \quad = \frac{\dot{W}_{net}}{\dot{m}_{air}\ c_p\,(T_4 - T_3)}$$

$$= \frac{\dot{W}_{net}}{\dot{m}_{air}\ c_p T_2\,(T_4/T_2 - T_3/T_2)}. \tag{4.9}$$

The actual temperature ratio across the compressor in equation 4.9 is given by

$$T_3/T_2 \quad = 1 + \frac{r^{(\gamma-1)/\gamma} - 1}{\eta_{comp}}.$$

The thermal efficiency of the open-cycle gas turbine is the ratio of the net power to the energy input by the combustion of fuel and is

$$\eta_{th} = \frac{\dot{W}_{net}}{\dot{m}_f\ LCV}. \tag{4.10}$$

but using equation 4.2

$$\dot{m}_f\,LCV = \dot{m}_{air}\,c_p\,(T_4 - T_3)$$

it is clear that within the approximations we are adopting the thermal efficiency of the open-cycle gas turbine and the cycle efficiency of the closed-cycle gas turbine are equal, $\eta_{th} = \eta_{cycle}$. When the thermal efficiency was introduced in section 3.3 it was specific to the jet engine and all the power was transferred into the increase in kinetic energy of the flow through the engine, assuming that the jet nozzle was loss free.

In calculating the thrust and propulsive efficiency it was assumed that the mass flow of fuel is negligible compared to that of air. According to the approximations we have adopted the temperature rise in the combustion chamber is related to the fuel flow by equation 4.2. Suppose the turbine inlet temperature $T_4 = 1700$ K and the compressor produces a pressure ratio of 45 at an efficiency of 90% with air entering it at $T_2 = 288$ K. The compressor outlet temperature is therefore 912.5 K and the temperature rise in the combustor is 787.5 K, requiring an enthalpy rise of 786 kJ/kg. For a fuel with $LCV = 43$ MJ/kg, the fuel flow per unit mass flow of air is therefore 0.018. This is the mass flow of air through the core; were the bypass ratio to be 6, for example, the ratio of fuel to air would be down to 3 parts in 1000 for the whole engine.

Exercise
4.3*   a) For the example of Exercises 4.1 and 4.2, take off on a standard day, calculate the thermal efficiency and the net work per kg of air flowing. Assume isentropic efficiencies of 90% for the compressor and turbine.   (**Ans:** $\eta_{th} = 0.492$, $W_{net} = 387$ kJ/kg)

The calculation **could** be repeated for the following:

    b) $T_2 = 308$ K, $T_4 = 1700$K (as for a take off on a hot day), $PR = 45$,
(**Ans:** $\eta_{th} = 0.475$, $W_{net} = 342$ kJ/kg)

    c) $T_2 = 259.5$K, $T_4 = 1575$K (representative of top of climb at 31000ft, $M = 0.85$), $PR = 45$,
(**Ans:** $\eta_{th} = 0.498$, $W_{net} = 374$ kJ/kg)

    d) $T_2 = 259.5$K, $T_4 = 1450$K (representative of initial cruise conditions), $PR = 40$,
(**Ans:** $\eta_{th} = 0.478$, $W_{net} = 313$ kJ/kg)

    e) $T_2 = 259.5$K, $T_4 = 1450$K (representative of initial cruise conditions), $PR = 35$,
(**Ans:** $\eta_{th} = 0.475$, $W_{net} = 326$ kJ/kg)

and   f) $T_2 = 259.5$K, $T_4 = 1450$K (representative of initial cruise conditions), $PR = 40$ when for the compressor and turbine $\eta_c = \eta_t = 0.85$.   (**Ans:** $\eta_{th} = 0.375$, $W_{net} = 234$kJ/kg)

Discuss these results of a–f, commenting on the effect of inlet temperature, turbine inlet temperature, pressure ratio and component isentropic efficiency.

Notes
1)   The answers here show the highest work output for conditions at take off on a standard day – note that this is actually work per unit mass flow. In fact the power from any gas turbine falls rapidly with altitude because as the density of the air drops the mass flow rate of air through the engine falls.

2)   These efficiencies are really quite high. For comparison, a high quality diesel engine in a large truck has an overall efficiency only just about 40%.

3)   The gas turbine is not alone in suffering a reduction in power output and efficiency as the inlet air temperature rises. In road vehicles we normally pay no attention to the maximum temperature in the engine (assuming that the engine has been developed to stand the worst conditions to which we are likely to expose it) but the peak temperature will rise with inlet temperature, assuming the same fuel input. Gasoline engines and diesel engines experience a marked reduction in power output as the inlet air density drops, which is most obvious at high altitudes; one of the Interstate freeways in the USA exceeds 12000 feet and there the effect is pronounced.

Figure 4.3(b) shows the cycle efficiency for a gas turbine as a function of pressure ratio, with the curves denoting different values of the temperature ratio T_4/T_2. It is again assumed that η_{comp} = η_{turb} = 0.90. It is clear how important both the temperature ratio and pressure ratio are in determining the efficiency of the cycle. (For the ideal gas turbine, in which η_{comp} = η_{turb} = 1.00, the efficiency is independent of T_4/T_2.) Put simply, for realistic cycles (where the components are not assumed to be isentropic) the cycle efficiency continues to rise as the mean temperature of heat input rises relative to T_2, the ambient temperature, and in practical gas turbines it is essential for T_4, the temperature of the gas entering the turbine, to be high[4].

The cycle efficiency depends on pressure ratio and for a given value of T_4/T_2 there is a pressure ratio corresponding to the peak value of η_{cy}. The pressure ratio at which the peak cycle efficiency occurs depends on the isentropic efficiencies of the compressor and turbine. For the higher values of T_4/T_2 the variation of η_{cy} with pressure ratio is very slight (i.e. the curves are fairly flat) to a considerable distance either side of the maximum. A pressure ratio for maximum efficiency occurs because increasing pressure ratio raises the mean temperature of heat addition, but the extra power into the compressor has to be supplied by the turbine and irreversibilities in these take a larger share of the available power. As will be discussed in Chapter 5, the value of T_4/T_2 appropriate for cruise is approximately 5.5.

By comparing Fig. 4.3(a) and 4.3(b) it may be noted that peak efficiency occurs at a substantially higher pressure ratio than that for the maximum net power output per unit mass flow rate of air: for T_4/T_2 =5.5, for example, the pressure ratios for maximum power and maximum efficiency are around 12 and 32 respectively. Thus for engines for which maximum thrust is the goal (the normal requirement for military engines) the pressure ratio might be around 12 for T_4/T_2=5.5, whilst for engines for which efficiency is most important (most civil engines) the pressure ratio would be nearer 40. The reason for the maximum power occurring at lower pressure ratio than maximum efficiency is as follows. At low pressure ratios the power output increases with pressure ratio because the efficiency rises so rapidly that more of the heat is converted to work. As the pressure ratio increases, the temperature at compressor outlet also increases and if the turbine inlet temperature is held constant the permissible heat input (equivalent to amount of fuel burned) reduces. At higher pressure ratios the heat input decreases more rapidly with pressure than the increase in efficiency (the conversion of heat into work) and where the two effects have equal magnitude defines the pressure ratio for maximum power output per unit mass flow rate of air.

---

[4] Because here we are not dealing with a real closed cycle, but with an open system in which the heat input is replaced by a combustion process, the effect of temperature at the end of the combustion process should really be handled more carefully. Combustion produces flame temperatures typically in excess of 2300 K, which are higher than the temperature allowed into the turbine (typically no more than 1500 K for prolonged operation). The dilution of the combustion products to lower the temperature is associated with a rise in entropy and a loss in the capability of turning thermal energy into work.

The inlet temperature T_2 is determined by the altitude and the forward speed and at 31000 feet and $M = 0.85$ this is $T_2 = 259.5$ K. As discussed later in section 5.2, the turbine inlet temperature is fixed by metallurgical considerations (i.e. what temperature can the metal stand at a given level of stress), by cooling technology and by considerations of longevity – lower temperatures lead to longer life. With $T_4/T_2 = 5.5$ the peak cycle efficiency is about 0.474 and occurs at a pressure ratio of about 40; the variation in efficiency to either side of this pressure ratio is small. As the pressure ratio increases it becomes harder to design a satisfactory compressor and the isentropic efficiency tends to fall, so there is some advantage in staying to the lower side of the peak in η_{cy}. Furthermore, by putting the cruise at a pressure ratio of, say, 40 allows the maximum climb condition to occur in an acceptable range of pressure ratios up to, say, 45.

There is an additional issue concerning pressure ratio, which will be discussed more fully in Chapter 6. For an aircraft cruising at a Mach number of 0.85 the pressure at inlet to the compressor is raised by a factor of 1.60, compared to the surrounding atmosphere, by the forward motion. At outlet from the engine (i.e. at nozzle outlet) the pressure is not raised in this way, but remains at the atmospheric pressure. The effect of forward speed is therefore to raise the compression ratio of the whole engine by 1.60; raising the effective pressure ratio from 40 to about 64 has only a small effect on the cycle efficiency, which we can neglect.

Figure 4.3(c) also shows η_{cycle} versus pressure ratio but this time for a fixed temperature ratio $T_4/T_2 = 5.5$ and various values of compressor and turbine efficiency. It can be seen that the cycle efficiency is very sensitive to the component efficiencies: at a pressure ratio of 40 a reduction in compressor and turbine efficiencies from 90% to 85% would lower the cycle efficiency from 47% to 37%, corresponding to about 21% less power for the same rate of energy input in the form of fuel.

Taking equal turbine and compressor efficiencies is an oversimplification, and for the pressure ratios now being considered this becomes significant. As the pressure ratio increases the efficiency of the turbine tends to rise, whilst that of the compressor falls by a similar amount. The broken curve in Fig. 4.3(c) explores this, with equal magnitude changes of component isentropic efficiency of opposite sign in turbine and compressor. Although the alteration in η_{cy} is significant, the trends found with equal values of isentropic efficiency are not altered.

4.4 THE EFFECT OF WORKING GAS PROPERTIES

The analysis up to now, and in most of what follows up to Chapter 10, treats the working fluid in the gas turbine as a perfect gas with the same properties as air at standard conditions. This is done to make the treatment as simple as possible, and it does yield the correct trends. In a serious design study, however, the gas would be treated as semi-perfect and the products of combustion in the stream through the turbine would be included. In the semi-perfect gas approximation c_p, R, and $\gamma = c_p/c_v$ are functions of temperature and composition but *not* of pressure. With this approximation, which is sufficiently accurate for all gas turbine applications, the useful relation between gas properties $p/\rho = RT$ is retained. As is discussed in more detail in Chapter 11, R is virtually independent of temperature and composition for the gases occurring throughout a gas turbine, and it may be taken to be constant at 0.287 kJ/kgK. The specific heat capacity may be obtained from $c_p = \gamma R/(\gamma - 1)$.

Table 4.1 compares the results of Exercise 4.3 (based on a gas with $\gamma = 1.40$, giving $c_p = 1.005$ kJ/kgK) with results for the same pressure ratio and compressor and turbine entry temperatures but using accurate values for γ and c_p based on local temperature and composition. In these more accurate calculations (under the heading 'Variable γ') the mass flow rate of fuel is added to the exhaust gas. The columns show the overall cycle efficiency and the net work produced per unit mass of air through the engine.

Table 4.1   Comparison of cycle efficiency and net work based on constant gas properties and those related to local conditions

Exercise number	T_2 K	T_4 K	$\eta_c = \eta_t$	p_3/p_2	p_2 bar	$\gamma = 1.40$		Variable γ	
						η_{cy}	W_{net} kJ/kg	η_{cy}	W_{net} kJ/kg
4.3a	288	1700	0.90	45	1.00	0.492	387	0.477	496
4.3b	308	1700	0.90	45	1.00	0.475	342	0.466	454
4.3c	259.5	1575	0.90	45	0.46	0.498	374	0.485	466
4.3d	259.5	1450	0.90	40	0.46	0.478	313	0.469	387
4.3e	259.5	1450	0.90	35	0.46	0.475	326	0.464	397
4.3f	259.5	1450	0.85	40	0.46	0.375	234	0.385	305

The table shows that treating the gas with equal properties throughout gives quite good estimates for the efficiency, reflecting the variation with operating condition and, most significantly, with component efficiency. With the simple gas treatment, however, the work output is underestimated by around 20%, though the trends are similar to those calculated with the more realistic assumptions for the gas properties. It would be possible to improve the constant-property gas model by having different values of c_p for the compressor and turbine (a much better approximation for the flow in the turbine would be to take $c_p = 1.244$ kJ/kg and $\gamma = 1.3$). The mass flow through the turbine should be about 2% larger than through the compressor because of the fuel, but the effect of cooling air and combustor pressure loss need to be included. These complications are, however, unwarranted for the present purpose but are revisited in Chapter 20.

The underestimate of power predicted assuming a perfect gas with equal properties for the flow through the turbine and compressor is likely to lead to a core being designed which is about 20% bigger (in terms of the mass of air which passes through it) than that which would result if the correct gas properties were used. In fact the perfect gas design is not likely to give such an oversized core for a number of reasons. The model has neglected the cooling flows (which may take 20% of the air passing through the compressor). For an aircraft, air must be bled off the compressor to pressurise the cabin and to bring about de-icing of the wing and nacelle, whilst electrical and hydraulic power is taken from the engine. There is also drop in pressure in the combustor (perhaps 5% of the local pressure) is also neglected. Lastly the 90% isentropic efficiencies for compressor and turbine used in the design calculations are quite ambitious and these might not be achieved; the net power has been shown to be strongly affected by compressor and turbine isentropic efficiency and this too means that the actual core might be larger than a design shown here using real gas properties and $\eta_c = \eta_t = 0.90$.

4.5 THE GAS TURBINE AND THE JET ENGINE

High cycle efficiency therefore depends on having a high temperature ratio and a pressure ratio appropriate to this. The net power generated by the core could all be used to accelerate the flow through it, in which case we would have produced a simple turbojet, as in Fig. 5.1. The problem that this creates is that with the high turbine inlet temperatures necessary to give high efficiency there is such a large amount of net work available that if all of this were put into the kinetic energy of the jet, the jet velocity would be very high. As a result the propulsive efficiency η_p would be low and therefore so too would be the overall efficiency. The way around this is to use the available energy of the flow out of the core to drive a turbine which is used to move a much larger mass flow of air, either by a propeller (to form a turboprop) or a fan in a duct (as in the high bypass ratio engine used now on most large aircraft).

    The noise power generated by jets is approximately proportional to the eighth power of jet velocity, so there is an incentive, in addition to propulsive efficiency, to reduce jet velocity.

Exercise
4.4   **a**) A simple (no bypass) turbojet engine flies at 256.5 m/s. The compressor and turbine are as described in Exercises 4.1, 4.2 and 4.3. Find the jet velocity which would be produced with the temperatures given as appropriate for initial cruise (31000 feet, $M = 0.85$) in part d of Exercise 4.3. Assume that all the net work is used to increase the kinetic energy of the flow.
    Calculate the propulsive efficiency η_p and the overall efficiency $\eta_o = \eta_p \eta_{cycle}$ at this flight speed. Explain why the propulsive efficiency and overall efficiency is low. Indicate ways in which η_o could be raised at this flight speed.   (**Ans:** $V_j = 832$ m/s, $\eta_p = 0.471$, $\eta_o = 0.225$)
    b) Recalculate the efficiencies if the flight speed were 600 m/s ($M = 1.99$ at 31000 feet).
      (**Ans:** $V_j = 993$ m/s, $\eta_p = 0.753$, $\eta_o = 0.360$)

SUMMARY CHAPTER 4

The gas turbine may be envisaged as a core with various loads fitted to it; one possible load is a nozzle to create a simple turbojet, another is a turbine driving a bypass fan. In a closed cycle gas turbine a fixed amount of air is compressed, heated, expanded in a turbine and cooled; the efficiency of such a cycle is the cycle efficiency η_{cy}. Most gas turbines, and all jet engines, are open cycle, replacing the heating process by combustion of fuel and eliminating the cooling process by instead taking in fresh air to the compressor. For the open-cycle gas turbine the measure of performance is thermal efficiency η_{th}. Both η_{cy} and η_{th} vary in a similar way with pressure ratio and the ratio of turbine inlet temperature to compressor inlet temperature, T_4/T_2.

Isentropic efficiencies are the conventional method of relating actual compressor and turbine performance to the ideal; nowadays isentropic efficiencies of around 90% are expected, perhaps a few per cent lower for the compressor and a similar amount higher for the turbine.

The cycle efficiency, the ratio of the net power to the equivalent heat input rate, is strongly dependent on the isentropic efficiencies. It is also a strong function of T_4/T_2. For any given value of T_4/T_2 the pressure ratio to give maximum net power would be lower than the pressure ratio for maximum cycle efficiency.

To obtain high cycle efficiency it is essential to operate at high values of T_4/T_2. This invariably means that there is a high level of power output, and if this were used to accelerate only the core flow it would lead to jet velocities unacceptably high for subsonic aircraft. The solution is to choose a combination of high temperature ratio and pressure ratio for the core, but to use the available power output via a turbine to drive a fan, supplying a relatively small increment in kinetic energy to a large bypass stream.

The elementary description of the gas turbine has made it possible to show what is needed to obtain large power output and high cycle efficiency and, equivalently, high thermal efficiency. Some important limitations of the present treatment are:

- no cooling flows (air taken from the compressor to cool or shield the turbine blades);
- a perfect gas with properties of ambient air ($\gamma=1.40$) has been assumed throughout;
- the mass of fuel in calculating power from the turbine has been neglected;
- pressure drop in the combustor, which may be 5% of pressure, is omitted;
- simple assumptions have been made for compressor and turbine efficiency;
- electrical power off-takes and air for cabin pressurisation have been neglected.

Despite these shortcomings the trends predicted are correct and that the magnitudes for efficiency and available power are plausible. The manner in which this can be used effectively is explored in the next chapter.

APPENDIX

A brief summary of thermodynamics from an engineering perspective

This is not the place to give a detailed account of thermodynamic theory and principles, but a brief summary in the nature of revision may be helpful. For those who find the treatment here insufficient it is recommended that one of the very many texts be consulted. (There are so many books on engineering thermodynamics available that it is invidious to select only one, but some guidance does seem in order. The text by Van Wylen and Sonntag, 1985, gives an excellent treatment.) In the interest of brevity and simplicity details will be omitted wherever possible and no attempt is made to include every restriction and caveat. A feature of engineering thermodynamics is the common use of a closed control surface to enclose the process or device under consideration; when using a control surface attention is directed at what crosses the surface, both material flows of gas and liquid as well as work and heat being transferred across it.

The **first law** is a formal statement of the conservation of energy. For a *fixed* mass m of a substance this can be written in differential form as

$$\mathrm{d}Q - \mathrm{d}W = m\,\mathrm{d}e$$

where Q is heat transferred to the substance, W is work extracted from the substance and e is the energy per unit mass. We will neglect changes in energy associated with chemical, electrical and magnetic changes. If potential and kinetic mechanical energy are also neglected for the moment, energy e is restricted to the internal energy denoted by u, the specific internal energy. The work here can be done by, for example, gas expanding and pushing back the atmosphere, or it could be work driving a shaft or electrical work.

The application we have for the first law in the gas turbine and jet engine is for steady flow of gas through a device (either a whole engine or a component like a compressor or turbine). To facilitate this we draw an imaginary control surface around the device or the process we are considering. The control surface is closed, so any matter entering or leaving the device must cross the control surface; properties entering are given subscript 1 whilst those leaving have subscript 2. We specify here that there is a steady mass flow $\dot{m}$ into and out of the control surface. We also have heat transfer rate *into* the control surface $\dot{Q}$ and work transfer rate *out from* the control surface $\dot{W}_s$. The first law is now written for the closed control surface as

$$\dot{Q} - \dot{W}_s = \dot{m}\{(h_2 + V_2^2/2) - (h_1 + V_1^2/2)\}. \tag{A1}$$

The terms on the right hand side are the energy transported across the control surface by the flow. The kinetic energy is recognisable as $V^2/2$ and h is the enthalpy. For a flow process it is appropriate to use the specific enthalpy $h = u + p/\rho$, where u is the internal energy per unit mass,

p is the pressure and ρ is the density. Enthalpy allows for the displacement work done by flow entering and leaving the control surface. The work $\dot{W}_s$ in equation A1 is often referred to as shaft work (to distinguish it from the so-called displacement work done by flow entering and leaving the control surface) even though the work may be removed without a shaft, for example electrically.

For an ideal gas of fixed composition the enthalpy is a function of only the temperature. The formal expression is

$$dh = c_p\, dT$$

where c_p is the specific heat capacity at constant pressure. If c_p is constant, as for a perfect gas, equation (A1) can be written

$$\dot{Q} - \dot{W} = \dot{m}\{(c_p T_2 + V_2^2/2) - (c_p T_1 + V_1^2/2)\}. \tag{A.2}$$

It is frequently useful to refer to the stagnation enthalpy h_0, defined by

$$h_0 = h + V^2/2.$$

and the corresponding stagnation temperature

$$T_0 = T + V^2/2c_p.$$

The **second law** leads to the property entropy. Entropy is useful because in an ideal process for changing energy from one form to another the net change of entropy of the system and the environment is zero. Since zero entropy increase gives the ideal limit, the magnitude of entropy rise gives a measure of how far short of ideal the real process is.

Entropy, as a property, can be expressed in terms of other properties, for example $s = s(T,p)$. It is defined for an ideal process (normally referred to as a reversible process) by

$$ds = dQ/T \tag{A3}$$

where s denotes the specific entropy (entropy per unit mass). By using the first law in differential form the change in entropy in equation (A3) can be rewritten as

$$ds = dh/T - dp/\rho T$$

and this can be integrated for a perfect gas to give

$$s - s_{ref} = c_p \ln(T/T_{ref}) - R \ln(p/p_{ref}) \tag{A4}$$

where s_{ref} is the entropy at the reference pressure p_{ref} and temperature T_{ref}.

As the energy of a gas is increased the temperature must also increase; if pressure is held constant equation (A4) shows that entropy also rises. But if temperature is held constant and the pressure is reduced, entropy also rises. The implications of equation (A4) can be understood a little better by considering the adiabatic (no heat transfer) flow of gas along a pipe or through a throttle; in both cases there is a drop in pressure but, because there is no work or heat transfer, no change in energy and therefore no change in temperature. The fall in pressure therefore leads to a rise in entropy. In fact the magnitude of the entropy rise is a measure of the pressure loss. An entropy rise in the absence of heat transfer is evidence of a dissipative process; the effect is irreversible, since reversing the process would not reduce the entropy. Ideal processes are therefore often referred to as reversible, whereas processes which lead to loss (that is, processes for which the entropy rises by more than follows from equation (A3)) are often referred to as irreversible.

   Compressors and turbines have a small external area in relation to the mass flow passing through them. As a result the heat transfer from them is small and a good approximation is normally to treat both compressors and turbines as adiabatic. When ideal (hypothetical) machines are used as a standard to compare with real machines, it is normal to specify them to be adiabatic. The ideal compressor or turbine will have no dissipative processes – the flow is loss free and is normally described as reversible. A reversible adiabatic flow will experience no rise in entropy and such ideal processes are normally referred to as isentropic.

   An ideal compressor or turbine, producing a reversible and adiabatic change in pressure, produces no change in the entropy of the flow passing through. With no change in entropy equation (A4) leads immediately to

$$0 = \ln(T_2/T_1) - \frac{R}{c_p} \ln(p_2/p_1),$$

and since $c_p = \dfrac{\gamma R}{\gamma-1}$, the familiar expression relating temperature and pressure changes results

$$T_2/T_1 = (p_2/p_1)^{(\gamma-1)/\gamma}.$$

CHAPTER 5

THE PRINCIPLE AND LAYOUT OF JET ENGINES

5.0 INTRODUCTION

This chapter looks at the layout of some jet engines, using cross-sectional drawings, beginning with relatively simple ones and leading up to the large engines for one of the most recent aircraft, the Boeing 777. Two concepts are introduced. One is the multi-shaft engine with separate low-pressure and high-pressure spools. The other is the bypass engine in which some, very often most, of the air compressed by the fan bypasses the combustor and turbines.

Any consideration of practical engines must address the temperature limitations on the turbine. The chapter ends with some discussion of cooling technology and of the concept of cooling effectiveness.

5.1 THE TURBOJET AND THE TURBOFAN

Figure 5.1 shows a cut-away drawing of a Rolls-Royce Viper engine. This is typical of the simplest form of turbojet engine, which were the norm about 40 years ago, with an axial compressor coupled to an axial turbine, all on the same shaft. (The shaft, the compressor on one end and turbine on the other are sometimes referred to together as a spool.) Even for this very simple engine, which was originally designed to be expendable as a power source for target drones, the drawing is complicated and for more advanced engines such drawings become unhelpful at this small scale. Simplified cross-sections are therefore more satisfactory and these will be shown for more advanced engines. A simplified cross-section is also shown for the Viper in Fig. 5.1, as well as a cartoon showing the major components.

More recent engines have two or three spools so that the compression and expansion are split into parts. For flight at sustained speeds well in excess of the speed of sound a turbojet engine remains an attractive option and a two-shaft example, the Rolls-Royce Olympus 593, is shown in Fig. 5.2. Four of these engines are used to propel the Concorde at around twice the speed of sound. The low-pressure (LP) compressor and LP turbine are mounted on one shaft to form the LP spool. The LP shaft passes through the high-pressure (HP) shaft on which are mounted the HP compressor and the HP turbine. The compression process is split between two spools for reasons to do with operation at speeds below the design speed, including starting; this is discussed in some detail in Chapter 12.

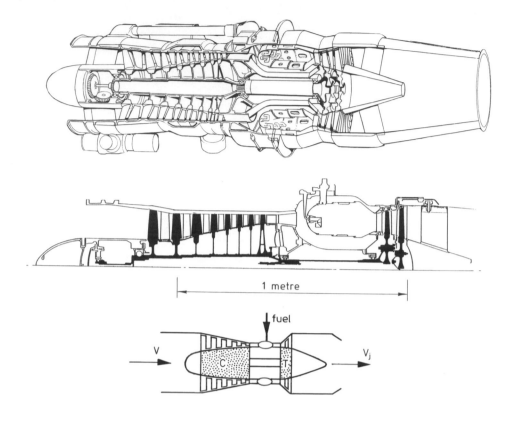

Figure 5.1. The Rolls-Royce Viper Mark 601 single-shaft turbojet shown as a cut-away, in simplified cross-section and as a schematic

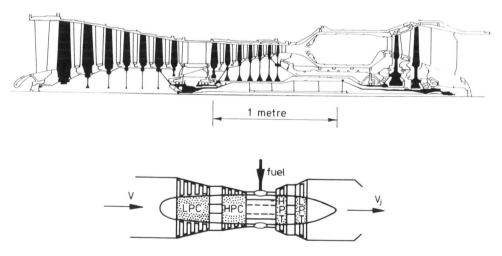

Figure 5.2. The Rolls-Royce Olympus 593 shown as a simplified cross-section and as a schematic

The arrangement of twin spools in Fig. 5.2 does not alter the problem revealed in Exercise 4.4: the jet velocity is too high to give good propulsive efficiency at all but very high flight speeds. The way to raise the propulsive efficiency at subsonic flight speeds is to go to a bypass engine, sometimes known as a turbofan. An early turbofan, the Pratt & Whitney JT8D-1, which was manufactured in large numbers to propel the Boeing 727 and 737, is shown in Fig. 5.3. In this engine some of the air compressed by the LP compressor is passed around the outside of the engine and does not go through the combustor, i.e. it is bypassed around the core. These early turbofan engines have a bypass ratio (the mass flow of air bypassed around the core divided by the mass of air going through the core) typically between 0.3 and 1.5. They have been widely used in older civil aircraft and a similar arrangement can be seen in many military engines.

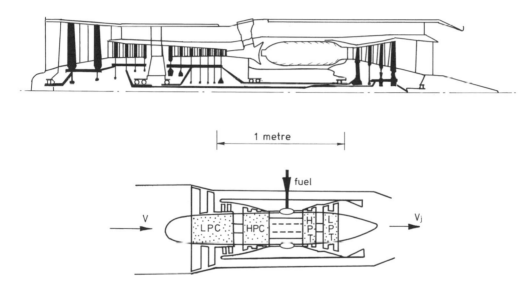

Figure 5.3. The Pratt & Whitney JT8D-1 shown as a simplified cross-section and as a schematic

5.2 THE HIGH BYPASS RATIO ENGINE

The arrangement which is normal for modern airliners is the high bypass ratio engine, with a bypass ratio of 5 or more. Three examples are shown in Fig. 5.4, all roughly contemporary and intended to propel the Boeing 777: (a) is the Trent developed by Rolls-Royce from the RB211, (b) the Pratt & Whitney 4084 and (c) the GE90 developed by General Electric.

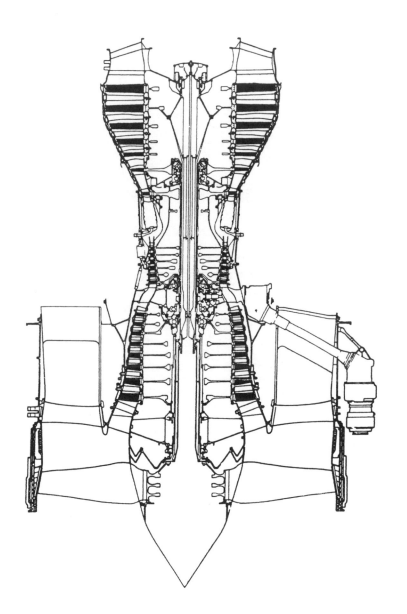

Figure 5.4(a) The Rolls-Royce Trent 884
(Fan tip diameter 2.79 m)

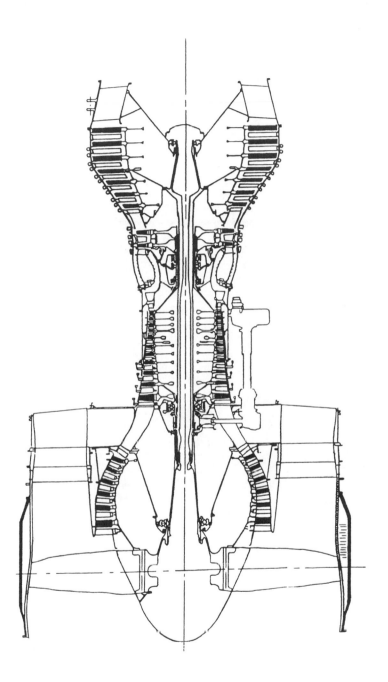

Figure 5.4(b) The Pratt & Whitney 4084
(Fan tip diameter 2.84 m)

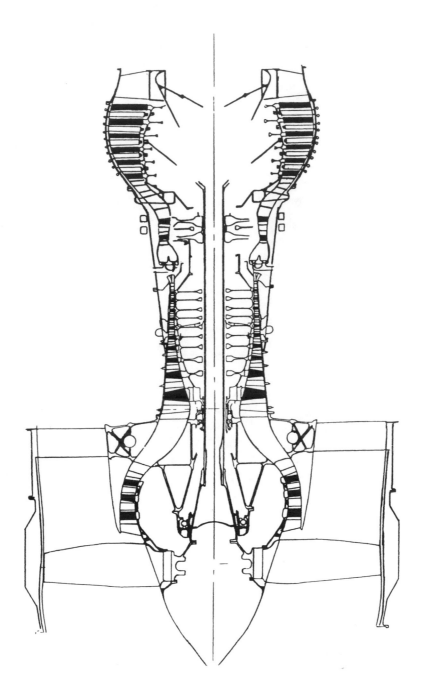

Figure 5.4(c) The General Electric GE90
(Fan tip diameter 3.12 m)

The Trent and the P&W 4084 have a bypass ratio of about 6, whereas the GE90 has a bypass ratio of about 9. The Trent is noticeably different with its three concentric shafts, LP, IP (intermediate pressure) and HP, whereas the P&W 4084 and GE90 have only two, the LP and HP. Of these western companies only Rolls-Royce uses three shafts; the configuration has aerodynamic advantages, particularly at part speed, at some cost in mechanical complexity. For high bypass ratio engines the front compressor is always known as the fan. These are highly specialised compressor stages for which the relative flow into the rotor is supersonic near the tip (for current engines the fan tip speed may be as high as 450 m/s at take off) and subsonic near the hub.

The flow through a bypass engine is divided into the bypass stream and the core stream, the latter going through the HP compressor and combustor. The part of the engine which handles the core stream is generally referred to as the core. In a two-shaft engine the fan is on the LP shaft driven by the LP turbine and the core is on the HP shaft. In fact this is an oversimplification of the way most two-shaft engines are designed, since they usually have some 'booster' stages just behind the fan on the LP spool to compress the core air before it enters the core itself; typically at cruise the fan and booster might have a pressure ratio of about 2.5 and the HP compressor a pressure ratio of 16 to give an overall pressure ratio of around 40. For the Trent the core air is compressed in three separate sections. At cruise the pressure ratio is about 1.6 in the fan (near the hub) and a little over 5 in the IP compressor and in the HP compressor to give about 40 overall.

In the simplified treatment here it is assumed that all of the power from the HP turbine is used to drive the HP compressor. In fact a relatively small proportion is taken to drive fuel pumps, generate electricity and provide hydraulic power for the aircraft. Similarly it is assumed that all of the air compressed in the core passes through the turbine; in fact some is bled off to pressurise the cabin and de-ice (by warming) some surfaces of the wing and nacelle. Most of the power from the LP turbine is used in compressing the bypass flow and only a small proportion is used to raise the pressure of the core.

At this stage of the design there is no need to consider whether to have a two-shaft engine with booster stages (the Pratt & Whitney and General Electric solution) or separate IP and HP compressors in a three- shaft engine (the Rolls-Royce solution). (As noted above, the advantages of having separate IP and HP shafts appear mainly during off-design operation, including starting.) It suffices for the present treatment to have only the fan on the LP shaft and the remainder of the core flow compression on a single HP shaft.

Exercises

5.1* At the start of cruise the temperature and pressure of the air entering the engine may be taken to be 259 .5K and 46.0 kPa. Assume a pressure ratio of 1.6 for flow through the fan which enters the core and a pressure ratio of 25 in the core compressor itself. Assume isentropic efficiencies of 90% in each component. Find the temperature rise across the fan of the flow entering the core compressor, the temperature entering the core compressor and thence the temperature at exit from the core compressor.   (**Ans**: 41.4 K; 300.9 K; 805.2K)

The power produced by the core turbine must equal the power into the core compressor; since we are assuming a perfect gas with the properties of air and treating the combustion as equivalent to a heat transfer, the temperature drop in the core turbine must equal the temperature rise in the core compressor . If the temperature of the gas leaving the combustor (i.e. entering the HP turbine) is 1450K, find the temperature at core turbine outlet and thence the pressure at outlet from the core turbine. Assume a turbine efficiency of 90%. (Neglect any pressure drop in the combustor.) 45

(**Ans**: 945.7 K, 333 kPa)

Explain why the pressure ratio across the core turbine is less than the pressure ratio across the corresponding compressor. Then show (but do not work out the numbers) how the difference will increase as either turbine entry temperature is increased or compressor entry temperature is reduced.

5.2 Find the temperature drop and pressure ratio across the LP turbine, taking the temperature into it as calculated in Exercise 5.1, and then the temperature and pressure at outlet from the LP turbine. Assume a bypass ratio (i.e. the ratio of the mass flow of the bypass stream to the mass flow through the core) equal to 6. Take the pressure ratio and efficiency across the fan for the bypass flow to be 1.6 and 90% (as for the fan flow going through the core).   (**Ans**: 290.0 K,4.30; 655.7K, 77.5 kPa)

Note: This exercise assumes that the value of bypass ratio and the fan pressure ratio are both specified. The actual choice of these is carried out in Exercise 7.1.

5.3 TURBINE INLET TEMPERATURE

Turbine inlet temperature is important because increasing its value makes the pressure ratio across the core turbine smaller in relation to the pressure rise of the core compressor and thereby increases the power available from the LP turbine. Increasing this temperature also increases the thermal efficiency, provided that the pressure ratio increases by an appropriate amount. These trends were shown in the idealised treatment of the gas turbine cycle presented earlier in sections 4.2 and 4.3.

In Exercise 5.1 the temperature entering the turbine was given as 1450 K, a plausible value for cruise operation for long periods of time. For take off the temperature into the turbine would be higher, and 1700 K is a representative value for the same engine. This high temperature would *not* be sustainable for the long periods typical of cruise without excessive amounts of cooling. In fact the temperature of the gas stream entering the turbine is higher than the melting temperature of the material from which the blades are made, about 1550 K.

It will be recalled from section 4.3 that it was the *ratio* of the turbine inlet temperature to the compressor inlet temperature T_4/T_2 which was crucial. Reducing the compressor inlet temperature, as occurs at high altitude, has a similar effect to increasing the turbine inlet temperature on the ground. The temperature ratios at take off, top of climb and cruise, for example, are less different than might be imagined, as the example in Table 5.1 from a recent engine shows:

Table 5.1   Representative compressor and turbine inlet temperatures

	engine inlet T_2	turbine inlet T_4	T_4/T_2
Take off (standard day, sea level)	288.15 K	1700 K	5.90
Top of climb (31000ft, M = 0.85)	259.5 K	1575 K	6.07
Start of cruise (31000 ft, M = 0.85)	259.5 K	1450 K	5.59

Note for second edition. These turbine inlet temperatures would now be rather low for an all-new engine for a large aircraft and values 100 K higher might be more realistic. They have been retained to avoid reworking all the exercises and many of the graphs.

It is apparent from this table that it is conditions at top of climb when the temperature ratio is greatest; at this condition the engine is working hardest in a non-dimensional sense, an issue discussed further in Chapter 8. It can also be seen that the low compressor inlet temperature at cruise means that T_4/T_2 is only about 5% down on the take off value even though the turbine inlet temperature is 250 K lower. The values in this table will be taken as appropriate for our design.

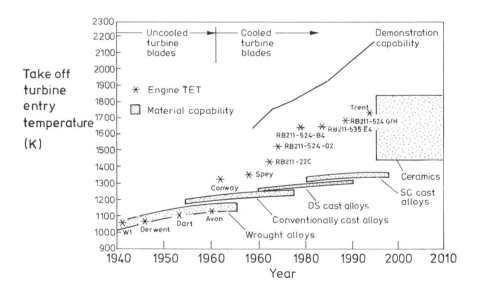

Figure 5.5. Turbine entry temperature[1] for Rolls-Royce engines since 1940:
a figure drawn in about 1993.

---

[1]For cooled turbines the temperature needs to be defined carefully. What is shown here (and used throughout the book) is the stator outlet temperature (SOT). This is the temperature after a hypothetical complete mixing of the cooling air with the hot gas downstream of the first turbine stator row (the HP nozzle guide vane).

The ability to operate at higher temperatures has been crucial in improving the performance of jet engines. The higher temperatures are possible in part because of better materials. More important, however, has been the use from the early 1960s of compressor air to cool the blades – it is the improvements in the way that cooling air is used that have brought the biggest gains. Figure 5.5 shows how the turbine inlet temperature used by one company, Rolls-Royce, has increased over more than 50 years; over the last 20 or so years there has been an average increase of about 8 K rise per year in turbine entry temperature. The rise in gas temperature shown in Fig. 5.5 is therefore all the more remarkable when it is appreciated that the turbines of newer engines are now operating in service for periods in excess of 20000 hours. The life of hot blades are primarily limited by creep, by oxidation or by thermal fatigue. Creep is the continuing and gradual extension of materials under stress at high temperature. A rule of thumb for blades limited by creep is that turbine blade life is halved (at a given level of cooling technology and material) for each 10 K rise in temperature of the metal. Thermal fatigue is not a function of the length of time the engine runs but how many operating cycles it goes through, in other words how many times the engine is started, accelerated and stopped with the attendant raising and lowering of the turbine temperature.

Originally the blades for high temperature use were forged, but better creep performance could be obtained when the blades were cast. (In addition the cooling passages can be cast inside the blade.) Then it was found that a better blade could be made by arranging for the crystals to form elongated in the direction of the span, so-called directionally solidified blades. A still better blade was obtained by casting each blade as a single crystal, and this is now the norm for the HP blades.

It is only possible for turbine blades to operate at such high temperatures because relatively cool air (nearly 900 K at take off) is taken from the compressor and fed to the inside of the turbine blades. Examples of this are shown in Fig. 5.6. Inside the blade there are complicated passages, with roughness elements to encourage heat transfer to the cooling air, which have the effect of lowering the metal temperature. More modern blades have even more complicated internal configurations to increase the heat transfer and to achieve more gradual variations in metal temperature. The air which has passed inside the blade then emerges from many small holes in the surface. These are positioned so that a relatively cool film of air exists around the blade outer surface, thereby shielding the surface from the very hot gas. This process is known as film cooling. Between 15 and 25% of the core air may be taken from the compressor and used in this way; for the same level of technology, the amount of air used for cooling must rise as turbine inlet temperature is increased. To some extent choice of turbine inlet temperature is a balance between engine performance and turbine life. There is also a balance between turbine inlet temperature and cooling air requirements, since cooling air reduces the efficiency and thrust of an engine. Turbine cooling is an expensive technology to develop and, since raising the temperature can have such a big effect on engine performance (as demonstrated in Exercise 5.1), turbine cooling technology is one of the areas where competition and product differentiation are most intense.

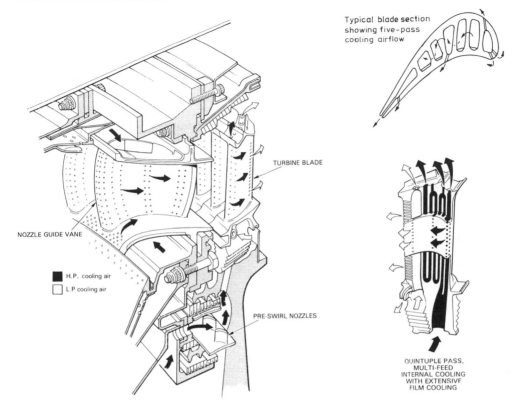

Figure 5.6. Arrangements for cooling an HP turbine and rotor blade. It should be noted that the rotor blade configuration is unique to Rolls-Royce, with a shroud on the tip.

(From *The Jet Engine*, 1986)

One way of assessing the performance of blade cooling is to use the cooling effectiveness defined by

$$\varepsilon = \frac{T_g - T_m}{T_g - T_c}$$

where T_g is the temperature[2] of the hot gas stream, T_m is the temperature of the metal and T_c is the temperature of the cooling air. The level of effectiveness is determined by the sophistication of the cooling technology, but it is also increased when the amount of cooling air is increased. A value of between 0.6 and 0.7 is currently 'state of the art'. The expression for effectiveness also brings out the serious consequences of a rise in cooling air temperature, which is normally close to compressor delivery temperature. If the compressor efficiency turns out to be lower than

---

[2]The temperature of the hot gas T_g and the temperature of the cooling air T_c are both *stagnation* temperatures, as are all the fluid temperatures used in this chapter. Stagnation temperature is discussed in Chapter 6.

expected, but the effectiveness is unchanged, the metal temperature will be raised, with potentially serious effects on the life. In this case the effectiveness may have to be increased by use of more cooling air, but this has detrimental effects on the thermal efficiency, fuel consumption and thrust.

It is not only the turbine which has a temperature limit on it. Modern engines have such high pressure ratios that the temperature of the air is sufficiently high at the back of the compressor (i.e. before entering the combustion chamber) that it is difficult to stand the high stresses in the compressor rotor. If the compressor rotor is made of titanium alloy the upper temperature allowable is around 870 K, whereas if nickel based alloys can be used, temperatures of around 990 K are possible. Nickel alloys are much heavier, so titanium based ones are preferred whenever possible. There is clearly a balance between allowable stress, weight and temperature; where the balance is struck must depend on the application and the relative desirability of low weight and of low fuel consumption.

In Fig. 5.5 a region is shown for ceramics. Ceramics appear very attractive for turbines, offering very high temperature performance as well as reduced density. Unfortunately the date of application to turbine blades tends to move to the right by about 12 months every year and Fig. 5.5, which was produced in the early 1990's, looks far too optimistic. Fundamentally the problem is that ceramics are vulnerable to defects and lack the safe ductile characteristics of metals. Some of the advantages of ceramics have been achieved by using thermal barrier coatings (TBC) on top of metal blades. The TBC reduces the temperature of the metal, since the conductivity is low, and provides a barrier against oxidation. For the same level of cooling flow and cooling technology the use of TBC has allowed an increase in turbine entry temperature of about 100 K.

Exercise

5.3 Assuming a cooling effectiveness of 0.65 calculate the metal temperature of the HP turbine rotor when the temperature of the gas sensed by the rotor T_g = 1600 K and T_c = T_{03} compressor delivery temperature. The overall pressure ratio is 40, the compressor isentropic efficiency is 0.90 and the inlet temperature T_{02} = 288 K   (**Ans**: T_m = 1136 K)

If the compressor efficiency is reduced to 0.85 but the other quantities, including cooling effectiveness, are unchanged, find the turbine metal temperature and estimate the reduction in creep life.   (**Ans:** T_m = 1159 K, creep life reduced by factor 4.6)

If the metal temperature were to be kept constant despite the reduction in compressor efficiency, what would the temperature of the gas T_g have to be reduced to?   (**Ans:** T_g = 1533K)

Note: Assuming the effectiveness stays constant, the cooling air temperature has a greater effect on metal temperature than the temperature of the gas from the combustor into the turbine. For an engine a decrease in compressor efficiency would reduce net power from the core and reducing the turbine entry temperature to restore the turbine metal temperature would reduce the power further.

SUMMARY CHAPTER 5

To make efficient use of high temperature ratios and pressure ratios in the core of the engine a bypass stream is normally used. Modern subsonic civil aircraft engines normally have bypass ratios of 5 or more. The fans on the front normally have supersonic tip speeds, but subsonic flow near the hub. The fan is driven by the LP turbine, with a separate LP shaft passing through the HP compressor and HP turbine.

The temperature of the gas entering the turbine is as high as the metal and the cooling arrangements will allow. At most operating conditions it is close to or above the melting temperature of the turbine blade material. During cruise the turbine entry temperature is typically about 250 K lower than at take off; this is desirable to prolong the life of the turbine but it also keeps the non-dimensional turbine inlet temperature T_4/T_2 nearly constant. The highest temperature ratio is encountered at top of climb and at this condition the non-dimensional variables in the engine, such as pressure ratio and non-dimensional rotational speed, will be greatest.

The pressure ratios now employed are sufficiently high that the temperature of the gas leaving the compressor is near to the limit possible with current materials.

With a turbine inlet temperature for initial cruise (at 31000 ft) of 1450 K it is sensible to take an overall pressure ratio of 40 and use this as the design condition. This may be divided into 1.6 for the core flow through the fan and 25 in the core itself. A pressure ratio of 40 for cruise would give a pressure ratio of about 45 at maximum climb and nearly this at take off.

There are aspects of the engine which require some understanding of the way gases flow at high speed. The velocities inside the engine are typically near to the speed of sound and at such speeds the pressure *changes* are a substantial fraction of the *absolute* pressure. These pressure changes bring significant variations in density and the flow is said to be compressible. Compressible fluid flow is the subject of the next chapter.

CHAPTER 6   ELEMENTARY FLUID MECHANICS
OF COMPRESSIBLE GASES

6.0 INTRODUCTION

In treating the gas turbine it is essential to make proper acknowledgement of the compressible nature of the air and combustion products. Compressible fluid mechanics is a large and highly developed subject, but here only that which is essential to appreciate the treatment and carry out the designs is given. There are also special approaches for handling the compressible, high-speed flow inside ducts which need to be introduced, and that is the purpose of this chapter. The most important book dealing with this topic is Shapiro (1953), but a more accessible account is given, for example, by Munson, Young and Okiishi (1994).

6.1 INCOMPRESSIBLE AND COMPRESSIBLE FLOW

For **liquids** the changes in density are normally negligible and it is possible to treat the flow as *incompressible*. Thus the equation for steady frictionless flow along a streamline,

$$VdV + dp/\rho = 0,$$

can be integrated directly, assuming the density is constant, to give Bernoulli's equation

$$\tfrac{1}{2}V^2 \;\; + \;\; p/\rho \;\; = \;\; p_0/\rho, \text{ a constant.}$$

p_0 is the *stagnation or total pressure* and corresponds to that pressure obtained when the flow is brought to rest in a frictionless or loss-free manner. The term $\tfrac{1}{2}\rho V^2$ is known as the *dynamic pressure* or *dynamic head*. A pitot tube records the stagnation pressure whereas a pressure tapping in a wall parallel to the flow records static pressure. The use of stagnation pressure is something like a book-keeping exercise – it indicates the pressure which is available to accelerate the flow. This is illustrated in Fig. 6.1. An analogy which is sometimes helpful can be drawn between the hydraulic system and a mechanical system: static pressure is analogous to potential energy and dynamic pressure is analogous to kinetic energy.

In dealing with gases at low speed (more precisely, at low Mach number) the density changes little and it is possible to use Bernoulli's equation as a reasonable approximation. The inaccuracy becomes significant when the dynamic pressure becomes a substantial fraction of the absolute pressure of the gas; this occurs when the Mach number exceeds about 0.3. For most of the processes inside jet engines the Mach number is nearer 1.0 than 0.3 and **Bernoulli's** equation is quite inappropriate and **MUST NOT BE USED.** (This is also true for the external aerodynamics of the aircraft itself.) To deal with compressible flow we have to start from a different approach.

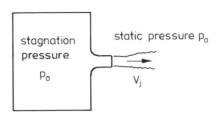

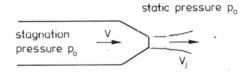

For INCOMPRESSIBLE FLOW $\quad V_j{}^2 = (p_0 - p_a)/\frac{1}{2}\rho$

For COMPRESSIBLE FLOW $\quad M_j{}^2 = \dfrac{2}{\gamma - 1}\,[(p_0/p_a)^{(\gamma-1)/\gamma} - 1]$

Figure 6.1. Schematic representation of stagnation conditions

6.2 STATIC AND STAGNATION CONDITIONS

The steady flow energy equation for the flow of gas with no heat transfer and no external work transfer can be written

$$h_1 + V_1{}^2/2 \quad = \quad h_2 + V_2{}^2/2$$

where h is the enthalpy, which can be rewritten in a way more convenient for us as

$$c_p T \quad + \quad V^2/2 \quad = \quad c_p T_0. \tag{6.1}$$

Here T_0 is the *stagnation* temperature, being that temperature which the gas would attain if brought to rest without work and heat transfer, not necessarily in an ideal or loss-free manner.

The specific heat can be rewritten $c_p = \gamma R /(\gamma - 1)$ and on rearranging and inserting this into the equation above the stagnation temperature may be written

$$T_0 /T = 1 + \frac{\gamma - 1}{2} \frac{V^2}{\gamma R T} .$$

(6.2)

The speed of sound is given by $a = \sqrt{\gamma R T}$ and the Mach number is given by $M = V/a$, so the equation for the stagnation temperature becomes

$$T_0 /T = 1 + \frac{\gamma - 1}{2} M^2.$$

(6.3)

Again, this equation between stagnation temperature T_0 and static temperature T does not imply ideal or loss-free acceleration or deceleration. To go further and find the corresponding relation between stagnation pressure and static pressure does require some idealisation.

At this point we suppose that the acceleration or deceleration of the gas between the static state p and T and that at stagnation p_0 and T_0 is reversible and adiabatic (i.e. *isentropic*), for which we know that

$$p/\rho^\gamma = \text{constant} \qquad \text{and} \qquad p/T^{\gamma/(\gamma-1)} = \text{constant}.$$

(6.4)

It then follows immediately that $p_0/p = (T_0/T)^{\gamma/(\gamma-1)}$

or
$$p_0 /p = \left(1 + \frac{\gamma - 1}{2} M^2\right)^{\gamma/(\gamma-1)}.$$

(6.5)

In incompressible flow it was often possible to use gauge pressures, but in all work with compressible flow it is important to remember to use the **absolute** pressures and temperatures .

It should be explained that until this section in the book a distinction between static and stagnation temperatures has not been drawn, or else the description of conditions has been made in such a way that it was not necessary to distinguish. From now on we must be more careful. In calculating such quantities as the speed of sound or the density of the air through which the aircraft is travelling it is the **static** quantities which are relevant. In carrying out the cycle analyses, for example finding the work or the heat in a gas turbine, it is the **stagnation** quantities which should be used. Returning to the book-keeping analogy, the use of stagnation properties like stagnation temperature or pressure gives a measure of the total amount of energy or capacity to accelerate the flow which is available. Stagnation properties are not the 'real' ones, except at places where the velocity may be zero like on the nose of an aerofoil, but the static properties are real. The density of the air, the speed of sound and the rate of chemical reactions depend on the local static properties.

Stagnation pressure and temperature change with the frame of reference. For a stationary atmosphere (i.e. one with no wind, $V = 0$) the static and stagnation properties are equal, but an observer in an aircraft travelling through the stationary atmosphere would not see the static and stagnation properties as equal. The observer travelling at velocity V, Mach number M would perceive the stagnation temperature given by

$$T_0 = T + V^2/2c_p$$

$$= T\{1 + \frac{\gamma-1}{2} M^2\}, \tag{6.6}$$

and the stagnation pressure, calculable via the isentropic relation, as

$$p_0 = p\{1 + \frac{\gamma-1}{2} M^2\}^{\gamma/(\gamma-1)}. \tag{6.7}$$

The effect of forward speed is to increase the overall pressure ratio of the engine. At its inlet the engine responds to the stagnation temperature and pressure. For flight at a Mach number of 0.85 the inlet stagnation pressure is related to the ambient pressure by equation (6.5) giving $p_{02}/p_a = 1.604$. The flow at nozzle exit, however, responds to the ambient static pressure p_a so that as a result of the forward motion the pressure ratio across the engine is increased by 1.604.

Exercises

6.1 Expand the expression for p_0/p in equation 6.7 using the binomial expansion to express the relation in the form

$$p_0 - p = \frac{1}{2}\rho V^2\{1 + a M^2 + b M^4 + \ldots\}.$$

What is the highest Mach number at which the incompressible expression for stagnation pressure can be used if the error is not to exceed 1%? What is the error in calculating the stagnation pressure by Bernoulli's equation at $M = 0.3$ **(Ans:** 0.20, 2.25%**)**

6.2* If the aircraft cruises at Mach number of 0.85 at an altitude of 31000 ft find the stagnation temperature and pressure of the flow perceived by the aircraft. (These values were assumed in Exercise 5.1.) **(Ans:** 259.5 K, 46.0 kPa**)**

6.3 An aircraft flies at Mach 2 at 51000 feet ($p_a = 11.0$ kPa, $T_a = 216.7$ K) propelled by a simple turbojet engine (i.e. no bypass). If the inlet is effectively isentropic find the stagnation temperature and pressure into the compressor. The engine compressor has a pressure ratio of 10 with an isentropic efficiency of 90%: find the temperature and pressure at compressor exit. **(Ans:** 793.3 K; 0.861 MPa**)**
  In the combustor the velocities are low (so the stagnation and static pressures are equal) but the absolute stagnation pressure falls by 5%. At turbine entry the stagnation temperature is 1400 K and the turbine has an efficiency of 90%. Find the stagnation temperature and pressure downstream of the turbine. **(Ans:** 996.7 K; 0.212 MPa**)**

If the final propelling nozzle is isentropic, find the velocity of the jet: for this assume that the expansion is to the static pressure after the nozzle, which is equal to that of the surrounding atmosphere.

Calculate the gross and net thrust per unit mass flow, the propulsive efficiency and, from the temperature rise in the combustor, the overall efficiency.

(**Ans:** V_j = 1069 m/s; F_G = 1069 N/kg/s, F_N = 479 N/kg/s, η_p= 0.711, η_o= 0.464)

6.3 THE CHOKED NOZZLE

One other concept peculiar to compressible flow must be addressed at this stage, the maximum flow which can pass through a given cross-sectional area. Figure 6.1 shows flow from a reservoir or large volume entering a nozzle for which the minimum area occurs at the throat. We will assume here that the flow is one-dimensional (in other words, the flow is uniform in the direction perpendicular to the streamlines) and that it is adiabatic (no heat flux to or from the air). The mass flow rate per unit area of the throat is given by

$$\dot{m}/A = \rho V \tag{6.8}$$

and in steady flow the maximum value of $\dot{m}/A$ evidently occurs at the throat. For adiabatic flow the steady flow energy equation, equation 6.1, leads to

$$V^2 = 2c_p(T_0 - T)$$

or
$$V = \sqrt{2c_p(T_0 - T)} \tag{6.9}$$

where T is the local temperature of the air and T_0 denotes the stagnation temperature, which is uniform throughout the flow.

For isentropic flow $\quad T/\rho^{(\gamma-1)} = T_0/\rho_0^{(\gamma-1)} = $ constant

or
$$\rho = C T^{1/(\gamma-1)} \tag{6.10}$$

where $\quad C = 1/(T_0/\rho_0^{(\gamma-1)}) = $ constant.

The mass flow rate can then be written

$$\dot{m}/A = C T^{1/(\gamma-1)} \sqrt{2c_p(T_0 - T)}$$

$$= C \sqrt{2c_p(T_0 T^{2/(\gamma-1)} - T^{(\gamma+1)/(\gamma-1)})} \tag{6.11}$$

The maximum mass flow rate per unit area can be found by differentiating $\dot{m}/A$ with respect to T, noting that T_0 is a constant. (It is actually easier, of course, to square $\dot{m}/A$ before differentiating – the turning point will be the same.) When this is done the maximum occurs when

$$T = T_0 \frac{2}{\gamma+1} .$$

Substituting this into equation 6.9 yields

$$V = \sqrt{2c_pT \left(\frac{\gamma+1}{2} - 1\right)}$$

which simplifies to

$$V = \sqrt{c_pT(\gamma-1)} = \sqrt{\gamma R T} = a \qquad (6.12)$$

the speed of sound.

In other words when the mass flow rate through an orifice or nozzle is at its maximum, the velocity in the throat is sonic. At this condition the nozzle or orifice is said to be choked. A maximum occurs in the mass flow rate per unit area because the density decreases as the velocity increases. As the pressure ratio across the nozzle increases the velocity increases, but beyond Mach 1 the rate of fall in density exceeds the rate of increase in velocity. To accelerate a flow beyond sonic velocity it is first necessary to reduce the area to a throat and then increase the area downstream. Such nozzles are familiar at the back of rockets used to launch satellites, missiles etc. and are referred to as a convergent–divergent (often abbreviated to con–di) nozzles or sometimes as Laval nozzles.

For most operating conditions of a jet engine the final propelling nozzles will be choked. (It is also the case that the turbines in a gas turbine are effectively choked.) This has important consequences for the way the engine behaves, as will be discussed in Chapter 12. Although the propelling nozzles are choked, it is not normally worthwhile fitting a con–di nozzle on subsonic civil aircraft; the nozzle exit plane forms the throat and some expansion of the jet takes place downstream of the nozzle. For a high-speed military aircraft, however, the jet Mach number may be sufficiently high that a con–di nozzle is installed. Such con–di nozzles are made so they can be varied in shape and throat area to match the operating condition.

6.4 NORMALISED MASS FLOW PER UNIT AREA

Equation (6.11) for $\dot{m}/A$ can be expressed in terms of the local flow Mach number M and after some algebraic manipulation, the mass flow rate per unit area is given in non-dimensional form by the expression

$$\frac{\dot{m}\sqrt{c_pT_0}}{A\,p_0} = \frac{\gamma}{\sqrt{\gamma-1}} M \left(1+\frac{\gamma-1}{2}M^2\right)^{-(\gamma+1)/2(\gamma-1)} .$$

$$(6.13)$$

This function is so widely used that it is convenient to denote it by the symbol $\bar{m}$ so that

$$\bar{m}(M, \gamma) = \frac{\dot{m}\sqrt{c_p T_0}}{A\, p_0} . \tag{6.14}$$

For a given gas (i.e. a given value of the ratio of specific heats) $\bar{m}$ is a function of Mach number only, and this is shown for air ($\gamma = 1.40$) in Fig. 6.2. Note that at the throat, when $M = 1$,

$$\bar{m}(M = 1, \gamma) = \frac{\gamma}{\sqrt{\gamma - 1}} \left(\frac{\gamma + 1}{2} \right)^{-(\gamma+1)/2(\gamma-1)} . \tag{6.15}$$

Thus at the throat

$$\text{for } \gamma = 1.4, \quad \bar{m} = 1.281,$$

whereas

$$\text{for } \gamma = 1.30, \quad \bar{m} = 1.389.$$

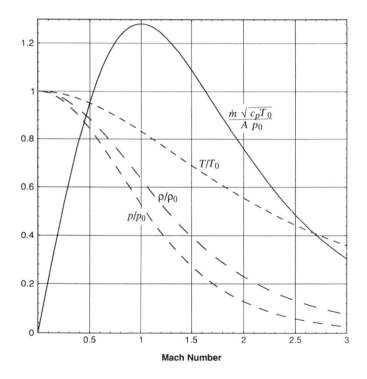

Figure 6.2. One-dimensional flow of a perfect gas, $\gamma = 1.4$

Equation 6.15 shows very clearly that the mass flow through a choked nozzle is proportional to the stagnation pressure of the flow and inversely proportional to the square root of the stagnation temperature. The *only* way to alter the mass flow through a choked nozzle of given size is to vary either the stagnation pressure or stagnation temperature. *Varying conditions downstream of the nozzle has no effect on mass flow rate or on conditions upstream of the throat.* This can be understood by realising that pressure disturbances travel at the speed of sound; if the flow is supersonic, the velocity exceeds the speed of sound and the downstream conditions cannot 'inform' the upstream conditions to reduce or increase the mass flow.

Three other curves are also shown in Fig. 6.2. One curve is the ratio of static temperature to stagnation temperature

$$T/T_0 = \{1 + \frac{\gamma-1}{2} M^2\}^{-1}.$$

The other two curves give the corresponding ratios of local static to stagnation pressure and density for isentropic flow along the nozzle, $p/p_0 = (T/T_0)^{\gamma/(\gamma-1)}$ and $\rho/\rho_0 = (T/T_0)^{1/(\gamma-1)}$ respectively.

Exercises

6.4 Verify that $\dot{m}\sqrt{c_p T_0}/A\,p_0$ is non-dimensional. If for a choked con-di nozzle the area of the throat is equal to 1 m^2, find the areas of the nozzle at sections where the Mach numbers are 0.5, 0.9 and 2.0. Treat the flow as isentropic, i.e. p_0 – constant as well as adiabatic, i.e. T_0 – constant.

(**Ans:** 1.340 m^2; 1.009 m^2; 1.688 m^2)

Note: The difference in area for $M = 1.00$ and $M = 0.90$ is very small indeed. Close to sonic velocity very small area changes produce large variations in Mach number and pressure.

6.5 A high bypass engine is 'throttled back' whilst the aircraft continues to fly at Mach 0.85. Neglecting any losses in stagnation pressure upstream and downstream of the fan, calculate the stagnation pressure ratio across the fan at which the bypass nozzle will just unchoke. (**Ans:** 1.18)

6.6 For the simple turbojet engine of Exercise 6.3 (no bypass stream) find the area of the throat of the propelling nozzle per unit mass flow through the engine. There is a divergent section downstream of the throat – find the final area if the expansion is isentropic to the static pressure of the surrounding atmosphere. (**Ans:** 0.00369 m^{2}kg^{-1}s; 0.010 m^{2}kg^{-1}s)

SUMMARY CHAPTER 6

Bernoulli's equation is **not** valid for the flow of gases unless the Mach number is below about 0.3; this will not normally be the case in jet engines. Stagnation pressure must therefore be defined using

$$p_0 = p\{1 + \frac{\gamma-1}{2} M^2\}^{\gamma/(\gamma-1)}.$$

Stagnation temperature and pressure provide a convenient way of accounting for the effects of velocity - they are the temperature and pressure of a gas where it to be brought to rest adiabatically and, in the case of pressure, reversibly. For the flow of a compressible gas through a converging duct or nozzle there is a maximum mass flow rate per unit area. This occurs when the velocity is sonic in the section of duct where the cross-sectional area is a minimum, normally called the throat. To accelerate a flow to supersonic velocity it is necessary to first converge the stream and then diverge it downstream of the minimum area where the velocity is sonic.

Mass flow per unit area may be made non-dimensional in a form which, for fixed gas properties, is a unique function of Mach number and ratio of specific heats,

$$\frac{\dot{m}\sqrt{c_p T_0}}{A\, p_0} \;=\; \bar{m}(M, \gamma)\,,$$

and this non-dimensional form will also be useful in connection with engine performance.

The flow rate through a choked nozzle is determined by the upstream stagnation pressure and temperature and is independent of the conditions downstream of the throat. For jet engines the propelling nozzles are normally choked; effectively so too are most turbines.

The understanding of the way in which a choked orifice works is indispensable to the treatment of jet engines, as will become apparent in the later chapters. It will also be found that the non-dimensional expression above for the mass flow will be useful in expressing the mass flow through the engine.

CHAPTER 7

SELECTION OF
BYPASS RATIO

7.0 INTRODUCTION

In this chapter we set about selecting the bypass ratio and with it the fan pressure ratio. The overall pressure ratio and the temperature ratio T_{04}/T_{02} have been selected in Chapter 5. Once we have selected the bypass ratio and fan pressure ratio the mass flow of air needed to give the required thrust for cruise can be obtained; the thrust required was found in Chapter 2.

7.1 OVERALL SPECIFICATION

The bypass ratio is defined as the mass flow of air passing outside the core divided by the mass flow through the core

$$bpr \; = \frac{\dot{m}_b}{\dot{m}_c} \; .$$

The choice of bypass ratio has a major effect on the efficiency, because for a given core the bypass ratio determines the jet velocity. The bypass ratio also greatly affects the appearance, size and weight of the engine: the pure turbojet has a small diameter relative to its length, whereas high bypass ratio engines have their overall diameter comparable to their length. Previous engines (the RB211, CF6 and JT9D generation) had bypass ratios of around 5 and we may assume that a bypass ratio at least as great as this will be employed in a new engine for the New Large Aircraft. For a number of practical reasons bypass ratios in excess of about 10 are not attractive at the present time. To achieve this it would probably be necessary to install a gear box between the LP turbine and the fan, to allow the turbine to run faster; this requires considerable development and will very probably be heavy.

For a civil transport engine a major requirement is low fuel consumption, equivalent to high overall efficiency. The overall efficiency is the product of propulsive and thermal efficiency,

$$\eta_0 = \eta_p\eta_{th},$$

and we have now investigated this enough to be able to put the ideas and constraints together to be able to choose rationally a suitable bypass ratio. In Chapter 4 we showed that to obtain a high thermal efficiency η_{th} we must operate with a high turbine inlet temperature (this was taken to be around 1450 K for cruise) and high pressure ratio (which is likely to be around 60, taking account of the forward motion of the aircraft). This combination of temperature and pressure would lead to a high jet velocity if all the available power were used to accelerate the core flow. In Chapter 3, however, we have shown that very high jet velocities give low propulsive efficiency η_p, and an efficient engine will generate its thrust by accelerating a large mass flow of air by only a small amount. The combination of constraints on η_{th} and η_p give rise to the high bypass engine for subsonic propulsion.

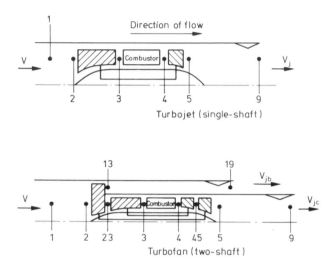

Figure 7.1. Standard numbering schemes for stations in jet engines.

It is necessary to introduce a system for designating the various stations through the engine and Fig. 7.1 shows this for a simple turbojet and for a simplified high bypass engine; for the latter only some of the stations are labelled; the label designations may not seem the ideal system of numbering, but it is used here because it is close to the internationally recommended one! (If a booster compressor were installed on the LP shaft downstream of the fan then station 23 would be downstream of the booster stages.)

When the overall pressure ratio of the engine and the turbine inlet temperature are fixed the available power per unit mass flow rate through the core is determined. In Exercise 5.1 the conditions at outlet from the core, that is at outlet from the HP turbine and inlet to the LP turbine, were determined for cruise at 31000 ft. At this station the stagnation temperature and pressure are $T_{045} = 945.7$ K and $p_{045} = 333$ kPa. What we now have to do is to decide how much of the pressure shall be expanded in the LP turbine and how much in the nozzle of the core stream. Equivalently how much power is extracted in the LP turbine to be used on the bypass stream and how much kinetic energy is given to the core jet. The process is illustrated by the temperature–entropy diagram in Fig. 7.2; what we need to do is decide on the pressure p_{05} in the jet pipe of the core.

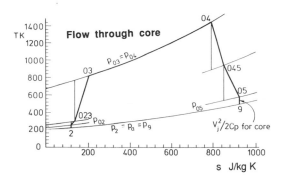

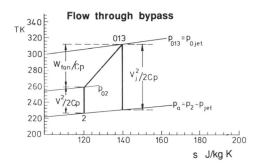

Figure 7.2. Temperature–entropy diagrams for the high bypass ratio engine.
Core and bypass jet velocities equal.
Cruise at $M = 0.85$ at 31000 ft. Bypass ratio = 6, core pressure ratio = 32,
turbine inlet temperature = 1407 K, fan, compressor and turbine efficiencies = 90%.

7.2 JET VELOCITY AND THE SPECIFICATION OF FAN PRESSURE RATIO

In practice the velocity of the core jet V_{jc} is normally larger than the velocity of the bypass jet V_{jb}. This is because the LP turbine and the fan are both to some extent irreversible, whereas the core nozzle is virtually loss free. It can be shown that larger thrust is obtained if V_{jb}/V_{jc} is approximately equal to the product $\eta_t\eta_f$ of the LP turbine and fan isentropic efficiencies. The calculations are made easier if the jet velocities are assumed equal and, as will be shown later, the errors introduced by this in estimating thrust or fuel consumption are small.

The most natural way to explore the effect of bypass ratio on sfc and thrust would be to select values of the bypass ratio and then evaluate the jet velocities from which thrust and sfc could be found. This is not difficult, but it does involve an iteration. For the present it is easier to avoid an iteration and this is possible if we choose the power output in the LP turbine, which then allows the core jet velocity to be found. Since the core velocity is taken to be equal to the bypass jet velocity, and since all the LP turbine power goes into the fan, the only remaining unknown is the bypass ratio.

The expansion in the LP turbine is taken to have an isentropic efficiency η_t , so if the power from the turbine, per unit mass flow rate through the core, is specified, the temperature drop is known and from it the pressure ratio across the turbine may be found from

$$T_{045} - T_{05} = \eta_t T_{045} \{ \ 1 - (p_{05}/p_{045})^{\gamma-1/\gamma} \}. \tag{7.1}$$

This fixes the pressure downstream of the turbine p_{05}.

The nozzle is presumed to be effectively isentropic and the *static* pressure p_9 at outlet from the nozzle is assumed equal to the atmospheric pressure p_a . The expansion in this is given by

$$T_{05} - T_9 = T_{05}\{ \ 1 - (p_a/p_{05})^{\gamma-1/\gamma} \} \tag{7.2}$$

which leads directly to the core jet velocity

$$V_j^2 \ = \ 2c_p (T_{05} - T_9) \ . \tag{7.3}$$

The power from the LP turbine goes entirely to the fan, a fraction to compress the flow entering the core and most of it to compress the bypass stream. The energy balance for the LP shaft, where $\dot{m}_c$ is the mass flow rate through the core, is

$$\dot{m}_c c_p (T_{045} - T_{05}) = \dot{m}_c c_p (T_{023} - T_{02}) + bpr \, \dot{m}_c c_p (T_{013} - T_{02}) \tag{7.4}$$

where $T_{013} - T_{02}$ is the stagnation temperature rise in the bypass stream through the fan and $T_{023} - T_{02}$ is the temperature rise for the core stream through the fan. We assume here that the pressure ratio for the core flow through the fan is $p_{023}/p_{02} \ = 1.6$, and Exercise 5.1 showed the temperature rise $T_{023} - T_{02}$ to be 41.4 K. Until the jet velocity required of the bypass stream has been selected we do not know the temperature rise or the pressure ratio of the bypass stream.

The stagnation temperature rise through the fan is that necessary to give the required bypass jet velocity V_j , chosen here to be equal to the core jet velocity. If the fan were ideal the stagnation temperature rise through it would be that for an isentropic compression, that is

$$T_{013s} - T_{02} = \ (V_j^2 - V^2)/2c_p$$

where V is the flight speed. In other words the static temperature of the bypass jet would be equal to the ambient temperature, but the stagnation temperature would be higher by the difference in the kinetic energy of the jet and incoming flow. The fan is not isentropic, but its efficiency is equal to η_f, so the actual temperature rise in the bypass stream is given by

$$T_{013} - T_{02} = \ (V_j^2 - V^2)/2c_p \, \eta_f \, . \tag{7.5}$$

The jet velocity has been specified by the core stream and the flight velocity is known so inserting equation 7.5 in 7.4 allows the bypass ratio to be found.

Knowing the temperature rise in the bypass stream of the fan allows the pressure ratio to be found since

$$(p_{013}/p_{02})^{\gamma-1/\gamma} - 1 = \eta_f(T_{013} - T_{02})/T_{02} = (V_j^2 - V^2)/2c_pT_{02}. \qquad (7.6)$$

Exercise

7.1* Use the results of Exercise 5.1 to provide the inlet conditions into the LP turbine at a flight Mach number of 0.85 at 31000 ft. Assume equal velocity for the core and bypass jets.

Assume that the inlet decelerates the flow isentropically and the nozzles expand the flow isentropically to ambient pressure. Take the isentropic efficiencies for the fan and for the LP turbine to be each 0.90. (Note that the inlet temperature given in Exercise 5.1 already took the effect of flight speed into account, i.e. the value of inlet temperature given corresponds to the stagnation value.)

Calculate the following:
      a) the jet velocity (where expansion is from turbine outlet to ambient *static* conditions;
      b) the propulsive efficiency;
      c) the gross thrust F_G and net thrust F_N from the whole engine (core plus bypass) per unit mass flow THROUGH THE CORE (i.e. comparing engines for same size core);
      d) the overall efficiency, given by (net thrust)× (flight speed) $\div c_p(T_{04}-T_{03})$;
  and   e) the sfc (see section 3.3) taking for the fuel $LCV = 43$ MJ/kg;
for zero bypass ratio and at least one other bypass ratio in the range from 1 to 12. (Better still, write a short programme or use a spread-sheet and work the values out for a number of cases in the bypass ratio range and plot these against bypass ratio.)

Compare the results with those shown in Fig. 7.3. Sample results at three *bpr* are given below.

(Ans:

bpr =0; V_j=932 m/s, η_p=0.432, F_G=0.932 kN kg^{-1}s, F_N=0.672 kN kg^{-1}s, η_0=0.268, *sfc*=0.788 kg/h/kg,

bpr =6; V_j=403 m/s, η_p=0.778, F_G=2.82 kN kg^{-1}s, F_N=1.023 kN kg^{-1}s, η_0=0.404, *sfc*=0.520 kg/h/kg,

bpr =10; V_j=355 m/s, η_p=0.839, F_G=3.91 kN kg^{-1}s, F_N=1.087 kN kg^{-1}s, η_0=0.428, *sfc*=0.492 kg/h/kg)

Approach to Exercise 7.1: Begin with zero bypass ratio – only one stream makes it simple to find the LP turbine power (which is just to raise the pressure of the flow through the core by 1.6) and the jet velocity. For the next case choose a suitably larger LP turbine power – whence conditions out of the turbine and the core jet velocity can be found. With the bypass jet velocity chosen equal to the core velocity, and the extra LP turbine power equal to the bypass flow times the enthalpy rise of the bypass flow, it is possible to determine the bypass ratio. With a little care the entire range of bypass ratios can be covered in a few suitable steps.

For convenience, to remove the need for an iterative solution, the drop in LP turbine temperature for the bypass ratios used to obtain the above answers are shown here:

bpr	LP turbine temp. drop
0	41.4 K
6	361.0K
10	376.2 K

The approach adopted in Exercise 7.1 has been used to generate Fig. 7.3. Raising the bypass ratio increases the thrust; increasing bypass ratio from 6 to 10 increases the gross thrust F_G by 39% whereas the increase in net thrust F_N is about 6%. At cruise conditions it is net thrust which is most relevant, but during take off, when the ram drag is small, net thrust is nearly equal to the gross thrust. For the same core a greater bypass ratio therefore gives a much larger take-off thrust; likewise for the same thrust at cruise, a higher bypass ratio gives a much greater thrust at take off or on the test bed.

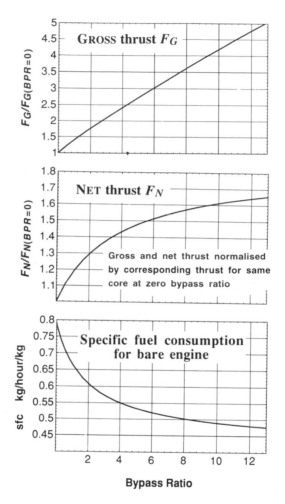

Figure 7.3. Predicted variation in thrust and sfc with bypass ratio for a constant core.
Bare engine. Aircraft cruising at $M = 0.85$ at 31000 ft.
Overall pressure ratio for core = 40, turbine inlet temperature = 1450 K. Pressure ratio for core flow through fan = 1.6, core and bypass jet velocities equal.

7.3 BARE AND INSTALLED SPECIFIC FUEL CONSUMPTION

The actual levels of sfc predicted in Exercise 7.1   for a bypass ratio of around 5 or 6 are quite similar to those of engines in service now, that is around 0.57 to 0.6. The absolute levels are, as already noted, sensitive to the isentropic efficiencies of the compressor and turbine, assumptions relating to cooling air, pressure loss and to the choice of c_p or γ for the gas. The fact that the levels predicted are similar does give confidence that the trends are correct. The results found in Exercise 7.1 and plotted in Fig. 7.3 show a lowering of sfc as bypass ratio is increased due to an improvement in propulsive efficiency but the amount of improvement declines as bypass ratio increases and V_j decreases, since propulsive efficiency must tend to a limit of 1.0 as the jet velocity tends to the flight speed. There are, moreover, practical snags associated with increasing the bypass ratio. As the bypass ratio increases, for a given size of core, the engine becomes larger and heavier. The larger nacelle involves increased drag, effectively reducing the net thrust, which has the effect of increasing the sfc. For a bypass ratio of 5 (the early RB211, CF6, JT9D generation) the increase in sfc associated with the nacelles is about 8%, for which the approximate breakdown is 1% in the inlet, 2% in the bypass duct and 6% on the outside. (This is explored in Exercise 7.2 below.) With more modern and better aerodynamic design of the nacelle its drag might be a little lower, but it is certain that the overall percentage will rise inexorably as the bypass ratio goes up. Also with a new engine, with a smaller and more powerful core, the bypass ratio can be higher for the same size of fan and the same nacelle drag - when this is taken into account using bypass ratio ceases to be the most useful term to describe an engine.

Exercise

7.2*   **a**) The results of Exercise 7.1 (or plots in Fig. 7.3) were for a 'bare' engine and neglected the drag and losses associated with the nacelle. It is plausible to assume that the increase in sfc due to the nacelle is linearly dependent on bypass ratio and is equal to 4% at a bypass ratio of 1 and 9% at a bypass ratio of 6. Make an estimate of the effect of the nacelle on the sfc by modifying the values plotted in Fig. 7.3. (The comparison between the bare and installed engine sfc is shown in Fig. 7.4.) Make a design choice for the bypass ratio somewhere in the range between 5 and 10.

(**Ans**: bpr = 0, sfc = 0.820 kg/h/kg; bpr = 6, sfc = 0.567 kg/h/kg; bpr = 10, sfc = 0.556 kg/h/kg)

b) In Exercise 2.2 the net thrust required from each engine was shown to be F_N = 75.1 kN for initial cruise at 31000 ft and M = 0.85. Use this with the chosen bypass ratio to find the mass flow of air, the gross thrust and the jet velocity at start of cruise. Use results from Exercise 7.1 for the bare engine (i.e. ignore installation effects on thrust). Sample results shown below.

(**Ans**:   bpr = 0: $\dot{m}_{air}$ = 111 kg/s; F_G = 103.5 kN; V_j = 932 m/s;
bpr = 6.0: $\dot{m}_{air}$ = 514 kg/s; F_G = 207 kN; V_j = 403 m/s;
bpr = 10.0: $\dot{m}_{air}$ = 760 kg/s; F_G = 270 kN; V_j = 355 m/s)

Note: The term *Specific Thrust* is used to denote the net thrust per unit mass flow through the engine, $F_N/\dot{m}_{air}$, and this characterises a wide range of engines better than the bypass ratio, see Chapter 15. For the bypass ratios of 0, 6 and 10 in the above exercise the specific thrusts are 677, 146 and 99 N/kg/s respectively. The units are those of velocity and it is easy to show that specific thrust is actually equal to the difference between jet and flight velocity if the mass flow rate of fuel is small. Industry, however, confuses the situation by expressing specific thrust in units of pounds thrust per pound per second of air (lb/lb/s is also equal to kg/kg/s) and quotes the values in seconds; in this barbaric system the specific thrusts would be 69,14.9 and 10.1 seconds respectively for the three bypass ratios. (Using pounds thrust per pound per second of air confuses pounds force with pounds mass, something which does not happen in the SI system, and is a potential source of confusion.)

The comparison of sfc for the 'bare' engine and that for the engine installed is shown in Fig. 7.4. The increase in bypass ratio from 6 to 10 now produces a reduction of 2.0% in sfc. The benefits of increased bypass ratio on the net thrust will be reduced as well. A consequence of the higher bypass ratio will be an engine somewhat heavier and a nacelle which is substantially heavier; such increases in weight will give rise to increased cruise drag and either reduced payload or reduced range.

It is possible to forget that the object of the engine maker is *not* to make the most efficient engine but the one which gives the largest yield to the airline. As the weight of the engine increases the same wings are able to lift less payload, so the revenue declines. Secondly, since for given wings the drag is proportional to the lift, an increase in engine weight leads to a small increase in the total drag and therefore in the fuel consumption. Thirdly, if the bypass ratio were to be very large, and the whole engine correspondingly large, the aircraft itself begins to be affected; the wings have to be higher off the ground and their aerodynamic performance is impaired by the engines. To make the wings higher off the ground requires a longer undercarriage, which is heavy, and may make the aircraft too high to be conveniently accommodated in existing airports.

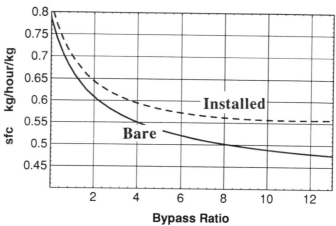

Figure 7.4 Predicted variation of sfc with bypass ratio for bare and installed engine.
Conditions as for Figure 7.3.
Assumed: $sfc_{installed} = \{1.04 + 0.01(bpr - 1)\}sfc_{bare}.$

For these reasons it is not possible to select the correct bypass ratio just on the basis of a calculation similar to that in Exercise 7.1, even if the calculation were more refined. As noted earlier, when the Boeing 777 was designed the engine manufacturers were divided; Pratt & Whitney and Rolls-Royce chose a bypass ratio of about 6 whilst General Electric selected a bypass ratio of about 9 for their engine, the GE90. By 2001 the various engines have been in service long enough to make some assessment. Numbers given by Boeing show that the fuel burn per seat-mile is lowest for the GE90, but it is less than 1% below that of the other two engines. More significantly the GE90 is considerably heavier, giving an empty aircraft which is about 3 tonne heavier than the lightest, which is the Rolls-Royce Trent. As a result the plane with the lighter engines can carry almost 6% extra payload or, for the same sfc, operate over increased maximum range.

Where the GE90 engine has been most successful is in giving markedly lower noise levels. The higher bypass ratio translates into a smaller fan pressure ratio, which in turn allows a lower fan tip speed and gives a lower jet velocity; lower fan speed and jet velocity are both expected to lower the noise. In the time since the Boeing 777 entered service there has been a marked tightening of the noise regulations, mainly local airport regulations such as the London Heathrow Quota Counts. (The topic of noise is addressed very briefly in the appendix to this book.) The lesson of high bypass ratio from the GE90 has been quickly assimilated, so the Rolls-Royce Trent 500 engine, being developed for the Airbus A340-500 and -600, has a bypass ratio of about 8.5. With the passage of time the technology improves and the power output from a given size of core (i.e. the mass flow through the core) increases. In other words the optimum bypass ratio for lowest fuel consumption is likely to increase with time. Nevertheless, it is still commonly believed that with current technology levels, a bypass ratio of 8 is greater than that for lowest fuel burn. At the time of writing the second edition the engines are being designed and developed for the Airbus A380: the GP7200 engine, from the alliance of General Electric and Pratt & Whitney, and the Trent 900, from Rolls-Royce. To minimise the noise the makers sought to minimise the jet velocity and the fan speed, requiring the largest possible bypass ratio and fan diameter. Both have settled on a bypass ratio of around 8 with a fan diameter of 2.95 m. The choice of fan diameter is for a reason which might not be guessed. Spare engines have to be capable of being moved by air freight and the normal way to do this is in a Boeing 747 freighter. A fan diameter of 2.95 m is the largest for which the engine can just fit into the aircraft.

7.4 THE EFFECT OF UNEQUAL BYPASS AND CORE JET VELOCITY

In Exercises 7.1 and 7.2, as well as in Figs. 7.3 and 7.4, it has been assumed that the jet velocity is equal for the core and bypass jets. Figure 7.5 shows the results of a numerical exploration for the same core used for Exercise 7.1 and Figs. 7.3 and 7.4. The lowest sfc and highest net thrust occur[1] when $V_{jb}/V_{jc} \approx 0.78$ but, compared to the values obtained earlier with the simple assumption that $V_{jb}/V_{jc} = 1.0$, the improvement is only around one per cent. Any serious design would, of course, allow for difference in jet velocity, but for the purpose of the exercises here, the simple assumption is entirely satisfactory in most cases.

---

[1]It is often supposed that the optimum velocity ratio is equal to the product of LP turbine efficiency and fan efficiency, which in this example would equal 0.81. This product of efficiencies is actually relevant to the optimum transfer of work to the two gas streams, to give the highest kinetic energy for the constraints imposed. What is wanted for the engine is the highest momentum from the two jet streams given the same constraints, and for this the bypass velocity should be slightly lower that the product of efficiencies.

Although the differences in thrust and fuel consumption are fairly small, Table 7.1 shows that the quantities in the bypass duct and in the core downstream of the LP turbine do differ significantly. The table is for a constant bypass ratio of 6 and the same inlet conditions to the LP turbine and to the fan as used in Exercises 7.1 and 7.2.

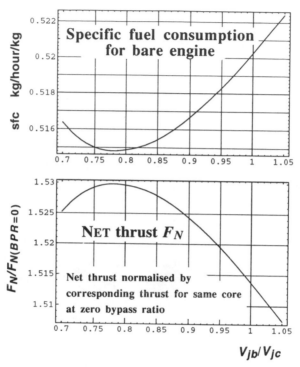

Figure 7.5. Predicted variation of sfc and thrust with ratio of bypass jet velocity to core jet velocity. Conditions as for Figure 7.3.

Table 7.1   Comparison with unequal and equal jet velocities
(Parameters as in Exercises 7.1 and 7.2)

		$V_{jb} = 0.78\,V_{jc}$	$V_{jb} = V_{jc}$
Core stream	V_{jc}	498 m/s	403 m/s
	p_{05}	62.2 kPa	48.3 kPa
	T_{05}	622 K	585 K
	$T_{045}-T_{05}$	324 K	361 K
	Nozzle area	0.72 m^2	0.92 m^2
Bypass stream	V_{jb}	388 m/s	403 m/s
	p_{013}/p_{02}	1.70	1.81
	T_{013}	306.5 K	312.9
	Nozzle area	2.42 m^2	2.31 m^2

SUMMARY CHAPTER 7

Decisions have to be made regarding the division of power between the bypass streams and the core. It is convenient to assume equal velocities for each in the current exercise and the error associated with this is not more than one per cent of net thrust or sfc.

The propulsive efficiency rises and the sfc falls as the bypass ratio rises. The rate of improvement with bypass ratio decreases as bypass ratio rise, and eventually the increased penalty associated with the drag of the nacelle outweighs the benefits predicted for the bare engine. In addition, an increase in bypass ratio engine leads to an increase in engine weight, so the optimum bypass ratio for generating profits will be lower than that for minimum installed sfc.

Minimising the noise has become crucial and has come to exercise a major effect on the engines for large aircraft. The requirements of the airports at which it is intended to be used may be more critical than the internationally agreed limits. The noise is reduced if the jet velocity is low and the fan large and relatively slow; this leads to aircraft which are not optimised for fuel burn or range and the ultimate limit on the size of the fan may be for logistic reasons not associated with performance.

It is still common to use the bypass ratio as the way to characterise engines. Certainly this gives a good idea of what the engine will look like, since the size of the core and the size of the fan are roughly proportional to the mass flows through them But in terms of propulsive efficiency or noise generation the bypass ratio is flawed as a descriptor. This is because, as the technology has advanced, the pressure ratio and turbine entry temperature of new engines have both increased, and as a result the core is producing more power for the same mass flow through it. This means, were the bypass ratio to be held constant while the technology is improved, the jet velocity of bypass stream would rise; alternatively, if the jet velocity were held constant the bypass ratio would be increased.

A better descriptor of engine performance (for propulsive efficiency and noise) is produced by the specific thrust. This is the net thrust per unit mass flow through the engine, which is equal to the difference between the average jet velocity and the velocity of flight. In consistent SI units the specific thrust is expressed in m/s.

The engine can now be regarded as defined in terms of its size, pressure ratio, bypass ratio and specific thrust. We can obtain considerable insight into the way engines behave by using dimensional analysis, the subject of the next chapter.

CHAPTER 8

DYNAMIC SCALING
AND DIMENSIONAL ANALYSIS

8.0 INTRODUCTION

It would be possible to calculate the performance of an engine in the manner of Exercise 7.1 and 7.2 for every conceivable operating condition, e.g. for each altitude, forward speed, rotational speed of the components. This is not an attractive way of considering variations and it does not bring out the trends as clearly as it might. An alternative is to predict the variations by using the appropriate dynamic scaling – apart from its usefulness in the context of engine prediction, the application of dimensional analysis is illuminating. The creation of groups which are actually non-dimensional is less important than obtaining groups with the correct quantities in them. The reasoning behind these ideas is discussed in Chapter 1 of Cumpsty (1989). For compatibility with the usual terminology the phrase 'dimensional analysis' will be retained here.

Using the ideas developed on the basis of dynamic scaling it is possible to estimate the engine performance at different altitudes and flight Mach numbers when the engine is operating at the same non-dimensional condition. From this it is possible to assess the consequences of losing thrust from an engine and the provision that needs to be made to cope with this at either take off or cruise.

8.1 ENGINE VARIABLES AND DEPENDENCE

Figure 8.1 shows a schematic engine installed under a wing. The only effects of the pylon are assumed here to be the transmission of a force between the engine and the wing and the passage of fuel to the engine. The primary control to the engine is the variation in the fuel flow. Until relatively recently the pilot controlled the fuel flow and based on this and the inlet air temperature and pressure any bleeds or variable vane settings were determined by a mechanical controller in the engine. Now the whole process can be controlled by an electronic controller which determines the fuel flow and the settings of any variables. The electronic controller does not alter the fundamental dependence of, for example, rotational speeds of the engine on the amount of fuel flow, the atmospheric conditions and the speed of flight. We can therefore say that any variable in the engine is expressed as a function of $\dot{m}_f$, p_a, T_a, and V.

Because the engine is self contained in this way it follows that fixing one set of engine non-dimensional parameters fixes all the others. For example, specifying the ratio of *stagnation* temperature at turbine inlet to *stagnation* temperature at compressor inlet T_{04}/T_{02} fixes the pressure ratios, the non-dimensional fuel flow and the non-dimensional rotational speed (or speeds). Likewise in a multi-shaft engine the *ratio* between the shaft speeds is also fixed by the temperature ratio. The engine condition could equally well be specified by one of the pressure ratios (for example, p_{03}/p_{02}) or a non-dimensional rotational speed.

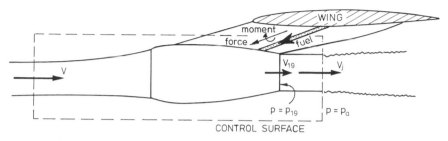

Figure 8.1. Engine mounted on wing pylon showing control surface for calculation of thrust. (Engine shown has mixed core and bypass streams through the final nozzle.)

We can simplify things further without significant loss of usefulness. For most operating conditions of an engine the final propelling nozzles for the core flow and the bypass flow will be choked. It will be recalled that when this occurs the flow upstream of the throat is unaware of the conditions downstream of the throat. This means that the engine is unaware of the static pressure at outlet from the nozzle which, to a good approximation, is the static pressure of the atmosphere. The engine is very much aware of the air pressure which is entering, and it is the *stagnation* pressure p_{02} and temperature T_{02} of the air entering the engine which determine conditions inside the engine. Put another way the engine responds to the inlet stagnation conditions but is unaware of the forward speed. (In this chapter when inlet stagnation conditions are calculated from the ambient static conditions and the forward speed it is assumed that any loss in stagnation pressure in the inlet is independent of flight speed.)

8.2 NON-DIMENSIONAL VARIABLES OF THE ENGINE

To illustrate non-dimensional scaling, consider the mass flow of air through the engine. This can be written as a function of the rotational speed N of one of the shafts

$$\dot{m}_{air} = f(N, \ p_{02}, \ T_{02}) \tag{8.1a}$$

or in terms of the turbine inlet temperature

$$\dot{m}_{air} = f(T_{04}, \ p_{02}, \ T_{02}) \tag{8.1b}$$

or in terms of the fuel flow rate

$$\dot{m}_{air} = f(\dot{m}_f, \ p_{02}, \ T_{02}). \tag{8.1c}$$

Which particular combination of variables is used depends on what is useful – in other words what information is being sought and what is available. The natural non-dimensional variable for the mass flow rate is $\bar{m}$ which was defined in section 6.4,

$$\bar{m} \;=\; \frac{\dot{m}_{air}\,\sqrt{c_p T_{02}}}{D^2 p_{02}}$$

where D is a characteristic diameter of the engine, typically the diameter of the inlet of the fan, and D^2 denotes a representative area. Case (8.1a) can then be written in non-dimensional form as

$$\frac{\dot{m}_{air}\,\sqrt{c_p T_{02}}}{D^2 p_{02}} = \mathrm{F}\{\;\frac{ND}{\sqrt{\gamma R T_{02}}}\;\} \tag{8.1a$'$}$$

where $ND/\sqrt{\gamma R T_{02}}$ is proportional to the rotor tip speed divided by a speed of sound based on the inlet stagnation temperature. Case (8.1b) likewise leads to the non-dimensional form

$$\frac{\dot{m}_{air}\,\sqrt{c_p T_{02}}}{D^2 p_{02}} = \mathrm{F}\{\;\frac{T_{04}}{T_{02}}\;\}. \tag{8.1b$'$}$$

Form (8.1c), in terms of the fuel flow rate, requires a little more consideration. The fuel flow is more properly to be interpreted as a flow of energy, equal to $\dot{m}_f\,LCV$, where LCV is the lower calorific value. A flow of energy is equivalent to rate of doing work, or force $\times$ velocity. The 'natural' force is the inlet stagnation pressure times inlet area $p_{02}D^2$ and the 'natural' velocity to use is the speed of sound based on inlet conditions $\sqrt{\gamma R T_{02}}$, (which is proportional to $\sqrt{c_p T_{02}}$, the normal expression used). The non-dimensional fuel flow is therefore

$$\frac{\dot{m}_f\,LCV}{\sqrt{c_p T_{02}}\,D^2 p_{02}}$$

so that equation (8.1c) becomes

$$\frac{\dot{m}_{air}\,\sqrt{c_p T_{02}}}{D^2 p_{02}} \quad=\quad \mathrm{F}\{\frac{\dot{m}_f\,LCV}{\sqrt{c_p T_{02}}\,D^2 p_{02}}\}. \tag{8.1c$'$}$$

It is, of course, equally true that there are functional relationships between all the non-dimensional variables; for example for the turbine inlet temperature in terms of fuel flow,

$$\frac{T_{04}}{T_{02}} \quad=\quad \mathrm{F}\{\frac{\dot{m}_f\,LCV}{\sqrt{c_p T_{02}}\,D^2 p_{02}}\}$$

or the rotation speed of the shaft in terms of fuel flow,

$$\frac{ND}{\sqrt{\gamma R T_{02}}} \quad = \quad \mathrm{F}\{\frac{\dot{m}_f \, LCV}{\sqrt{c_p T_{02}} \, D^2 p_{02}}\} \, .$$

8.3 NON-DIMENSIONAL TREATMENT OF THRUST

One of the principal variables we would like to scale between different conditions is the thrust of the engine. We need to distinguish between the *gross thrust* F_G and the *net thrust* F_N. The net thrust, it will be recalled from Chapter 3, is related to the gross thrust by

$$F_N = F_G - \dot{m}_{air} V$$

where V is the flight velocity and $\dot{m}_{air} V$ is often referred to as the *ram drag* or *inlet-momentum drag*. The net thrust is not a function of the engine operating condition alone because the flight speed is involved. It is therefore more convenient to work with the gross thrust in obtaining non-dimensional groups and convert this to net thrust when needed.

Retaining the assumptions that the mass flow of fuel is very small compared with the mass flow of air, the gross thrust is defined by

$$F_G = \dot{m}_{air} V_j \, .$$

With a convergent nozzle (i.e. not a convergent–divergent nozzle) some of the expansion takes place downstream of the nozzle throat, which is the exit plane of the nozzle; with only a small error it is nevertheless possible to calculate the jet velocity by assuming an isentropic expansion through the nozzle down to the atmospheric pressure, as was done in earlier chapters. The engine is unaffected by the atmospheric pressure p_a because the nozzle is choked, but the engine is affected by the inlet stagnation pressure p_{02} The dynamic scaling of the engine gross thrust is therefore done in terms of p_{02} but the dependence of V_j on p_a complicates this.

It is possible to apply conservation of momentum flux out of the nozzle and the pressure force acting on the nozzle, as in Fig. 8.1, to obtain

$$F_G = \dot{m}_{air} V_j = \dot{m}_{air} V_{19} + (p_{19} - p_a) A_N \qquad (8.2)$$

where $(p_{19} - p_a)$ is the pressure difference between the nozzle exit plane and the surrounding static pressure of the atmosphere, V_{19} is the velocity at the nozzle exit plane (less than V_j which is produced when the jet has expanded down to atmospheric pressure) and A_N is the nozzle area. For simplicity we neglect the common occurrence in bypass engines where there is a separate nozzle for the core and bypass. Also, because the mass flow through the bypass is so much larger than that of the core, the velocity, pressure and area used at nozzle outlet are those for the bypass with subscript 19. The pressure at the nozzle exit plane, p_{19}, is the static pressure corresponding to sonic flow of the bypass flow; it will be greater than the ambient pressure, so $p_{19} - p_a$ is positive. The area in the above equation is the nozzle outlet area because, as the jet expands in area downstream of the nozzle, the same ambient pressure will act over the sides of the jet as over the cross-section where the flow has expanded to atmospheric pressure.

Rearranging the expression for gross thrust gives

$$F_G + p_a A_N = \dot{m}_{air} V_{19} + p_{19} A_N .$$ (8.3)

The significance of using gross thrust plus $p_a A_N$ now emerges, for all the terms on the right hand side, $\dot{m}_{air}$, V_{19}, and p_{19}, are all wholly determined by the conditions *inside* the engine. In other words once the operating point of the engine is fixed they depend only on inlet stagnation pressure and temperature. Thus, in non-dimensional terms the combination of engine conditions is

$$\frac{\dot{m}_{air} V_{19} + p_{19} A_N}{D^2 p_{02}} = F\{\frac{ND}{\sqrt{\gamma R T_{02}}}\} \text{ or } = F\{\frac{T_{04}}{T_{02}}\} \text{ or } = F\{\frac{\dot{m}_f LCV}{\sqrt{c_p T_{02}} D^2 p_{02}}\}.$$ (8.4)

Equivalently, in terms of the gross thrust, the non-dimensional relationship is therefore

$$\frac{F_G + p_a A_N}{D^2 p_{02}} = F\{\frac{ND}{\sqrt{\gamma R T_{02}}}\} \text{ or } = F\{\frac{T_{04}}{T_{02}}\} \text{ or } = F\{\frac{\dot{m}_f LCV}{\sqrt{c_p T_{02}} D^2 p_{02}}\}.$$ (8.5)

It is clear that in the above expressions the term D^2 denotes merely a characteristic area, which can equally well be replaced by the nozzle area A_N.

The variation of gross thrust at fixed engine condition and with variation in inlet stagnation conditions may then be obtained from equation 8.5. Likewise the variation in mass flow rate of air with these same variables may be found. Knowing the gross thrust, the mass flow of air and the flight speed, the net thrust follows immediately. This incidentally exemplifies a feature of dimensional analysis which is normally ignored: making the non-dimensional groups is comparatively easy, the problem is deciding what are the relevant variables, which in the present example is $F_G + p_a A_N$ and not thrust alone.

To illustrate some of the functional dependence, Fig. 8.2 shows results from a high bypass engine already in service plotted in two different ways. In each case specific fuel consumption (sfc) is shown versus thrust, and curves are drawn for constant flight Mach number. The remaining curves are for constant turbine inlet temperature in one case, for constant speed of the HP shaft in the other. It can be seen that the lines of constant turbine temperature and constant speed are exactly parallel to one another, confirming that once one variable is chosen the other variables of the engine are also fixed. It may be noted from Fig. 8.2 that the sfc rises with flight Mach number for constant engine speed and T_{04}. This is because the ram drag is increased and the net thrust reduced as Mach number increases. It may also be noted that for the same Mach number the sfc is slightly higher at maximum thrust than at somewhat lower thrusts. The explanation for this is that most of the time during cruise the engine is not called upon to produce its maximum allowable thrust for the altitude, and as a result it is sensible to set the component maximum efficiency to occur at speeds and pressure ratios for typical cruise, which are lower than the maxima.

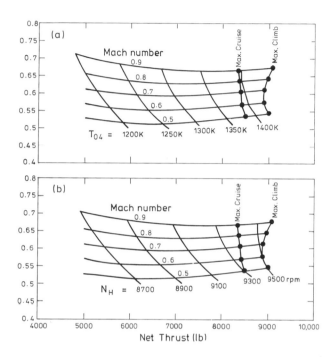

Figure 8.2. Performance maps for an existing high bypass ratio engine, showing sfc versus thrust for constant altitude flight at various flight Mach numbers. Case (a) shows the variation in terms of turbine inlet temperature, case (b) shows the same range in terms of HP rotational speed.

8.4 PRACTICAL SCALING PARAMETERS

Engineers tend to resent writing more than is needed. In the non-dimensional group for mass flow of air it is clear that c_p is constant and D^2 will be constant for a given engine. Therefore the group

$$\frac{\dot{m}_{air}\sqrt{T_{02}}}{p_{02}}$$

is often used to describe mass flow, even though it has units. Similarly for the fuel flow the calorific value will normally be constant and an abbreviated form of the group is used,

$$\frac{\dot{m}_f}{\sqrt{T_{02}} \; p_{02}} \; ,$$

again with dimensions. Notice, however, that the term involving the inlet stagnation temperature is in the numerator for the air flow but in the denominator for the fuel flow. This difference only emerges from a physical understanding of the processes involved *prior* to setting up the non-dimensional groups – recognising that the contribution of the fuel flow is really an energy input. A blind process following rules for dimensional analysis would treat the mass flow rate of fuel and the mass flow rate of air as equivalent and would not produce this difference.

The same practice of omitting constant terms is used for other variables so, for example, the non-dimensional speed is abbreviated to $N/\sqrt{T_{02}}$. This leads to the corrected speed

$$N/\sqrt{\theta},$$

where $\theta = T_{02}/T_{02ref}$, T_{02ref}, being a reference inlet temperature, typically 288.15 K. The magnitude of θ is usually fairly close to unity so corrected speed retains the advantage of being close to the actual speed, and therefore the intuitive sense of magnitude is retained. The same connection with the actual magnitude can be claimed for the corrected mass flow (units kg/s) derived from the non-dimensional expression,

$$\frac{\dot{m}_{air} \; \sqrt{\theta}}{\delta} \; ,$$

where $\theta = T_{02}/T_{02ref}$ and $\delta = p_{02}/p_{02ref}$.

Exercises

8.1   a) Verify that for constant flight Mach number the ratio of inlet stagnation temperature to ambient static temperature, T_{02}/T_a , and inlet stagnation pressure to ambient static pressure, p_{02}/p_a, are constant. Use the constancy of p_{02}/p_a to show that the gross thrust will scale as $F_G/D^2 p_{02}$ for a constant engine non-dimensional operating point and Mach number, and thence that $F_G/D^2 p_a$ is constant too. (In other words gross thrust is proportional to ambient pressure at constant Mach number and engine condition.)

The net thrust, F_N, is equal to the gross thrust minus the ram drag equal to $\dot{m}_{air}V$. Show that for a constant flight Mach number and engine condition $\dot{m}_{air}\sqrt{c_p T_a} /D^2 p_a =$ constant and hence that $\dot{m}_{air} V /D^2 p_a =$ constant. Use this to show that for a constant flight Mach number and engine non-dimensional operating point the net thrust, F_N, is also proportional to ambient pressure.

b) In Exercise 7.2b the *net* thrust required at the start of cruise at an altitude of 31000 ft and a flight Mach number of 0.85 allowed the mass flow of air at this condition, and thence the gross thrust, to be found. For the same flight Mach number calculate the mass flow of air, and the gross and net thrust from each engine at an altitude of 41000 feet (p_a = 17.9 kPa, T_a = 216.7 K) with the engine at the same non-dimensional operating condition.

(**Ans:** for bpr = 6; $\dot{m}_{air}$ = 328 kg/s, F_G = 129 kN, F_N = 46.8 kN)

If the lift–drag ratio of the aircraft is unaltered at 20, what is the maximum aircraft mass which can be propelled at M = 0.85 and 41000 ft with all four engines operating at the same non-dimensional conditions? (**Ans**: maximum mass = 381 tonne)

8.2* **a)** For the engine of Exercise 7.1 and 7.2, with a bypass ratio of 6, the stagnation temperature and pressure in the bypass duct downstream of the fan may be shown to be given by T_{013} = 312.8 K and p_{013} = 83.2 kPa at a cruise altitude of 31000 ft and M = 0.85. The mass flow rate through the bypass is 6/7 times the whole mass flow, found in Exercise 7.2b to be 514 kg/s. The bypass nozzle is convergent (no divergent section) so the Mach number is unity at the exit plane. (For an area with sonic velocity $\dot{m}\sqrt{c_p T_0}/A p_0$ = 1.281.) Find the area A_{Nb} of the bypass nozzle and the static temperature, static pressure and velocity at the bypass nozzle exit plane.

(**Ans:** A_{Nb} = 2.31m^2; T_{19} = 261 K; p_{19} = 43.9 kPa; V_{19} = 323.6 m/s)

b) From the curves shown in Fig.7.5 it is clear that an engine with bpr = 6 would realistically be designed to have the bypass jet velocity about 0.78 times the core jet velocity and the resulting parameters are summarised in the table in Chapter 7. For this combination of jet velocities the sum of the bypass and core nozzle areas A_N is 3.14 m^2. Use this value of A_N to evaluate the expression $(F_G + p_a A_N)/A_N p_{02}$ at cruise (note that the characteristic area D^2 in the equations in the text has been replaced by A_N). Recall that at this altitude and Mach number p_a = 28.7 kPa and p_{02} = 46.0 kPa.

(**Ans:** 2.06)

For the same engine non-dimensional working point (for example $N/\sqrt{T_{02}}$ or T_{04}/T_{02} constant) the group $(F_G + p_a A_N)/A_N p_{02}$ will be unaltered. Hence find the gross thrust for a sea-level static test when p_{02} = p_a = 101 kPa, T_{02} = T_a = 288 K. (**Ans:** $F_G = F_N$ = 336 kN)

c) Find the turbine inlet temperature and the total engine mass flow for the sea-level test.

(**Ans:** T_{04} = 1610 K; $\dot{m}_{air}$ = 1070 kg/s)

8.3 The sfc has been obtained at start of cruise in Exercise 7.1. What would the sfc be during a sea-level static test of the engine with bpr = 6 at the same non-dimensional operating point?
(Note that sfc is based on net thrust) (**Ans**: sfc = 0.269 kg/h/kg)

Note: Results in Exercises 8.2 and 8.3 show what would be obtained in a sea-level test bed with the engine operating at its non-dimensional conditions for cruise. Notice that the turbine operates at a substantially higher inlet temperature than it would be at cruise (1610 K versus 1450K) though its non-dimensional operating point is the same. During take off it is possible to allow the engine to operate at conditions which are sustainable only for short periods, and a turbine inlet temperature of 1700K is currently realistic. It should be noted that the thrust obtainable on the test bed is *very* much greater than that available at cruise, the difference increasing with bypass ratio.

It might be noticed from the table in Chapter 7 that the pressure ratio for the bypass stream across the fan is only 1.70 when the jet velocity of core and bypass are in the ratio 0.78. This means that on a static test bed the bypass nozzle would not be choked at this non-dimensional condition, since this requires a pressure ratio of at least 1.89. The fan would therefore see a different operating point (it would actually 'think' that the nozzle had become smaller) and the analysis is therefore not exactly correct. By the time the aircraft forward speed reaches a Mach number of about 0.4 the bypass nozzle chokes.

8.5 LOSS OF THRUST FROM AN ENGINE

Any aircraft in civil operation has to be able to fly safely even if one of the engines fails completely, the engine-out condition. A particularly critical requirement is that it must be able to complete a take off, climb and then land safely even if the engine fails at the worst possible moment, just as the plane is about to leave the ground. It must also be able to fly above mountains if an engine is lost at cruise altitude, taken here to be 31000 feet, and have adequate range to fly to an alternative airfield. To meet these contingencies the engine can operate at higher thrusts than those corresponding to the non-dimensional condition for the design at cruise; in practice this normally means allowing a higher turbine inlet temperature. In a complete analysis of engine-out operation the optimum condition for the remaining engines would be determined. Nevertheless we can learn a lot by examining engine-out behaviour with the remaining engines maintained at the same non-dimensional condition as prior to the loss of an engine. Because for high bypass engines the non-dimensional operating point does not alter greatly between cruise, climb and take off, as the temperature ratios in section 5.3 clearly show, the engine thrust at cruise can be indicative of thrust at take off with the appropriate scaling.

Loss of thrust at take off

If an engine were to fail just as the aircraft left the ground it would be crucial for the remaining engines to produce enough thrust for the plane to continue to climb and accelerate. The flight speed at lift-off may be assumed to be about 90 m/s, and at that condition the lift/drag ratio would be about 10. The minimum acceptable rate of climb is about 3% (i.e $\sin \theta = 0.03$), so the thrust must be at least $(1/10 + 0.03) \times$ weight. The maximum mass at take off is about 636 tonnes, so the total minimum thrust required is $0.13 \times 636 \ 10^3 \times 9.81 = 811$ kN. If the aircraft has four engines and one fails, the thrust required from each of the remaining three engines is therefore 270 kN.

The forward speed of the aircraft, even at 90 m/s, affects the stagnation conditions into the engine, and using the standard sea-level atmospheric values it can be shown that $T_{02} = 292$K and $p_{02} = 106$ kPa. For an engine with bypass ratio 6, capable of giving a net thrust of 75.1kN at cruise, it was found in Exercise 8.2b that the thrust on a sea-level test bed was 336 kN for the same non-dimensional condition of the engine. The mass flow rate on the test bed was found to be 1070 kg/s and, taking $\dot{m}_{air} \sqrt{T_{02}} / p_{02}$ to be constant (see section 8.2) it can be shown that at the moment of take off the mass flow in each engine is 1115 kg/s. The momentum of the incoming flow (the ram drag) is given by $\dot{m}_{air} V = 1115 \times 90 = 100.3$ kN.

The gross thrust at the condition when $V = 90$ m/s can be found by equating the expression $(F_G + p_a A_N)/A_N p_{02}$ (see section 8.3) with the stagnation pressure increased to 106 kPa, but the static pressure 101 kPa. The result is that the gross thrust for 90 m/s forward speed is 369 kN. The net thrust is then obtained by subtracting the ram drag, so from each engine at the cruise non-dimensional operating point $F_N = 268$ kN. This is virtually equal to the minimum value shown above to be necessary in the case of a total loss in thrust from one of the four engines at take off. At this condition the turbine inlet temperature would be 1632 K, well below the value assumed allowable for take off, 1700 K. With a four-engine plane there is evidently no need to have additional engine thrust to cope with the loss of an engine at take off.

The same argument can be applied to two- and three-engine aircraft. A plane which was sized to cruise at 31000 feet, $M = 0.85$ with three engines of the same size would be smaller than that forming the basis of the design here in Part 1, having approximately three quarters the mass. For this smaller aircraft the thrust required at take off would be about 608 kN, based on the same assumptions as those used for the four-engine plane. If a three-engine plane lost an engine it would require each of the two remaining engines to give a thrust of 304 kN, some 10 per cent more than the engine at the cruise non-dimensional condition. With the temperature into the turbine raised to the value for take off, 1700 K, the necessary thrust should be available; again in the case of the three-engine aircraft the loss of an engine at take off can be allowed for without including substantial additional thrust. On the same basis, for a two-engine aircraft the thrust required from the remaining engine, if one were lost at take off, would be 406 kN. This is significantly more thrust than that available at the cruise non-dimensional condition, 268 kN, and some additional engine thrust capacity (bigger or more powerful engines) would be necessary. In the case of a two-engine aircraft the aircraft designer normally includes a somewhat greater wing area. The larger wing area lowers the take-off speed, but more significantly it causes the two-engine plane to achieve maximum ML/D at a higher altitude. As shown above the lower density air at higher altitude reduces the net thrust from the engine, so the engines can be operating near their optimum condition. In other words, setting a higher initial cruise altitude for two-engine aircraft allows the same engine to operate near its design point at cruise and produce the necessary thrust at take off to cope with loss of an engine.

One of the more remarkable features of the high bypass engine is the very great increase in net thrust at low flight speeds. This means that for three- and four-engine aircraft no major compromise is needed to cope with the loss of an engine at take off and the total installed thrust can be fixed by the climb requirement (particularly top of climb, near to the cruise altitude). This can be fairly close to the condition for cruise, at which lowest fuel consumption is paramount. With the two-engine aircraft some adaptation is needed, resulting in higher cruise altitudes; without the increase in thrust at low forward speed which comes from the high bypass ratios, however, two-engine aircraft would be too severely compromised to make long range operation economical.

Loss of thrust during cruise

As discussed in Chapter 2, the aircraft will normally cruise at an altitude such that the lift /drag ratio is near its maximum. At this condition the engines should be sized such that they produce just enough thrust to equal the drag. If thrust is lost from one engine there is not normally enough additional thrust available in the remaining engines, by increasing the flow of fuel to raise the turbine inlet temperature, to compensate for the loss of an engine without bringing about a severe shortening of engine life. The normal procedure is for the aircraft to descend to a lower altitude where the higher density of the air allows the remaining engines to increase their thrust. To retain the maximum lift/drag ratio (more correctly VL/D) for the aircraft at the lower altitude the lift coefficient must be maintained close to its value at cruise, see Fig.2.2; for the present we will assume that the lift coefficient is held constant, so the flight speed is inversely proportional to the square root of atmospheric density. If the lift coefficient is constant it is a reasonable assumption that the lift/drag ratio is constant too. Figure 2.2 shows some increase in L/D as the Mach number is reduced, but this would be offset by the drag of the 'dead' engine and drag due to the rudder force needed to balance the moment of the asymmetric engine thrust on the aircraft. If L/D does remain constant the total thrust required for steady flight from the remaining engines is exactly equal to that at the cruise condition prior to the engine failure since aircraft weight is the same: this is the assumption adopted here.

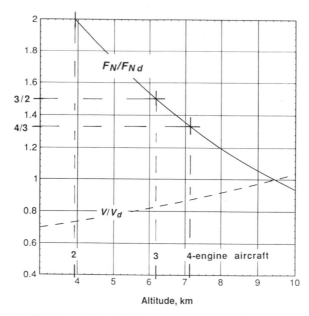

Figure 8.3. Predicted variation of flight speed and net thrust as function of altitude for engine at constant non-dimensional condition. Aircraft cruising at 31000 ft (9.45 km) altitude prior to shut down of one engine.
Bypass ratio = 6.0, conditions as in Exercise 7.1. Aircraft lift/drag ratio = 20.

We will also assume here that the engine non-dimensional operating condition remains at the same condition as it was during cruise, prior to the loss of thrust in one engine; this is not a necessary condition, but it allows the use of the non-dimensional scaling rules and the predictions are not far off the mark.

Figure 8.3 shows two curves and is based on cruise at 31000 feet and M = 0.85 prior to the loss of thrust in an engine; the flight speed and net thrust at the cruise condition are denoted by V_d and F_{Nd} respectively. The curve V/V_d shows the locus of the aircraft maintaining constant lift coefficient. The curve for the ratio of net thrust F_N/F_{Nd} is the result of using the results of section 8.3, holding the same non-dimensional condition for the engine, when the forward speed, the inlet stagnation temperature and pressure and the atmospheric pressure (acting on the exit of the nozzle) alter as a result of the reduction in altitude and flight speed. For a four-engine aircraft which loses thrust in one engine, the required value of F_N/F_{Nd} for steady flight is equal to 4/3, and this value is marked on the ordinate. The aircraft which was at 31000 ft (9.45 km) altitude therefore descends to 23400 ft (7.1 km). This altitude would allow the aircraft to fly over most mountain ranges. Using the results in section 8.3 it can also be shown that the range of the aircraft is reduced by only 7 per cent in this condition.

In Fig. 8.3 the conditions for loss of an engine in two- and three-engine aircraft are also shown, corresponding to F_N/F_{Nd} equal 3/2 and 2 respectively. A two-engine aircraft cruising at 31000 feet would, after losing power in one engine but keeping the other engine at the same non-dimensional condition, fly steadily at only 12800 ft (3.9 km) which is below many mountain ranges; some care is necessary in planning routes! Even for this condition the loss in range for a two-engine aircraft is only about 17 per cent.

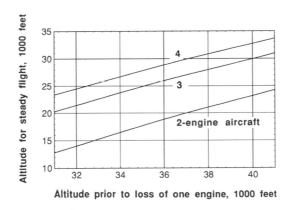

Figure 8.4. Steady altitude for flight when thrust from one engine is lost.
Non-dimensional conditions unchanged for remaining engines.
Lift/drag ratio constant, bypass ratio = 6.0 and conditions as in Exercise 7.1.

The variation in altitude for steady flight after losing thrust from an engine is shown in Fig. 8.4 for two-, three- and four-engine aircraft as a function of the cruise altitude prior to the loss of thrust. It is assumed that the Mach number at cruise, prior to loss of an engine, remains equal to 0.85; the altitude is increased to maintain maximum *L/D* as fuel is burnt off and the weight of the aircraft is reduced. It is clear from Fig. 8.4 that a four-engine aircraft can accommodate the loss of an engine at cruise with a reduction in altitude of only about 7000 feet. The reduction in range consequent on a loss of an engine is not affected by the altitude of cruise at which the loss occurred.

Although this section has assumed that the engine remains at the non-dimensional condition for which it is designed, that for cruise, this condition is not so far removed from what is probable. For example, in the most extreme case when a two-engine plane loses engine thrust while cruising at 31000 ft the new altitude for steady flight is 12800 feet and at that height the air temperature entering the engine is considerably higher than at 31000 feet. As a result the stagnation temperature entering the turbine is raised at the same engine condition from the cruise value, 1450 K, to about 1526 K, which should be compared with the value of 1575 K allowed during climb, a condition at which prolonged operation without rapid deterioration is expected.

SUMMARY CHAPTER 8

Making non-dimensional groups is much easier than choosing the relevant variables. Most engines operate with choked nozzles and the engine behaviour is then determined entirely by the fuel flow, the inlet stagnation temperature and the inlet stagnation pressure. Temperature ratios and pressure ratios in the engine, and non-dimensional speeds are determined, as well as the mass flow of air. To determine the gross thrust the ambient static pressure must also be included as a variable and to get the net thrust requires in addition the forward speed or Mach number.

The non-dimensional expression for the fuel mass flow is very different from that for the air mass flow, reflecting the different effects of fuel and air in the engine.

At a fixed Mach number the net thrust from a given engine is proportional to the ambient static pressure whereas the specific fuel consumption is inversely proportional to square root of ambient static temperature. Using non-dimensional variables, or dimensioned variables derived from them, it is possible to estimate performance at speeds and altitudes different from those at which engines were designed or tests were performed.

With a high bypass ratio engine the loss of an engine at take off creates a smaller problem than with low bypass engines (now gradually disappearing from airline service). For three- and four-engine aircraft this condition requires little if any compromise in the size of engines. With two engines some additional thrust capacity and larger wings are needed.

With loss of an engine during cruise the conditions for steady flight can be found. Losing thrust from one engine of a four-engine aircraft leads to a reduction in altitude of only about 7000 feet for steady flight and a loss of range of only about 7 per cent.

The engine has now been designed in outline and the next step is to see if a compressor and a turbine can be designed to meet the requirements which were assumed in the cycle analyses.

CHAPTER 9

TURBOMACHINERY:
COMPRESSORS AND TURBINES

9.0 INTRODUCTION

The compressor, which raises the pressure of the air before combustion, and the turbine, which extracts work from the hot pressurised combustion products, are at the very heart of the engine. Up to now we have assumed that it is possible to construct a suitable compressor and turbine without giving any attention to how this might be done. In this chapter an elementary treatment is given with the emphasis being to find the overall diameter of various components and the flow-path this entails, the number of stages of compressor and turbine, and suitable rotational speeds. The details of blade shape will not be addressed. Further information is obtainable at an elementary level in Dixon(1995) and at a more advanced level in Cumpsty(1989).

For the large engine that we are considering the most suitable compressor and turbine will be of the *axial* type. These are machines for which the flow is predominantly in the axial and tangential directions, and stand in contrast to radial machines for which the flow is radial at inlet or outlet.

Because the pressure rises in the direction of flow for the compressor there is always a great risk of the boundary layers separating, and when this happens the performance of the compressor drops precipitously and is said to stall. To obtain a large pressure rise (or, as it is more commonly expressed, pressure ratio) the compression is spread over a large number of *stages*. A stage consists of a row of rotating blades (the *rotor*) and a row of stationary blades (the *stator*). In a modern engine compressor there may be between 10 and 20 stages between the fan outlet and the combustor inlet. Each rotor or stator row will consist of many blades, typically anywhere between 30 and 100.

In the turbine the pressure falls in the flow direction and it is possible to have a much greater pressure ratio across a turbine stage than a compressor stage; quite commonly a single turbine stage can drive six or seven compressor stages on the same shaft. A turbine required to produce a larger drop in pressure than is appropriate will work less well; that is the efficiency will be lower than a turbine with a more moderate pressure drop. Nevertheless a turbine always produces a power output and the flow will be in the intended direction. In the case of a compressor, however, an attempt to get more pressure ratio than is appropriate may result in rotating stall or surge, either of which is quite unacceptable. Because the operation of a turbine is normally easier to understand than a compressor, the compressor will be described after the turbine. First, however, it is appropriate to consider how the blades work.

9.1 THE BLADES FOR AXIAL COMPRESSORS AND TURBINES

We need to distinguish between the directions in which the metal of the blades point, denoted by β here, and the direction in which the gas flows, denoted by α. In both cases the angles are measured from the *axial* direction. Relative to any blade row, compressor or turbine, we define the angles by

gas angle into blade $\quad \alpha_1 \qquad\qquad$ blade metal angle in $\quad \beta_1$

gas angle out of blade $\quad \alpha_2 \qquad\qquad$ blade metal angle out $\quad \beta_2$

The angle between the inlet gas and blade angles is defined as the incidence,

$$i = \alpha_1 - \beta_1,$$

which for the blade row concerned is normally treated as an independent variable. The corresponding angle between the flow of gas leaving the blades and the blade outlet angle is defined as the deviation, given by

$$\delta = \alpha_2 - \beta_2$$

and this is normally treated as a principal dependent variable. The flow does not turn quite as much as the blade, in other words the deviation δ is positive. Depending on the design of the blades and flow variables, principally the incidence and the inlet Mach number, the deviation can change.

Figure 9.1 shows schematic blade shapes and some properties of compressor and turbine blades as a function of incidence. The blades introduce a stagnation pressure loss and this, non-dimensionalised by the flow dynamic pressure, is shown in Fig. 9.1. It is important that the loss is not too great. Figure 9.1 also shows that the deviation is much larger for compressor blades than turbine blades, but in both cases the outlet flow direction (deviation) changes relatively little as the inlet flow direction (incidence) is altered.

It has been mentioned already that the rising pressure in the flow direction for the compressor blade makes the flow more difficult to control and limits the amount of turning and deceleration that can be achieved. It will be seen in Fig. 9.1 that the magnitude of the flow turning is much smaller for the compressor blades than the turbine: at zero incidence the compressor turns the flow by about 20° whereas for the turbine row it turns by about 63°. (More modern turbine blades often turn the flow more than 90°.) Another indication of the problems of compressors is the much narrower range of incidence for which the loss is small. When the loss starts to rise rapidly it is evidence of massive boundary layer separation; in the case of the compressor blades this coincides with a rapid reduction in the amount of flow turning produced.

The blades in Fig. 9.1 show another important aspect of their function. For compressor blades the flow is turned so that it is more nearly axial at outlet than at inlet, whereas for turbine blades it is the other way around, the flow is more nearly tangential at outlet. It is a common design choice to make the axial velocity V_x nearly constant through a blade row (and indeed through most of the compressor or turbine). It is usually possible to neglect the radial velocity in jet engines unless there is a radial (centrifugal) compressor; radial compressors are only used on small engines. In the case of a row of compressor blades the velocity at inlet to the row V_1 is given by $V_1 = V_x/\cos \alpha_1$ whilst at outlet the velocity is given by $V_2 = V_x/\cos \alpha_2$. It follows that $V_1/V_2 = \cos \alpha_2/\cos \alpha_1$ and since for a compressor $\alpha_2 < \alpha_1$ it also follows that $V_2 < V_1$. In other words the flow is decelerated in a compressor blade row. In contrast in the turbine the flow is turned away from the axial direction, $\alpha_2 > \alpha_1$, so the flow is accelerated in a turbine blade row.

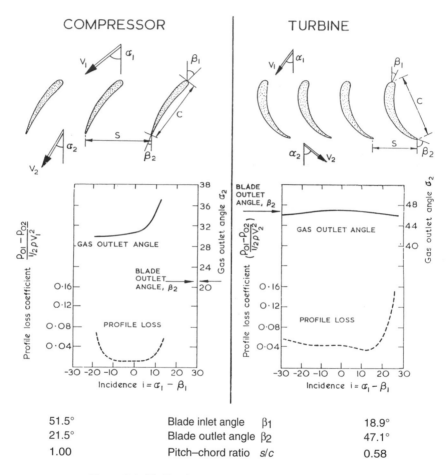

51.5°	Blade inlet angle β_1	18.9°
21.5°	Blade outlet angle β_2	47.1°
1.00	Pitch–chord ratio s/c	0.58

Figure 9.1. Blading for axial compressors and turbines.

The fan is a very special compressor stage. It is invariably the case in modern civil engines that the fan has the rotor row as the first row and a stator row behind (i.e. no upstream stator row). The flow out of the fan divides, with most of it going down the bypass duct to a propelling nozzle and a small fraction going into the core. The fan blades are long in relation to their axial extent; put another way, the ratio of the hub radius to the casing radius is small. A lower limit of about 0.35 can be put on this radius ratio, principally for mechanical reasons. The flow into the fan has a relatively high Mach number, and the mass flow per unit inlet area into the fan is about 88% of that required to choke the annulus – given the required mass flow of the engine this effectively determines the size of the engine. For reasons of efficiency, to prevent too much noise and to reduce the damage consequent on bird strike, the tip speed of the fan must not be allowed to become too high; a relative Mach number onto the tips of the fan of about 1.6 should be regarded as the upper limit. It is then possible to produce a pressure ratio of up to about 1.8 in a single stage fan with an efficiency of about 90%. Because the fan is such a specialised component we will not consider its design further here, but allow a choice of parameters so as not to go beyond those listed above.

Exercise

9.1* The engine thrust, bypass ratio and jet velocity were found in Exercises 7.1 and 7.2 for cruise at M = 0.85 at 31000 ft. Calculate the necessary pressure ratio across the fan. Assume that the core and bypass jet velocities are equal.   (**Ans**: For $bpr = 6$, $p_{013}/p_{02} = 1.81$)

     Using the mass flow calculated in Exercise 7.2 find the fan inlet diameter assuming that the mass flow per unit annulus area is 0.88 of that needed to choke the empty annulus and that the hub radius is equal to 0.35 of the fan tip radius.   (**Ans**: 2.71 m for $bpr = 6$)

9.2 THE EULER WORK EQUATION

For both compressors and turbines the work exchange is described by the Euler equation, which we derive here. Figure 9.2 shows a hypothetical rotor of very general shape rotating with an angular velocity Ω. The flow enters at radius r_1 with a velocity in the tangential direction $V_{\theta 1}$ and leaves at radius r_2 with tangential velocity $V_{\theta 2}$.

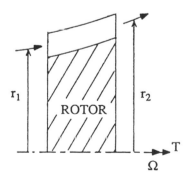

Figure 9.2. A hypothetical rotor for which flow enters at radius r_1 and leaves at radius r_2.
The torque created is T and the rotor rotates at Ω radian/s.

Consider an imaginary packet of fluid of mass $\delta m = \dot{m}\,\delta t$ which enters the rotor. This packet has a moment of momentum about the axis of rotation given by $\delta m\, r_1 V_{\theta 1}$. The corresponding moment of momentum for the same packet leaving at radius r_2 with velocity $V_{\theta 2}$ is $\delta m\, r_2 V_{\theta 2}$. Since the torque is equal to the rate of change of moment of momentum, this may be written in terms of the mass flow rate as

$$T = \dot{m}(r_2 V_{\theta 2} - r_1 V_{\theta 1}) \ . \tag{9.1}$$

The power is then given by

$$\dot{W} = T\,\Omega = \dot{m}\ \Omega(r_2 V_{\theta 2} - r_1 V_{\theta 1})$$

$$= \dot{m}(U_2 V_{\theta 2} - U_1 V_{\theta 1}) \tag{9.2}$$

where U_1 and U_2 are the speed of the blade row at inlet and outlet.

The power is also equal to mass flow rate times the change in stagnation enthalpy per unit mass, $\dot{W} = \dot{m}\ \Delta h_0$, and so by using equation (9.2) we get

$$\Delta h_0 = U_2 V_{\theta 2} - U_1 V_{\theta 1} \tag{9.3}$$

which is referred to as the Euler equation. In the case of a turbine rotor $V_{\theta 1} \gg V_{\theta 2}$ and the stagnation enthalpy falls, so the flow does work on the turbine. For a compressor $V_{\theta 1} \ll V_{\theta 2}$ and the stagnation enthalpy of the air rises so work is done by the compressor on the fluid. To assist the work exchange the designer often tries to maintain $r_2 \not< r_1$ for a compressor and $r_2 \not> r_1$ for a turbine; very often an adequate approximation is to take $r_2 = r_1$ and this is adopted here. With this restriction one can write

$$\Delta h_0 = U(V_{\theta 2} - V_{\theta 1}), \tag{9.4}$$

which leads to the natural non-dimensional form for the work coefficient

$$\Delta h_0/U^2 = V_{\theta 2}/U - V_{\theta 1}/U. \tag{9.5}$$

For the purpose of this very simplified treatment we will only carry out calculations at the mean radius (half way between hub and casing) and we will assume that this radius is constant across each compressor stage and across each stage of the LP turbine, though mean radius is not necessarily the same for the compressor and turbine.

9.3 FLOW COEFFICIENT AND WORK COEFFICIENT

As the air is compressed the density increases and to maintain the axial velocity at an acceptably high value it is necessary to reduce the area – this is very clear in the cross-sectional drawings of engines which show the radial length of the compressor blades decreasing progressively from front to back. Similarly in the turbine it is necessary to increase the area of the annulus as the gas expands and falls in density. There is no reason why the axial velocity should be constant, but this is often not far from the truth and it simplifies the present calculations. It has been found that compressors and turbines work most satisfactorily if the non-dimensional axial velocity, often called the flow coefficient V_x/U, is in a restricted range. For compressors the choice is normally $V_x/U \approx 0.4 - 0.7$, based on blade speed U at the mean radius. For turbines in the core $V_x/U \approx 0.5 - 0.65$ whilst for LP turbines $V_x/U \approx 0.9 - 1.0$.

Experience has allowed designers to choose combinations of work coefficient $\Delta h_0/U^2$ and flow coefficient Vx/U to give satisfactory performance. Figure 9.3 shows a plot for turbines

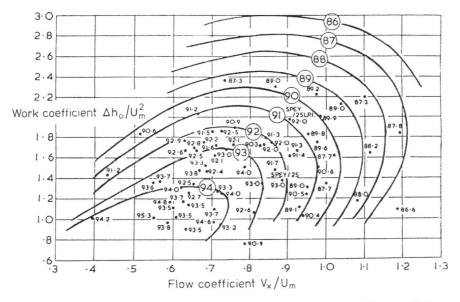

Figure 9.3 Variation of measured stage efficiency with stage loading and flow coefficient for axial-flow turbines (after Smith, 1965).

with the work coefficient and flow coefficient as axes and contours of efficiency superimposed. Very commonly the designer is unable to put the turbine working point at just the combination of V_x/U and $\Delta h_0/U^2$ which would give maximum efficiency for practical and/or geometric reasons. In the case of compressors there is no diagram equivalent to Fig. 9.3, but most stages are designed so that $\Delta h_0/U^2$ is in the range 0.35 to 0.5.

The temperature of the air rises in the compressor and so for the same velocity the Mach number decreases; stages near the front typically have a supersonic inlet flow whereas the stages near the rear are fully subsonic even though the blade speed and flow velocity may be nearly equal at front and rear.

Exercises

9.2* At cruise conditions assume that the fan gives a pressure ratio of 1.6 to the flow entering the core and that its efficiency is 90%. Find the stagnation temperature T_{023} of the air entering the core compressor. Using results from Exercise 7.2 find the cross-sectional area at entry to the core compressor if the mass flow is to be 85% of that necessary to choke the flow (this value, corresponding to a Mach number of about 0.6, has been found to be near the practical upper limit). Supposing that the ratio of the hub radius to casing radius at inlet to the core compressor is 0.70, find the casing and mean radii at inlet.

(**Ans:** $T_{023} = 300.9K$, area = 0.504 m^2, $r_t = 0.561$ m, $r_m = 0.477$ m)

The pressure ratio of the core compressor is 25, giving an overall pressure ratio of 40. Assume that the density at core compressor outlet is equal to 8.5 times that at inlet and that the axial velocity is uniform through the compressor. Find the cross-sectional area at outlet from the compressor. Then, assuming that the mean radius r_m is equal at inlet and outlet of the core compressor and that the outlet area may be calculated using $2\pi r_m h$, where h is the blade height, find the blade height at outlet from the compressor.

(**Ans:** $h = 19.7$ mm)

9.3* The relative Mach number at the tip of the first rotor of the core compressor is to be 1.10. Assume purely axial flow into the core with $V_x/U_t = 0.50$ at the tip, show that the resultant velocity into the tip is $U_t (1+0.5^2)^{1/2} = 1.118U_t$, where U_t is the blade speed at the rotor tip. Taking the **static** temperature T_{23} into the core compressor to be 287 K, find the blade tip speed and the rotational speed of the core shaft.

(**Ans**: $U_t = 334$ m/s, $\Omega = 94.8$ rev/s)

9.4* The enthalpy rise in each compressor stage is to be equal and is not to exceed $0.42U_m^2$, based on mean blade speed U_m, and the efficiency of the whole compressor is estimated to be 90%. The stagnation temperature entering the core compressor is derived in Exercise 9.2. Find the number of stages in the core compressor – with this number of stages, what is the average value of $\Delta h_0/U_m^2$ per stage?

(**Ans**: 15 stages, $\Delta h_0/U_m^2 = 0.418$)

Note: If the efficiencies of the first and the last stage were equal, the pressure ratios from each of these stages would not be equal even though the temperature rises were equal. This is because the pressure ratio across a stage is related to the temperature ratio; the same temperature rise gives a substantially smaller temperature ratio across the rear stages because the inlet temperature to these stages is much higher.

9.5* Exercise 9.3 has fixed the rotational speed of the core shaft so the mean blade speed for the turbine depends only on the mean radius. Previous experience and the wish to keep cost and weight down determines that there are to be only two core turbine stages. To maintain efficiency the non-dimensional work of the turbine, based on mean blade speed, $\Delta h_0/U_m^2$ is not to exceed 2.0. Find the necessary mean radius of the turbine, recalling that the power from the core turbine must equal the power into the core compressor.

(**Ans**: 0.597 m)

9.4 THE AXIAL TURBINE

It is conceptually easier to think of the turbine than the compressor, so we begin by considering the turbine. Figure 9.4 shows one and a half stages of a turbine: a stator row, downstream of this a rotor row and then a second stator. In practical turbines the rows are close together, the gap is perhaps 20% of the blade chord, and the rotor moves with a velocity in the tangential direction U which is normally not very different from the local speed of sound. In both the rotor and the stator rows the flow is more nearly tangential at outlet than inlet. The flow into the rotor is unsteady and likewise so is that into the stator – it is a flow of great complexity. Fortunately it was discovered a long time ago how to approximate this with sufficient accuracy for engineering purposes, and this method is invariably used in the gas and steam turbine industries.

The approach adopted is to use a frame of reference fixed to the blade row under consideration: for a stator row we use the *stationary* frame of reference and the velocities observed in this frame are conventionally described as *absolute*. For the rotor we use a frame of reference which moves with the rotor at speed U, and in this frame the velocity components are referred to as *relative*. The crucial assumption is that in the frame of reference fixed to a blade row the flow may be treated with sufficient accuracy as steady.

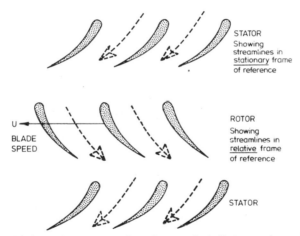

Figure 9.4. A schematic representation of one and a half stages of an axial turbine indicating the *steady relative* streamlines in each row.

The outlet conditions from the stator become the inlet conditions to the rotor and so we need a simple way of going from absolute to relative and back again. The approach always adopted is to use *velocity triangles* and these are shown in Fig. 9.5. We denote *absolute* velocities by V and *relative* velocities by V^{rel}. The flow enters stator row 1 with velocity V_1 inclined at angle α_1 to the *axial* direction. At this station the axial and tangential components are given by

$$V_{x1} = V_1 \cos \alpha_1 \quad \text{and} \quad V_{\theta 1} = V_1 \sin \alpha_1 \quad \text{respectively.}$$

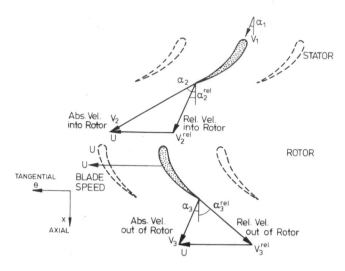

Figure 9.5. An axial turbine stage showing the velocity triangles into and out of the rotor.

At outlet from the stator the absolute velocity is V_2 and this can be resolved into its axial and tangential components just as for V_1. The relative velocity seen by an observer at inlet to the rotor, V_2^{rel}, may be obtained from the vector additions in the triangle drawn. It is, of course, the case that the *axial* velocity is equal in both the absolute and relative frames of reference, so $V_{x2}^{rel} = V_{x2}$, but the tangential velocities are different so that

$$V_{\theta 2}^{rel} = V_{\theta 2} - U.$$ (9.6)

From the components of velocity it is easy to obtain the various angles so, for example,

$$\tan \alpha_2 = V_{\theta 2}/V_{x2} \qquad \text{and} \qquad \tan \alpha_2^{rel} = V_{\theta 2}^{rel}/V_{x2}$$

or $\qquad \cos \alpha_2 = V_{x2}/V_2 \qquad \text{and} \qquad \cos \alpha_2^{rel} = V_{x2}/V_2^{rel}$ (9.7)

and so on for other possible combinations.

Conditions leaving the rotor are handled in the same way so, for example,

$$V_{\theta 3}^{rel} = V_{\theta 3} - U.$$ (9.8)

Inherent in equations is the simplification that the blade speed U is taken to be equal at inlet and outlet to the rotor, in other words the streamlines do not shift radially as they pass through the rotor blades. With this restriction to constant radius it is easy to show that the stagnation enthalpy is equal at inlet to and outlet from the rotor if the *relative* co-ordinate system is used, in other words

$$h_{02}^{rel} = h_{03}^{rel},$$

where $\quad h_{02}^{\text{rel}} = h_2 + (V_2^{\text{rel}})^2/2 \quad$ and $\quad h_{03}^{\text{rel}} = h_3 + (V_3^{\text{rel}})^2/2$. $\hfill$ (9.9)

This is analogous to the situation in stator rows for which the stagnation enthalpy in *absolute* co-ordinates is conserved. (But for the stator this is true even if the streamline does not remain at the same radius.) In the absence of losses it would also be true that the relative stagnation pressure would be conserved in the rotor and the absolute stagnation pressure in the stator; in fact losses do occur and in a turbine blade row between 3% and 6% of the exit dynamic pressure from the blades is typically lost.

The Euler work equation shows that the drop in *absolute* stagnation enthalpy through the stage, is given for the stage in Fig. 9.5 by

$$\Delta h_0 = U_2 V_{\theta 2} - U_3 V_{\theta 3}$$

$$= U(V_{\theta 2} - V_{\theta 3}) \tag{9.10}$$

with the restriction to streamlines at constant radius. This can be rewritten in terms of relative tangential velocity as

$$\Delta h_0 = U(V_{\theta 3}^{\text{rel}} - V_{\theta 2}^{\text{rel}}) \tag{9.11}$$

where V_θ and V_θ^{rel} are positive if in the same sense as the blade speed.

If the axial velocity is chosen to be constant equation (9.10) then simplifies to

$$\Delta h_0 = U V_x (\tan \alpha_2 - \tan \alpha_3) \tag{9.12}$$

which, it can be shown, is also equal to

$$\Delta h_0 = U V_x (\tan \alpha_3^{\text{rel}} - \tan \alpha_2^{\text{rel}}) \tag{9.13}$$

where α and α^{rel} are positive if they correspond to V_θ and V_θ^{rel} which are positive.

This work exchange takes place even when there are losses present. The effect of the losses is to make the pressure drop greater than it would be for the same temperature drop in an isentropic (loss-free) machine. If the design of the machine is unsatisfactory, however, these losses can become large, and it is for this reason that designs are normally restricted to appropriate ranges of $\Delta h_0/U^2$ and V_x/U.

The turbine stage can be visualised as a series of expansions in each blade row in which the velocity increases. If the high pressure existing in the combustion chamber were expanded in a single expansion, a very high velocity indeed (about 1450 m/s) would be produced, which it would be impossible to use efficiently. The 'trick' of the turbine is to make a series of smaller expansions, typically to velocities just over the speed of sound and then, by changing the frame of reference, to apparently reduce it on entry to the next blade row. This can be seen for the stage in Fig. 9.5: the velocity leaving the stator is high in the *absolute* frame of reference appropriate to the stator, but is much lower when seen by the rotor at entry. Likewise the velocity leaving the rotor is high in the *relative* frame of reference appropriate for it, but lower in the *absolute* frame of reference into the next stator. Each of the turbine blade rows takes in a flow which is not very far from axial and turns it towards the tangential, thereby reducing the flow area and increasing the velocity.

Exercises

Some of these do not have unique 'right' answers, they are open ended design exercises, for which one aims for the best solution.

9.6 Exercise 9.5 has fixed the non-dimensional work output from stages of the core turbine at mid-radius. Take $V_x/U_m = 0.55$ and suppose that in all stages the flow is purely axial into each stator row in the absolute frame of reference; this defines the outlet velocity triangle from the rotor as well as entry conditions to the stator. Find the flow direction out of the stator blades. Assume that the axial velocity is uniform from inlet to outlet. Draw velocity triangles and sketch corresponding blade cross-sections to achieve the desired loading. (For this purpose assume that the flow direction is the same as the blade outlet direction.)   (**Ans**: $\alpha_2 = 74.6°$)

9.7* The length of the turbine blades in the radial direction has not yet been chosen. Assume that the flow is choked at outlet from the first stator blades. Knowing the mass flow, the stagnation temperature (which is 1450 K at cruise assuming there is no drop in temperature in the stator) and the pressure (assumed equal to that at compressor exit) the area of the passage required can be found. Approximate the flow area A by treating the blade height h as short relative to the mean radius r_m, so
$$A = 2\pi r_m h \cos \alpha_2.$$
The mean radius was found in Exercise 9.5. Use this to find h at inlet to the first rotor.
                                    (**Ans**: $h = 38$ mm)
Sketch an outline of the core turbine in the axial–radial plane. Assume that the area increases in inverse proportion to the density based on stagnation conditions calculated using
$T_0/\rho_0^{\gamma-1} = $ constant. For the core turbine the aspect ratio based on the axial projection of chord (aspect ratio is the span of the blade divided by its chord) should not be less than about one and should not exceed about 2.5.

9.8* The rotational speed of the LP shaft is fixed by the requirement of the fan tip speed: assume that the LP rotational speed is 53 revs/s (3180 rpm). If the LP turbine blade speed can be high, which means making the mean radius large, we can have fewer stages for the same loading coefficient. If one is not careful, however, the flow path needed to get the large mean radius becomes very unsatisfactory. For the LP turbine we must choose how many stages, and a figure up to six would be acceptable, though the cost and weight of having more than four is serious.

Use as guide lines that $\Delta h_0/U_m^2$ should not exceed 2.5 and V_x/U_m should not exceed about 1.0 for an LP turbine to find the mean radius of the LP turbine and the number of stages required. (From Exercise 7.1 the temperature drop in the LP turbine is about 361 K for $bpr = 6$ and around 376.2 for a $bpr = 10$.) Assume that the flow is purely axial in the absolute frame of reference into each stator row, since this makes these estimates much easier - this assumption is usually not far from the case.   (**Ans:** for $bpr = 6$; with 5 stages $r_m \approx 0.51$ m, with 4 stages $r_m \approx 0.572$ m)

Make some sketches of possible layouts, allowing the aspect ratio based on axial projections of chord to rise in this case to around 4.

9.5 THE AXIAL CORE COMPRESSOR

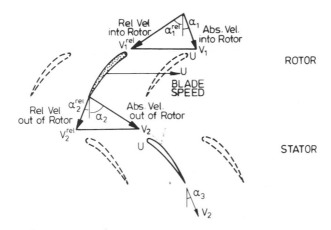

Figure 9.6. An axial compressor stage showing the velocity triangles into and out of the rotor.

Figure 9.6 shows a compressor stage and the corresponding velocity triangles. At entry to the rotor the *absolute* velocity is V_1 inclined at α_1 to the axial. In the *relative* co-ordinates, corresponding to what an observer on the rotor would perceive, the velocity is V_1^{rel} which is inclined at α_1^{rel} to the axial direction. For the flow into the rotor,

$$V_{x1} = V_{x1}^{rel} \quad \text{and} \quad V_{\theta 1} = V_{\theta 1}^{rel} - U \qquad (9.14)$$

and likewise for the flow out of the rotor. The same trigonometric expressions as those for the turbine link the velocities, for example

$$\tan \alpha_1 = V_{\theta 1}/V_{x1} \qquad \text{and} \qquad \tan \alpha_1^{rel} = V_{\theta 1}^{rel}/V_{x1} .$$

For the conditions in Fig. 9.6 the work input, which is equal to the stagnation enthalpy rise per stage, is given by Euler's equation and is

$$\Delta h_0 = U_2 V_{\theta 2} - U_1 V_{\theta 1}$$

$$= U(V_{\theta 2} - V_{\theta 1})$$

with the restriction to constant radius streamlines. If the axial velocity is constant

$$\Delta h_0 = UV_x(\tan \alpha_2 - \tan \alpha_1) \qquad (9.15)$$

or equivalently $\qquad \qquad \Delta h_0 = UV_x(\tan \alpha_1^{rel} - \tan \alpha_2^{rel}). \qquad (9.16)$

Whereas for a turbine the flow deflection may be $90°$ or more, for a compressor it is rarely more than about $45°$. As a result $\Delta h_0/U^2$ is several times smaller for a compressor stage than a turbine stage.

It has been remarked already that the compressor is less 'forgiving' than a turbine. With a turbine an excursion outside the normal range of values for $\Delta h_0/U^2$ and V_x/U is likely to lead to some loss in efficiency. In a compressor operation outside the normal bounds is likely to lead to the machine not working at all as expected. The seeds of the problem are to be seen in Fig. 9.1, which shows the fairly narrow range over which the compressor blade deflects the flow as expected and the rapid rise in deviation when this is exceeded. (Deviation is the angle between the flow outlet direction and the outlet direction of the blade itself.) What tends to happen is that the deviation rises to very high values if normal operating limits are exceeded. For the stator this would mean that α_1 would be much larger than the design intent, and likewise for the rotor α_2^{rel}, so that the work drops sharply.

It is almost impossible to reduce the velocity of a flow to less than about 50% of its entry value because the boundary layers tend to separate. Once separated the 'wake' can expand to fill the area, thus effectively removing any method of increasing the area and reducing the flow velocity further. In a compressor the 'trick' is to carry out the deceleration in a large number of steps, each one raising the pressure by a small amount. Thus in the rotor in Fig. 9.6 the flow is decelerated to a velocity just above that likely to cause boundary layer separation. The frame of reference is then changed to the *absolute* one appropriate for the stator and the velocity is thereby apparently raised, allowing deceleration to take place in the stator.

Exercises

9.9 The entry conditions to the core compressor were determined, including the hub and casing radii, in Exercise 9.2. The number of stages was also determined, so the enthalpy rise per stage is known. We will assume that the mean radius remains constant through the compressor and that the axial velocity is uniform from inlet to outlet and equal to $0.5U_m$. Suppose that the absolute flow is axial into each stage, and that the work input is equal in each stage.

Find the flow direction α_2 into the stator of a typical stage at mid-radius and sketch the blades, assuming zero incidence and deviation.   (**Ans**: $\alpha_2 = 39.8°$)

You could repeat this for the first stage for a section near the hub and another near the casing, assuming that the axial velocity is uniform along the span from hub to casing. If you did you would find that the rotor blades showed most alteration along the span, with significantly more curvature (camber) near the hub than the casing, but the whole more inclined towards the tangential direction near the casing.

9.10 Sketch the core compressor from inlet to outlet, holding the mean radius constant. The aspect ratio (based on the blade span and the projection of the chord in the axial direction) can be up to 3 for the first stage, but should decrease uniformly to around 1 at the last stage - this is known to give a reasonable compromise between length, robustness, good aerodynamics and a modest number of blades.

SUMMARY CHAPTER 9

Compressors and turbines consist of stages made of rows of stationary blades (stators) and rotating rows (rotors). The pressure rise of a compressor stage is much less than the pressure fall of a turbine stage because of the generally favourable pressure gradient in the turbine and the adverse pressure gradient in the compressor.

Compressor blades generally operate satisfactorily over a narrower range of incidence; when the incidence becomes too large, massive boundary layer separation can lead to a large increase in loss and a reduction in turning (i.e. deviation).

Satisfactory operation of both compressors and turbines is possible only in a fairly narrow range of V_x/U and $\Delta h_0/U^2$. Acceptable values of these non-dimensional parameters are frequently given in terms of the conditions at mean height along the blades. Practical constraints often make it impossible to stay within the desired limits; because the turbine is more 'forgiving' it is generally in this component where the compromises are most evident. Although there is no need to maintain the axial velocity exactly constant through a multistage compressor or turbine, this is a reasonable first approximation. This requires the blade height to reduce along the length of the compressor and to increase along the turbine.

The work exchange in a turbine or compressor is given by the Euler work equation

$$\Delta h_0 = U_2 V_{\theta 2} - U_1 V_{\theta 1}.$$

The practical way to consider turbine or compressor blade rows is to adopt a frame of reference fixed to the rotor (the *relative* frame) or the stator (the *absolute* frame), whichever is being studied. The easy way to carry this out is to use velocity triangles, and it is strongly recommended that these be drawn whenever change in frame of reference is carried out.

Considerable simplification is possible if the blade speed is equal at inlet and outlet to the rotor (i.e. the streamline radius does not shift) and also if the axial velocity is equal at inlet and outlet – both these approximations are often reasonably good for the core.

For a turbine the deviation is small and normally does not vary much with flow coefficient. For a compressor the deviation angle may be 20% of the turning of the blades and is inclined to rise sharply if the compressor is asked to operate beyond the normal incidence.

CHAPTER 10

OVERVIEW OF THE
CIVIL ENGINE DESIGN

The emphasis of Part 1 of the book has been overwhelmingly towards the aerodynamic and thermodynamic aspects of a jet engine. These are important, but must not be allowed to obscure the obvious importance of a wide range of mechanical and materials related issues. In terms of time, cost and number of people mechanical aspects of design consume more than those which are aerodynamic or thermodynamic. Nevertheless this book is concerned with the aerodynamic and thermodynamic aspects and it is these which play a large part in determining what are the *desired* features and layout of the engine. Nevertheless an aerodynamic specification which called for rotational speed beyond what was possible, or temperatures beyond those that materials could cope with, would be of no practical use.

An aircraft engine simultaneously calls for high speeds and temperatures, light weight and phenomenal reliability; each of these factors is pulling in a different direction and compromises have to be made. Ultimately an operator of jet engines, or a passenger, cares less about the efficiency of an engine than that it should not fall apart. Engines are now operating for times in excess of 20000 hours between major overhauls (at which point they must be removed from the wing), and this may entail upward of 10000 take-off and landing cycles. In flight shut-downs are now rare and many pilots will not experience a compulsory shut-down during their whole careers. The speeds of some of the components of an engine, particularly the fan blades, are comparable with the shells from some guns; it is vital that these very rarely come apart and that if they do they are contained inside the engine. To make it possible to contain the fan blades in the event of one becoming detached much engineering effort is expended in design and testing the containment ring around the fan and this ring adds considerable extra weight to the engine.

Any design is a compromise. The aerodynamicist would like thin blades, but for mechanical reasons they may have to be thicker than he would choose. The mechanical designer has to find a way to hold the bearings and to supply and remove the lubricating oil – these tend to give fairly massive structures which conflict with the aerodynamicist's wishes. This is a conflict that the mechanical designer is likely to win! The positioning of the bearings has a marked effect on the strength and integrity of the engine and on its weight – bearings which are not in the right axial place need heavy structural elements to transmit the loads to them.

A huge amount of work has gone into the specification of the layout of engines and the mechanical design which follows. If you look at drawings for engines of different companies you will see the differences in such important things as the positions of the bearings and the load bearing struts. As so often there is not a 'right' solution – there are a range of solutions, some of which are more elegant than others. Decisions taken at an early stage without much thought for the consequences may make other things more difficult at a later stage. Different companies have arrived at families of engine style for which they have evolved workable solutions to the problems inherent in the design. By keeping to the familiar style it is possible to avoid some of the problems of a new design. One of the fascinating aspects of this is that a very elegant solution to one problem, for example where to put a bearing, may cause a problem in another aspect, for example in the aerodynamic specification of the HP turbine.

The first ten chapters have taken the simplest possible approach to the design of civil transport engines. By sketching the design, as suggested in Exercise 10.1, a strong similarity will emerge with the engines shown in Fig. 5.4. This, of course, is no coincidence, since the assumptions and empiricism used are close to those adopted by the engine manufacturers. One difference is the layout of the compression system; in the design here all the core compression after the fan is carried out in the core compressor on the HP shaft, whereas the real engines either have booster stages on the LP shaft or else separate IP and HP shafts. The reasons for this are associated with off-design performance, including starting, and this is addressed in the following two chapters. Chapter 11 looks at the performance of the components of the engine, mainly the compressor and turbine, whilst Chapter 12 examines how the engines behave when they go off-design, for example when T_{04}/T_{02} is greater or less than the value at the design point.

The early civil engines evolved out of military engines intended for fighter aircraft. Over the years the designs of military and civil engines have diverged, especially with high bypass ratios for civil engines. Chapters 13, 14 and 15 look at the requirements for military combat engines which are more varied and complicated than those for the civil aircraft. Chapter 16 considers the design point operation of the engine for a possible new fighter and Chapter 17 the off-design operation. The study of military engines closes in Chapter 18 with a short section on the turbomachinery requirements.

In considering the military engine several of the complicating factors are included which were omitted in Chapters 1 to 10. Chapter 19 therefore returns to the engine for the New Large Aircraft with these additional effects included. Chapter 20 is a brief overview of the whole scene, emphasising that any straightforward treatment outlined in a short book is likely to be a serious oversimplification of what goes on in practice.

Exercise

10.1 Sketch an engine, having in mind the core compressor and turbine you have specified. An important component which we have neglected in this course is the combustor. Use the engine cross-sections in Fig. 5.4 to obtain some idea of this.

Try to avoid making the flow go through tortuous ducts to change radius. Try to put in bearing supports. To get everything to match is extremely difficult and you will not have the time to try many schemes. The main thing is to get a feel for the difficulty involved and the scope for different solutions. (It is recommended that this is done on squared paper.)

Part 2

Engine Component Characteristics

and Engine Matching

CHAPTER 11

COMPONENT

CHARACTERISTICS

11.0 INTRODUCTION

Up to this point consideration has been given only to the design point of the engine. This is clearly not adequate for a variety of reasons. Engines sometimes have to give less than their maximum thrust to make the aircraft controllable and to maintain an adequate life for the components. Furthermore all engines have to be started, and this requires the engine to accelerate from very low speeds achieved by the starter motor. The inlet temperature and pressure vary with altitude, climate, weather and forward speed and this needs to be allowed for.

To be able to predict the off-design performance it is necessary to have some understanding of the way the various components behave and this forms the topic of the present chapter. It is fortunate that to understand off-design operation and to make reasonably accurate predictions of trends it is possible to approximate some aspects of component performance. The most useful of these approximations is that the turbines and the final propelling nozzle are perceived by the flow upstream of them as choked. Another useful approximation is that turbine blades operate well over a wide range of incidence so that it is possible to assume a constant value of turbine efficiency independent of operating point. These approximations make it possible to consider the matching of a gas turbine jet engine – how the various components operate together at the conditions for which they are designed (the design point) and at off-design conditions – and this will form the topic of Chapter 12.

The present chapter will consider only the major components: the fan and compressor, the combustor, the turbine and the propelling nozzle. Because of its simplicity it is convenient to begin by considering the nozzle, but prior to this the issue of gas properties will be addressed.

11.1 GAS PROPERTIES IN THE AIRCRAFT GAS TURBINE

In the treatment of the engines for the New Large Aircraft in Chapters 1–10 the specific heat capacity of the gas at constant pressure c_p and ratio of specific heat capacities $\gamma = c_p/c_v$ were assumed to be equal for the air and for the products of combustion and to be constant regardless of temperature and pressure. This is a major over-simplification which will be corrected somewhat in the present and later chapters. A very extensive collection of thermodynamic properties relevant to the gas turbine is given in tabular form in Banes, McIntyre and Sims (1967).

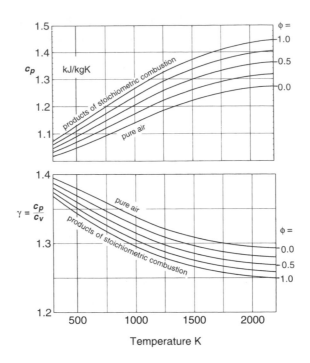

Figure 11.1. Variation in specific heat at constant pressure c_p and specific heat ratio γ with temperature for air and for combustion products of kerosene C_nH_{2n}. ϕ is the equivalence ratio

For a gas turbine the range of temperatures of interest extends from about 216 K to about 2200 K and the range of pressures from about 20 kPa to 45 MPa. In this range the effect of pressure on the values of c_p and γ is of the order of 0.1% and is therefore negligible for the level of precision required here. The effect of temperature and composition is not negligible, as Fig. 11.1 shows. Curves are shown for different levels of equivalence ratio ϕ, this being the ratio of the fuel–air ratio to the fuel–air ratio for stoichiometric combustion. If all of the reactants are consumed in the combustion and there is no excess fuel or oxygen, the combustion is said to be stoichiometric. For a kerosene fuel which is typical of what might be burned in an aircraft gas turbine, with the empirical formula C_nH_{2n}, the mass of fuel per unit mass of air in stoichiometric combustion is 0.0676. For a gas turbine combustor at maximum thrust the equivalence ratio is typically about 0.4, indicating that only 40% of the oxygen has been burned, whereas for an afterburner on a military engine at maximum thrust the equivalence ratio is about unity.

The curves in Fig. 11.1 were derived assuming that the component gases behave as perfect gases (i.e. there is no pressure dependence) and dissociation of the gas molecules as well as the formation of additional species, such as oxides of nitrogen, is neglected. These assumptions are reasonably good at the pressures and maximum temperatures involved: at

2000 K and a pressure of 100 kPa the molar concentration of atomic oxygen at equilibrium is about three orders of magnitude below that of molecular oxygen. The concentration of atomic nitrogen is many orders of magnitude below that of atomic oxygen. Of the oxides of nitrogen the highest concentration is that of NO at about 0.8%, smaller than the concentration of argon. Although such a small concentration of NO is not significant in the energy release in the combustion process, it is highly significant as a pollutant.

The curves in Fig. 11.1 show that any simple choice of gas properties is going to lead to inaccuracy if applied to compressions and expansions in which there are substantial temperature changes. For the air in the compressor γ falls from just below 1.40 to about 1.35 over the range of temperature involved, whereas for the turbine, with $\phi = 0.4$, γ on entry is around 1.28 increasing to about 1.32 at exit. For simplicity and consistency the values for γ of 1.40 and 1.30 will be used from this chapter forwards for the compressor and turbine respectively. The ratio of specific heats is related to the specific heat at constant pressure by the relation

$$c_p = \gamma R / (\gamma - 1)$$

where R is the gas constant with units kJ/kgK. For combustion of hydrocarbons the value of R hardly changes, rising from 0.2872 for pure air to 0.2877 for the products of stoichiometric combustion; it is also effectively independent of temperature. For the present work we will take $R = 0.287$ kJ/kgK throughout. Specifying γ therefore effectively determines c_p. For pure air with $\gamma = 1.40$ we obtain $c_p = 1.005$ kJ/kgK and for the products of combustion with $\gamma = 1.30$ we get $c_p = 1.244$ kJ/kgK. These are the values which will be used throughout.

11.2 THE PROPELLING NOZZLE

The bypass and core streams may have separate nozzles or the two flows may be mixed prior to the contraction forming the nozzle. In either case the flow is assumed to be uniform when it enters the nozzle; non-uniformity contributes to the discharge coefficient being less than unity. (Discharge coefficient is the ratio of the actual mass flow to the mass flow assuming an expansion to the exit static pressure with no loss and with uniform flow over an area equal to the geometric area of the nozzle.) For a nozzle which is just choked with a 5° convergence, the discharge coefficient is about 0.97, whilst the velocity coefficient is about 0.998, indicating levels of imperfection small enough to be neglected in the present treatment. For high speed propulsion it is not uncommon to use a convergent–divergent nozzle, but for subsonic aircraft the cost and weight is not justified.

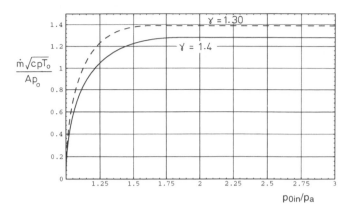

Figure 11.2. Variation in non-dimensional mass flow rate through a convergent nozzle with ratio of inlet stagnation pressure to atmospheric pressure at outlet.

The variation in non-dimensional mass flow through a convergent nozzle with pressure ratio (upstream stagnation pressure ÷ downstream static pressure) across the nozzle is shown in Fig. 11.2 for air ($\gamma = 1.40$) and for exhaust gases, for which $\gamma = 1.30$ is more appropriate. It was assumed here that the flow is reversible to the throat, which is the exit plane in the case of a simple convergent nozzle. The mass flow increases with pressure ratio until the pressure ratio reaches the value at which the nozzle chokes. As is shown in Chapter 6, choking of a convergent nozzle occurs when the inlet stagnation pressure is equal to at least $((\gamma+1)/2)^{\gamma/\gamma-1}$ times the downstream ambient static pressure, or 1.89 in the case when $\gamma = 1.40$. When the nozzle is choked, i.e. the Mach number is unity at the throat, the non-dimensional mass flow rate is constant, in other words

$$\bar{m}_{choke} = \left(\frac{\dot{m}\sqrt{c_p T_0}}{A p_0} \right)_{M=1.0} = \frac{\gamma}{\sqrt{\gamma-1}} \left(\frac{\gamma+1}{2} \right)^{-\frac{\gamma+1}{2(\gamma-1)}} .$$

For pure air, such as through the bypass nozzle, taking $\gamma = 1.4$ gives

$$\bar{m}_{choke} = 1.281.$$

For the flow in a turbine or in the core nozzle $\gamma = 1.3$ is a better approximation and then

$$\bar{m}_{choke} = 1.389.$$

The mass flow and all conditions *upstream* of the throat are unaffected by conditions downstream of the throat once the flow is choked. The relationship between $\bar{m}$ and p_0/p shown in Fig. 11.2 also applies to a convergent–divergent nozzle where A is then the throat area and p is the pressure at the throat. (The variation in area for an isentropic convergent–divergent nozzle is shown in Fig. 6.2 as a function of Mach number; using the curves of p/p_0 also in Fig. 6.2 the dependence of area on local static pressure can be deduced.) In almost all cases of interest for aircraft propulsion the nozzle will be choked; the exception is the bypass stream of a high bypass engine with negligible forward speed.

Although conditions downstream of a choked throat do not affect the conditions upstream in the engine this does not mean that what happens downstream of the throat is unimportant, since it greatly affects the thrust. This is illustrated in Exercise 11.1.

Exercise

11.1 Consider a nozzle discharging into an atmosphere with ambient static pressure p_a. The flow at nozzle exit plane (exit area A) has uniform static pressure and velocity, p_9 and V_9; in general p_9 is not equal to p_a. Some distance downstream of the exit plane the pressure in the jet is equal to p_a and at that condition the jet velocity is V_j. The mass flow in the jet remains effectively constant during this process and is equal to $\dot{m}$. By considering a suitable control volume and applying conservation of momentum, show that the gross thrust is given by

$$F_G = \dot{m}\, V_j \;=\; \dot{m}\, V_9 + (p_9 - p_a)\, A.$$

For which of the following conditions is the second term on the right hand side of this equation non-zero: an unchoked nozzle, a convergent nozzle which is just choked, a convergent nozzle which has a pressure ratio significantly greater than that to choke it, and a fully expanded convergent–divergent nozzle?

Show that the gross thrust may be written in non-dimensional form as

$$\frac{F_G}{\dot{m}\,\sqrt{c_p T_{09}}} = \frac{V_9}{\sqrt{c_p T_{09}}} + (p_9/p_{09} - p_a/p_{09})\left(\frac{\dot{m}\,\sqrt{c_p T_{09}}}{A\, p_{09}}\right)^{-1}.$$

Noting that the conditions at the nozzle exit plane are fixed for a choked convergent nozzle (p_9/p_{09}, non-dimensional velocity and mass flow all constant) explain why the rate of increase in thrust with p_{09}/p_a is so small for large values of pressure ratio.

Note: The very weak dependence of $F_G/\dot{m}\sqrt{c_p T_{09}}$ on p_{09}/p_a for large values of p_{09}/p_a means that at high speeds, ($M > 1.8$, say) when the inlet ram pressure is high, there is little benefit in having a large pressure ratio in the engine. Instead the designer should go for getting the highest mass flow through the engine possible, and this strategy is adopted for some modern high performance combat engines.

Fig. 11.3, based on Exercise 11.1, shows the non-dimensional gross thrust $F_G/\dot{m}\,\sqrt{c_p T_{09}}$ versus pressure ratio for two nozzles in the case of air ($\gamma = 1.40$). One is a simple convergent nozzle, the other a convergent–divergent nozzle of sufficient size at each pressure ratio to allow full reversible expansion. The expansion is assumed reversible inside the convergent nozzle but not downstream of its exit plane. The nozzle is choked for pressure ratios $p_{09}/p_a > 1.89$, but the effect of irreversibility in the flow downstream of the convergent nozzle does not become apparent until the pressure ratio exceeds about 3. This is substantially higher than is common for modern civil transport engines, but much lower than that for high-speed engines.

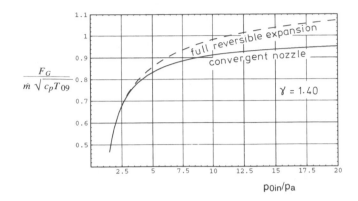

Figure 11.3. Variation in non-dimensional gross thrust from nozzles as a function of
the ratio of inlet stagnation pressure to atmospheric pressure at outlet.

A modern fighter engine may have a pressure ratio of 16 across the nozzle, and at this condition the loss in thrust associated with a convergent nozzle is more than ten per cent. From a plot like Fig.6.2 it can be seen that the exit areas can become large for a fully reversible expansion inside the nozzle if the pressure ratio is large; with a pressure ratio $p_{09}/p_a = 16$, for example, the exit area would be more than 2.5 times the throat area.

For test purposes the nozzle may be replaced by a variable area throttle, as for example, during the testing of a compressor. Again the non-dimensional mass flow rate is independent of pressure ratio once the pressure ratio is large enough to choke the throat or narrowest part. In this case, however, the flow is strongly non-uniform across the area of the throttle, so $\bar{m}_{choke}$ based on the throttle open area would not have the same magnitude as that for a well-shaped choked nozzle with nearly uniform flow. So far as the compressor is concerned, however, the throttle behaves just like a nozzle.

11.3 THE FAN

The fan is simply a specialised form of compressor. For civil applications it is a single stage producing a pressure ratio of no more than about 1.8. In military applications the pressure ratio may be as high as 4, with two or three stages. The fan should pass as much flow per unit area as possible, so the blades should be long and the hub radius small; the ratio of the hub to casing radius is small at fan entry, typically less than 0.4. In modern civil fans there are no stator blades (usually referred to as inlet guide vanes or IGVs) upstream of the rotor, but for military fans inlet guide vanes are commonly used.

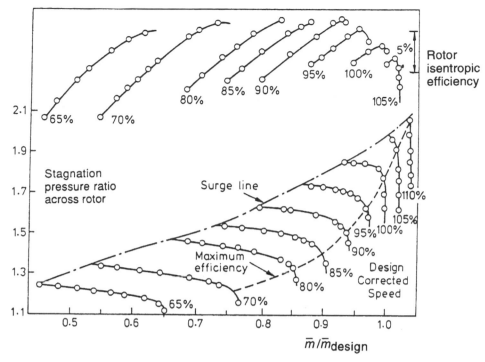

Figure 11.4. Characteristics of a modern civil fan showing pressure ratio and isentropic efficiency versus mass flow rate (non-dimensionalised by inlet conditions) for lines of constant non-dimensional rotational speed.

Figure 11.4 shows the characteristics of a civil fan plotted in the conventional way: the ratio of outlet to inlet stagnation pressure, p_{013}/p_{02}, versus non-dimensional mass flow rate for different non-dimensional speeds. (The numbering system for stations used throughout and illustrated in Fig.7.1 is retained here; p_{013} is the average pressure of the bypass stream whilst $\dot{m}$ is the mass flow rate of the core and bypass streams entering the fan.) As noted in Chapter 8 it is common to drop from the non-dimensional terms those quantities which are constant for a given machine, typically properties of air (R, c_p and γ) and area A. The mass flow is then expressed as

$$\frac{\dot{m}\ \sqrt{T_{02}}}{p_{02}}$$

and the rotational speed as

$$\frac{N}{\sqrt{T_{02}}}\ .$$

Results are often presented in terms of *corrected* mass flow and speed, where $\dot{m}$ and N are referred to standard inlet conditions using $\delta = p_{02}/p_{02\text{ref}}$ and $\theta = T_{02}/T_{02\text{ref}}$, typically $p_{02\text{ref}} = 1.01$ bar and $T_{02\text{ref}} = 288$ K.

The corrected mass flow and speed are then

$$\dot{m}_{corr} = \frac{\dot{m}\ \sqrt{\theta}}{\delta} \quad \text{and} \quad N_{corr} = N/\sqrt{\theta}.$$

In this system the units are retained (for example, kg/s and rev/min) and since the terms δ and θ are often near to unity the physical magnitudes are often helpful.

The curves for pressure ratio versus mass flow at constant rotational speed show properties common to all compressors. As the mass flow is reduced the pressure rise increases. Reducing the mass flow is equivalent to reducing the axial velocity into the fan or compressor and for constant rotational speed a reduction in axial velocity gives an increase in incidence. The effect of increased incidence can be appreciated in several different but compatible ways: an increase in the force on the blades; an increase in the change in whirl velocity ΔV_θ across the rotor; a greater increase in streamtube area passing through the rotor and therefore a larger decrease in velocity and increase in static pressure rise. All these are compatible with the pressure ratio increasing as the mass flow is reduced. Both the pressure ratio and the mass flow rate increase rapidly with speed, so the constant speed lines are well separated. If incidence were held constant the mass flow would be almost proportional to rotational speed and the rise in pressure almost proportional to the square of rotational speed.

The constant speed lines of pressure ratio on Fig. 11.4 vary in shape as the speed increases. At low rotational speeds the lines gradually curl over to be almost horizontal as flow rate is reduced, but for the higher speeds the lines become vertical at low pressure ratios, corresponding to the rotor blades choking. The mass flow at which this choking takes place increases with rotational speed because the increase in speed leads to an increase in stagnation pressure relative to the rotor blades. As the pressure ratio is reduced for these choked speeds the efficiency falls rapidly. Efficiency is generally strongly dependent on both mass flow and speed.

On the plot of pressure ratio against mass flow in Fig. 11.4 are two lines in addition to the lines of constant rotational speed. The highest is the stall or surge line, which denotes the maximum pressure rise which the fan can produce at any rotational speed; attempts to operate above and to the left of this line result in either a collapse of pressure ratio (stall) or a violent oscillatory flow (surge), either of which is unacceptable. The other line is the locus of maximum efficiency as rotational speed is altered. Not shown on Fig.11.4, but derived in exercise 11.4, is the *working line* produced by a nozzle at the rear of the bypass duct (with a fixed nozzle on the core section too); this is roughly parallel to the surge line and tends to move away from the locus of maximum efficiency as speed is reduced (see section 19.1). For static operating conditions the bypass nozzle is not normally choked for a civil fan, but for cruise conditions the high inlet stagnation pressure to the fan, relative to the ambient static, does give a choked bypass nozzle and the working line corresponding to this is the lowest line on the graph.

Also shown in Fig. 11.4 are curves of isentropic efficiency versus mass flow rate for different speeds. The isentropic efficiency for a compressor or fan is given by

$$\eta_{isen} = \frac{T_{013is} - T_{02}}{T_{013} - T_{02}} \tag{11.1}$$

where $T_{013is}/T_{02} = (p_{013}/p_{02})^{\gamma - 1/\gamma}$.

Figure 11.4 shows the conventional form for displaying compressor or fan performance, but comparable information would be conveyed if the temperature ratio T_{013}/T_{02} or the ratio of temperature rise to inlet temperature $(T_{013}-T_{02})/T_{02}$ were shown instead of pressure ratio.

The characteristic map of a military fan is generally similar to Fig. 11.4, though the maximum pressure ratio is likely to be about 4 and the efficiencies are significantly lower.

11.4 THE CORE COMPRESSOR

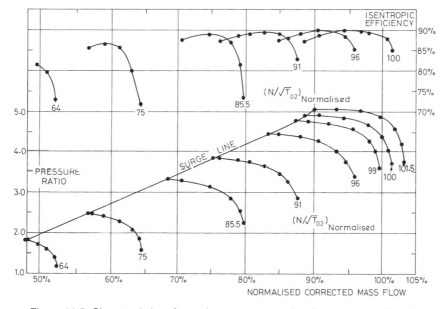

Figure 11.5. Characteristics of a modern compressor showing pressure ratio and isentropic efficiency versus corrected mass flow rate for lines of constant non-dimensional rotational speed.

Figure 11.5 shows the performance map of a multistage compressor in the form of pressure ratio p_{03}/p_{023} versus non-dimensional mass flow, $\bar{m}$, based on inlet conditions to the core compressor,

$$\bar{m}_{23} = \frac{\dot{m} \sqrt{c_p T_{023}}}{A\, p_{023}},$$

using the numbering system of Fig.7.1. The mass flow has been expressed as a percentage of the design value. The maximum pressure ratio of this compressor is around 5 and would be suitable for a three-shaft high bypass ratio engine. A pressure ratio of 6 is approaching the upper limit which can be achieved without incorporating variable stators for reasons which are discussed later in this section. Variable stators are arranged so that as the speed of the compressor falls the stators in the front few rows of the machine are turned to be more nearly tangential, i.e. their stagger angle is increased.

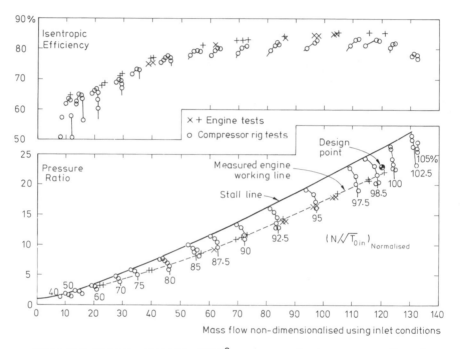

Figure 11.6. Characteristics of the GE E^3 compressor showing pressure ratio and isentropic efficiency versus mass flow rate (non-dimensionalised by inlet conditions) for lines of constant non-dimensional rotational speed.

Figure 11.6 is plotted in the same variables for a compressor with a pressure ratio of about 25. This is the General Electric E^3 (Energy Efficiency Engine) compressor, which, in modified form, is the core of the GE90 engine. For this compressor there are variable stators on the first 6 rows of stator blades (including the inlet guide vanes ahead of the first rotor). Notice again that at the highest speeds the compressor speed lines show choking at lower pressure ratios. Two types of data are shown on this figure, results from a compressor tested on a rig with an external motor and variable throttle (shown with open symbols) and data obtained in engines (shown with crosses). The data obtained on the rig is that which makes possible the speed lines (loci of constant speed operating lines). Obtaining speed lines on an engine requires the engine geometry to be modified between points and this is costly and time consuming so for the engine the data is measured along the working line.

Definitions of efficiency – isentropic and polytropic efficiency

In Figs.11.4, 11.5 and 11.6 the efficiencies shown are the *isentropic* efficiencies, defined in equation 11.1, and sometimes called overall or adiabatic efficiency. In dealing with multistage compressors and turbines it is useful to introduce an efficiency defined slightly differently, the *small-stage* or *polytropic* efficiency η_p. Using polytropic efficiency makes some algebra easier, but it also removes a bias in the isentropic efficiency when comparing machines of different pressure ratio.

Consider a small stagnation pressure rise δp_0 which will be accompanied by a temperature and enthalpy rise δT_0 and δh_0. The familiar thermodynamic equation states that

$$T_0 \delta s_0 = \delta h_0 - \delta p_0/\rho_0$$

where $\delta h_0 = c_p \delta T_0$ and by definition $\delta s_0 = \delta s$. For an ideal isentropic compression process, for which $\delta s = 0$, the enthalpy and pressure changes are related by

$$\delta h_{0i} = c_p \delta T_{0i} = \frac{\delta p_0}{\rho_0} = \frac{RT_0}{p_0} \delta p_0,$$

which may be integrated between states 1 and 2 to give the familiar

$$T_{02i}/T_{01} = (p_{02}/p_{01})^{(\gamma-1)/\gamma}.$$

The actual temperature rise for a real compression will be larger than the ideal and can be written in terms of the polytropic efficiency as

$$\delta h_0 = c_p \delta T_0 = \frac{\delta p_0}{\eta_p \rho_0} = \frac{RT_0}{\eta_p p_0} \delta p_0.$$

which integrates to give

$$\frac{T_{02}}{T_{01}} = \left(\frac{p_{02}}{p_{01}}\right)^{\frac{(\gamma-1)}{\eta_p \gamma}} \qquad \text{for a compressor} \qquad (11.2)$$

and

$$\frac{T_{02}}{T_{01}} = \left(\frac{p_{02}}{p_{01}}\right)^{\frac{\eta_p(\gamma-1)}{\gamma}} \qquad \text{for a turbine.} \qquad (11.3)$$

Exercise

11.2 Show that the polytropic or small-stage efficiency for a compression process between stagnation states 1 and 2 can be written as

$$\eta_p = \frac{\gamma - 1}{\gamma} \frac{\ln(p_{02}/p_{01})}{\ln(T_{02}/T_{01})} \tag{11.4}$$

and that the overall, adiabatic or isentropic efficiency is given by

$$\eta_{isen} = \frac{(p_{02}/p_{01})^{(\gamma-1)/\gamma} - 1}{(p_{02}/p_{01})^K - 1} \qquad \text{where } K = \frac{\gamma-1}{\eta_p \gamma} . \tag{11.5}$$

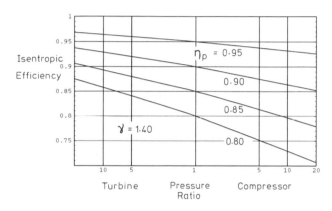

Figure 11.7. Variation in *isentropic* efficiency with pressure ratio for various levels of *polytropic* efficiency.

The *polytropic* efficiency may be related to the *isentropic* efficiency and this is shown in Fig. 11.7 for air for both a compressor and turbine. For compressors the overall isentropic efficiency is lower than the polytropic efficiency, whereas for turbines it is higher; in turbines this effect is referred to as the *reheat factor*. The reason for the lower isentropic efficiency for the compression process can be explained with the aid of the temperature–entropy diagram in Fig. 11.8, which shows a compression between stagnation states 01 and 03 carried out two ways: in one step or in two smaller equal steps. As drawn the overall isentropic efficiency for the single-step compression is 90%; an efficiency of 90% has also been used for each step of the two-step compression and in this case the final temperature $T_{03'}$ can be seen to be higher, implying a lower overall efficiency. In other words when the compression process is broken down into steps or stages the isentropic efficiency of each step must be higher than that over the whole. This is because the temperature of the gas is raised in the first step before it enters the next and as a result the temperature rise in the second step is higher than in the first even if the pressure ratio and the efficiency were equal for both steps. In the limit, when the increment in pressure rise across a step tends to zero, the isentropic efficiency of the step approaches the polytropic limit. For axial compressors the pressure rise across each stage is normally small and in this case the stage isentropic efficiency and stage polytropic efficiency are almost equal; furthermore if all the stages had equal efficiency the polytropic efficiency of the overall machine would equal that of each stage. This is illustrated in Exercise 11.3. The polytropic efficiency is therefore not only more convenient algebraically in many instances, but it offers a better comparative measure of the performance of machines having different pressure ratio.

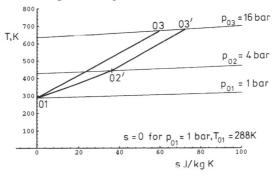

Figure 11.8. Temperature–entropy diagram for compression between 1 and 16 bar achieved in two ways: process 01–03 is in a single step with $\eta_{isen} = 0.9$; process 01–02'–03' is in two steps each with $\eta_{isen} = 0.9$.

Exercises

11.3 **a**) Suppose that a compressor has ten stages each with a pressure ratio of 1.3. Find the overall pressure ratio. If the polytropic efficiency of each stage is 90%, find the overall temperature ratio across the whole machine. Confirm that the polytropic efficiency across the whole machine is 90% and find the isentropic efficiency across a stage, $\eta_{isen,\,st}$, and across the whole machine, $\eta_{isen,\,ov}$.

b) Suppose that the stage pressure ratios remain 1.3, but that for the first five stages the efficiency is 90% and for the last five stages it is 80%. Find the overall temperature ratio and thence the polytropic and isentropic efficiencies for the whole machine.

(**Ans:** a) 13.79, 2.300, $\eta_{isen,ov} = 0.858$, $\eta_{isen,st} = 0.896$; b) $\eta_{p,\,ov} = 0.846$, $\eta_{isen,ov} = 0.784$)

11.4 For the fan in Fig. 11.4 draw the working line for a choked nozzle such that the line passes through the 100% speed line at the mass flow for maximum efficiency. (Three pressure ratios, one of which is at the design point, will suffice to draw the working line.) For simplicity take $\eta_p = 0.90$ all along the working line.

Note: the working line through a choked nozzle does not give a good match between the working line and the locus of maximum efficiency.

Non-dimensional mass flow based on outlet conditions

In the conventional presentation of compressor performance the non-dimensional mass flow used is that based on inlet conditions, $\overline{m}_{23}$ for the core compressor. Results can, however, be presented in terms of the outlet pressure and temperature

$$\bar{m}_3 = \frac{\dot{m}\sqrt{c_p T_{03}}}{A_{out} p_{03}} \qquad = \frac{\dot{m}\sqrt{c_p T_{023}}}{A_{in} p_{023}} \frac{p_{023}}{p_{03}} \frac{\sqrt{T_{03}}}{\sqrt{T_{023}}} \frac{A_{in}}{A_{out}}$$

$$= \bar{m}_{23}\left(\frac{p_{023}}{p_{03}}\right)^L \frac{A_{in}}{A_{out}}$$

where the index $L = 1 - (\gamma - 1)/2\eta_p\gamma$ and η_p is the polytropic efficiency. For reasonable value of efficiency, L is nearly equal to unity; in fact for $\gamma = 1.4$ and $\eta_p = 0.9$, $L = 0.84$, showing that the ratio of non-dimensional mass flow rates is almost proportional to the pressure ratio. (The manipulation carried out above illustrates how convenient the polytropic efficiency can be.)

Exercises

11.5 For the compressors shown in Fig. 11.5 and 11.6 calculate the polytropic efficiency corresponding to the highest isentropic efficiency at the design (100%) speed for each machine.

(Ans: 0.92, 0.88)

The compressors were each tested using a downstream throttle. Assuming that the throttle is choked and its area is constant (i.e. constant corrected mass flow through the throttle), draw a working line on Fig. 11.6. The working line should pass through a pressure ratio of 22.5 on the 100% speed line.

11.6 For the compressor in Fig. 11.6 redraw the 100%, 90% and 80% lines for constant speed in terms of the non-dimensional mass flow based on *outlet* conditions. Assume a constant value of polytropic efficiency on each speed line in calculating the outlet temperature – a simplification which will not seriously alter the form of the characteristic – and take the value for η_p derived in Exercise 11.6 for 100% speed, 0.86 at 90% speed and 0.83 at 80% speed.

Note: The non-dimensional mass flow rate based on outlet conditions can be equal at the different speeds. This is the case, for example, at a pressure ratio of 22.5 at 100% speed and about 12 and 7 for 90% and 80% respectively, corresponding to the working line drawn in Exercise 11.6.

Off-design operation of multistage compressors

It was remarked in Chapter 9 that the design of a compressor is inherently difficult because the pressure is increasing in the flow direction. If the pressure rise for a stage or for a complete compressor becomes too large for the speed of rotation or for the design the compressor can either surge or go into rotating stall. Shown on Figs. 11.5 and 11.6 are surge lines for the two compressors. (It is common to refer to the boundary where the flow breaks down into surge or rotating stall as the surge line regardless of which form of breakdown actually occurs.) Surge is an oscillatory motion of the air, which is usually violent; rotating stall is a non-uniform pattern with reduced flow rate and pressure rise. Either surge or rotating stall is an unacceptable operating condition which it is important to avoid, but this becomes progressively more difficult as the overall pressure ratio of the compressor is increased and this section attempts to explain why. Stall and surge in multistage compressors are more complicated than in a single-stage fan because the different stages from front to back may be operating under very different conditions at the same time. For example, the front stage may be close to stall whilst the back stage is choked, and this is a consequence of the matching of the stages at off-design conditions.

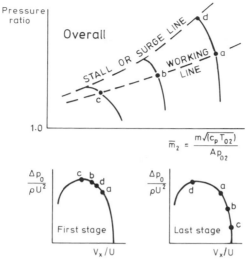

Figure 11.9. Schematic overall and stage characteristics for a multistage compressor
to illustrate different behaviour of front and back stages.

To understand the behaviour better it is helpful to consider the schematic example in Fig. 11.9, where the overall pressure ratio versus non-dimensional inlet mass flow $\bar{m}_2$ is shown. The compressor is assumed here to be made up of many stages of identical performance. For each stage the non-dimensional pressure rise $\Delta p_0/\rho U^2$ is the same unique function of V_x/U, U being the mean rotor blade speed and V_x the mean axial velocity[1], and the stage pressure rise–flow coefficient behaviour of the front and last stages are also shown in Fig. 11.9. At the design point the front and rear stages operate at the same V_x/U because the annulus height is reduced from front to back of the compressor to compensate for the rise in density. Reducing V_x/U leads to an increase in stagnation pressure rise until the stage starts to stall, when further reduction in flow leads to a reduction in pressure rise. It is assumed here, for sound reasons given in Chapter 9, that the pressure rise of a stage is proportional to U^2.

At the design point all the stages are operating at the same point, denoted by point 'a' in Fig. 11.9. On the pressure ratio versus $\bar{m}_2$ plot a working line is drawn through the design point and this working line might be produced by a throttle on a test-bed or by the turbine of an engine. Reducing the rotational speed, but staying on the working line through the design point, takes the compressor first to point 'b' and then to 'c'. To understand what is happening it is necessary to

---

[1] It can be shown that V_x/U is proportional to $(\dot{m}\sqrt{c_p T_0}\ /A\,p_0\,)\div(N/\sqrt{T_0}\,)$, so for constant non-dimensional rotational speed a reduction in V_x/U is equivalent to a reduction in non-dimensional mass flow.

look at the $\Delta p_0/\rho U^2$ versus V_x/U plots for the front and back stages. As the rotational speed is reduced the pressure rise from the stages will therefore decrease, for the same V_x/U, and with a reduction in pressure rise there is a corresponding reduction in density rise in the stage. Because the density rise is then smaller than that assumed in the design of the annulus, the decrease in annulus height from front to back is too much for compressors operating at reduced speed. This gives an increase in the axial velocity in the rear stages. This is illustrated for points 'b' and 'c'. Considering first point 'b' it can be seen that V_x/U has increased, compared to the design point 'a', for the rear stage and this tends to move the front stage to lower V_x/U. By the time point 'c' is reached the rear stage is virtually choked and in the front stage V_x/U has fallen to the point where the stage on its own would stall. (In fact front stages very often do have some rotating stall at low rotational speeds.)

Considering changes along a constant speed line, the effect of reduction in mass flow relative to the working line is shown by points 'd' on Fig. 11.9. Compared to point 'a', operation at point 'd' gives the first stage a lower value of V_x/U; this results in a higher pressure rise and therefore a greater increase in density than was envisaged at design. The greater increase in density means that the second stage has an even lower value of V_x/U than the first stage, with an even greater value of density rise. The effect is cumulative, so the last stage approaches stall whilst the front stage is only slightly altered. Similar effects occur at reduced speed, where the combination of throttling and the effect of speed changes described in the paragraph above must be considered together.

The strong effects of reduction in speed along the working line are revealed in Exercise 11.7 below. Although some rotating stall in the front stages is acceptable at reduced speed, the extent or degree of stall becomes too large when the design pressure ratio across the machine exceeds about 6 unless variable stators are fitted to the front stages. Even with several rows of variable stators the mismatch at low speeds becomes so severe that pressure ratios exceeding 20 have rarely been considered practical.

As noted above the matching problem at speeds below design arises from the tendency for the rear stage to choke. This can be alleviated at low speeds when efficiency is not crucial by bleeding off some of the air around the middle of the compressor and dumping it into the bypass duct. This has the effect of increasing the flow through the front stages, tending to unstall these stages, thereby improving their performance. If the front stages operate more efficiently (more rise in pressure, less rise in temperature) it is possible to pass more mass flow through the rear stages at the same value of corrected mass flow. All multistage compressors for aircraft engines use bleed for some low-speed conditions, including starting.

Exercise

11.7 For the E^3 compressor of Fig. 11.6 the non-dimensional mass flow along the working line will remain constant, based on compressor exit conditions, when tested with a choked throttle of constant area . This is because the throttle will be choked. The excursion in mass flow (and non-dimensional mass flow based on inlet conditions) is almost proportional to the pressure ratio at design to pressure ratio off design.

Suppose for simplicity that at design speed the *absolute* flow into the first rotor is axial, the *relative* inlet flow angle into the first rotor row is 60° at the mean radius and the incidence there is zero. If the axial velocity is proportional to mass flow at these conditions, what would be the relative flow direction and incidence at mean radius on the working line at 60% speed if there were no variable stators?

(**Ans:** $\beta = 81°$, $i = 21°$ – this incidence too large for the blades to work satisfactorily)

11.5 THE COMBUSTOR

The combustor is required to convert the chemical energy of the fuel into thermal energy with the smallest possible pressure loss and with the least emission of undesirable chemicals. To achieve this in a small volume it is necessary for the flow to be highly turbulent – the flow in a combustor is dominated by complex turbulent motions which still cannot be fully described quantitatively. Because the behaviour of the burning process depends on the turbulence, while at the same time the energy release brings about large alterations in the turbulence properties, it is not yet possible to carry out the detailed computations for the three-dimensional flow in the combustor that are now routine for other components of the engine. The design of the combustor therefore continues to rely to a considerable extent on insight and experience and for this book this means that the combustor does not lend itself to simple, useful calculations in the same way as the compressor and turbine. The combustor is therefore treated rather differently here than for the other components, with only one numerical exercise which does not address the major problems of combustor design.

The treatment is necessarily brief and more information is readily available elsewhere: Hill and Peterson (1992) give a very clear account of the fluid dynamic and thermodynamic aspects of combustion, whilst Kerrebrock(1992) gives a particularly good introductory account of the chemistry and of the progress to reduce pollution. For a much more detailed treatment the reader is referred to LeFebvre(1983) and Bahr and Dodds(1990). An excellent discussion of practical issues is given in *The Jet Engine* (1986).

Historically the combustor has always been one of the hardest components to get right, and problems with combustion held up the Whittle engine for many months; von Ohain avoided the problems in the engine flown in 1939 by burning hydrogen instead of kerosene. Burning hydrogen remains an attractive option so far as the design and operation of the combustor is concerned, since many problems are removed, but it is currently ruled out as a fuel by its cost and practical problems of its use. Hydrogen would need to be liquefied and this means a thick layer of insulation would be necessary, which in turn would mean that the wings were unsuitable for storing fuel. Finally the density of liquid hydrogen is low so that the energy stored per unit volume would be only about one third that of kerosene.

Chemical energy release

Inside the combustor of a gas turbine there is a very large energy release rate per unit volume, typically 100 times that for the boiler in a large steam power station. The combustor for a jet engine needs to be small to fit between the compressor and turbine without making the shaft unnecessarily long, since that would add to the weight and introduce problems of mechanical stiffness. (The stiffness of the engine is very important and the outer walls of the combustor are crucial in maintaining the integrity of the outer casing; because stiffness and strength are needed for these it is essential to maintain the structural walls at a temperature well below that of the combustion products.) The high energy release in a small volume is made possible by the high pressures (about 40 bar for a large civil engine at sea-level take off) and by the very high level of turbulence created in the combustor which mixes the fuel and air.

Fuel–air ratio and turbine inlet temperature

The combustor must deliver hot gas to the turbine at an average level which the turbine can tolerate with a level of uniformity which is tolerable. The crucial temperature is that to which the first turbine rotor is exposed and the stator outlet temperature (SOT) is the temperature of the flow into the rotor after hypothetical mixing of the cooling air with the hot gases. For a modern large civil engine SOT is not more that about 1850 K, but for combat engines being developed for entry into service in a few years time, this temperature can be as high as about 2300 K. (Stoichiometric combustion gives a temperature of about 2600 K.) The fuel–air ratio (the mass flow rate of fuel for a given mass flow rate of air through the core) can be shown to be 0.0272 in Exercise 11.8 for a compressor delivery temperature of 917.5 K and a turbine inlet temperature of 1700 K; for a turbine inlet temperature of 1850 K the ratio would rise to only 0.0319. (The maximum fuel–air ratio would be somewhat higher for a military engine at low flight Mach numbers. This is because the temperature leaving the combustor is somewhat higher for the military engine and the temperature of the air entering the combustor is lower in military engines because the pressure ratio is lower.)

A difficulty with hydrocarbon fuels is that they will not burn if the fuel–air ratio is far below stoichiometric, which, as discussed in section 11.1, is about 0.0676. The way around this is illustrated in Fig. 11.10. Fuel is injected as a fine sheet or spray and is broken up by a surrounding air blast. Most of the air from the compressor is diverted to avoid the region where the fuel is injected so that combustion starts in a relatively rich primary region with a fuel-air ratio of about 0.25 for take off, 0.1 at idle. Additional air is then fed in through holes in the combustor lining to complete the combustion process and reduce the temperature to the level acceptable to the turbine so that after dilution the effective overall air-fuel ratio is about 0.03. The air entering the dilution region also puts a layer of relatively cool air on the walls and modifies the exit temperature radial profile to be suitable for entry into the turbine. (It is desirable to have somewhat higher temperatures towards the outer radius of the turbine because the higher blade speed reduces the relative stagnation temperature more there. Furthermore the stress in the turbine rotor blades is highest near the root and to minimise creep it is desirable to have lower temperature near the root than the tip.)

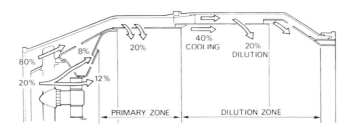

Figure 11.10. A schematic of a combustor showing the apportioning of air flow.
From *The Jet Engine* (1986).

As the combustor delivery temperature is raised there becomes progressively less air to cool the walls and to modify the radial profile - in the limit of stoichiometric combustion (when all the oxygen in the air is used) there is no spare air. Likewise as the combustor entry temperature rises the temperature of the whole combustor rises as well.

Flame speed and the stabilisation of flames

Air leaves the compressor at a speed on the order of 180 m/s, but this is reduced to around 50 m/s before entering the combustor. Laminar flames for hydrocarbon fuels, however, cannot travel at more than about 0.3 m/s, though this can be raised to around 5 – 8 m/s in turbulent flow. To prevent the flame from being blown away, i.e. to stabilise combustion, it is therefore necessary to set up local regions with much smaller velocity. One solution is to have the flame in the wake of a solid object, but it is now more common to set up a recirculation zone. The recirculation is normally created by swirling the flow around the fuel spray so that an enclosed region is formed. To create the swirl which will produce the recirculation, and the high turbulence to increase the flame speed, there has to be a drop in stagnation pressure in the combustor. The actual drop in pressure depends on the type of combustor, but a loss of 5% of the absolute stagnation pressure entering the combustor is representative.

Fuel injection, combustion rate and combustion efficiency

Most combustors inject a spray of liquid droplets but it is normally arranged that a flow of air passes close to the fuel injector and this 'air-blast' then further breaks up the fuel. The need to have a substantial velocity for the air-blast, typically around 100 m/s, requires a drop in pressure between the compressor and the centre of the burning regions and sets one lower limit on the pressure drop in the combustor.

Burning liquid droplets, even ones as small as 30×10^{-3} mm, requires the liquid to evaporate and then for the fuel and air to diffuse together to form locally a near stoichiometric mixture. This process is normally much slower than the main chemical process of combustion. The chemical process of combustion normally takes place very quickly, but the rate falls roughly as the square of the pressure. An additional problem for lower air pressure is in the preparation of the liquid fuel.

Varying the fuel pressure is the main method used to vary the fuel flow rate; when the fuel flow is low and the air pressure is low the break-up of the fuel into small droplets is less satisfactory than at high powers. The coarser spray and lower temperatures in the combustor at low thrust can lead to incomplete combustion. This in turn produces low combustion efficiency with high levels of carbon monoxide and unburned hydrocarbons in the exhaust. The efficiency of combustion (defined as the actual temperature rise divided by that for complete combustion) is very close to 100% for high thrust levels at sea level but decreases to perhaps 99.9% at cruise conditions for a civil airliner. After starting and during the run up to idle conditions the combustion efficiency can be very low. At very high altitude, or at normal cruise altitude when the engine is throttled back to give small thrust, the combustion efficiency can become so low that the process is no longer self-sustaining and the combustor can blow out.

An important requirement ,which has to be designed for, is the ability to re-light the combustor at altitude, even when the compressor is producing very little pressure rise. Combustion efficiency at starting is important for altitude restarts, when the engine can fail to pull away to idle if the combustor is not designed correctly.

Wall cooling and the use of annular combustors

The temperatures of the gas in the combustion region are high enough to lead to rapid failure of the combustor walls if they came in contact. Modern large engines are expected to have a combustor life of around 20000 hours. Moreover, since thermal stress and fatigue are critical, it is worth noting that they should be able to tolerate about 5000 cycles of take off, climb, cruise, landing and taxi-ing. The combustor walls are therefore shielded and cooled, as illustrated in Fig. 11.10. There is an obvious advantage in reducing the amount of surface area in relation to the flow area[1]. Early engines arranged the combustor as a series of discrete cans or tubes (referred to as can-annular or tubo-annular combustors) but the high bypass ratio engines designed since the 1960s have used an annular geometry, as illustrated in Fig. 11.11, with corresponding saving in surface area. The fuel is still supplied through a number of discrete injectors.

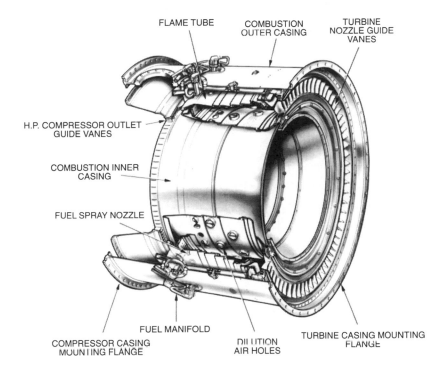

Figure 11.11. An annular combustor. From *The Jet Engine* (1986).

Modern combustors normally line the inner walls with thermal barrier coatings. These reduce the metal temperatures, for the same gas temperatures, and also inhibit oxidation of the surfaces. The modern combustor often has many small cooling holes, similar to the surface of turbine blades, but an alternative is to use metallic tiles, which are cooled from behind. The tiles carry very little load and can therefore operate at higher temperature than the load bearing walls.

---

[1] In this treatment many aspects of the design of the combustor are being neglected, one of which is the specification of overall size. The volume of the combustor is determined primarily by the requirement to be able to restart the engine at high altitude in the unlikely event of a flame-out. Increasing the volume lowers the gas speeds in the combustor and therefore makes it easier to ignite and burn the fuel.

Emissions – formation, regulation and control

Emissions, the creation of harmful or toxic gases during combustion, can be considered important for two different points of view. One relates to the effect on the environment, such as global warming, climate change and ozone depletion, and is primarily a problem during cruise when most of the fuel is burned. The other relates to the immediate surroundings of the airport. It was the effect of aircraft during starting, taxi-ing, taking-off and landing which were first apparent and gave rise to the first protests and then to legislation. At present legislation applies only to operation near the airport, even though the consequences of aviation emissions during cruise are potentially far more serious; fortunately steps taken to reduce emissions near the airport will lead to reductions for the rest of the flight.

The level of emission of a pollutant is expressed in terms of the emissions index (EI), which is the emissions in grams for each kilogram of fuel burned. The CO_2 and H_2O are unavoidable consequences of the burning of hydrocarbon fuel, which can only be reduced significantly by making the engine more efficient and reducing the drag of the aircraft. The oxides of sulphur SO_x are determined wholly by the amount of sulphur in the fuel after refining rather than by the engine; the level of sulphur is normally kept very low. Oxides of nitrogen, (NO_x), unburned hydrocarbons (UHC), CO and particulates (mainly soot, which is unburned carbon) depend on the performance of the combustion chamber and in an ideal one would be virtually zero. As the table below shows for a typical modern aircraft at cruise, the EI (in grams per kg of fuel) for UHC, CO and particulates are small, indicating very satisfactory operation.

Species	CO_2	H_2O	NO_x	SO_x	CO	UHC	Particulate
EI	3200	1300	9 -15	0.3 - 0.8	0.2 - 0.6	0 - 0.1	0.01 - 0.05

There is growing concern about the effect of CO_2, H_2O and NO_x introduced by air traffic into the upper atmosphere Currently about 2 - 3% of the man-made CO_2 is produced by aviation. Perhaps of greater concern, but certainly of greater uncertainty, is the effect of H_2O introduced into the upper atmosphere, particularly into the stratosphere, which is naturally dry. The effect of NO_x is complicated, affecting the ozone and the greenhouse effect. All of this is discussed at considerable length in the special report of the IPCC on Aviation and the Global Atmosphere, published in 1999. Kerrebrock(1992) explains how until the late 1970s the designers of combustors were so fully occupied with making the combustion stable and efficient and avoiding the burning of the walls that they could do little for the pollution aspects. The first target pressed by the US Environmental Protection Agency was visible smoke (unburned carbon particles) in the phase of operation below 3000 ft altitude that had a direct impact on the environment around the airport, and smoke had been greatly reduced before the ICAO[2] regulations for new aircraft

---

[2]ICAO, the International Civil Aviation Organisation, is the United Nations specialised agency that has global responsibility for the establishment of standards, recommended practices, and guidance on various aspects of international civil aviation, including environmental protection.

types were issued in 1981. Carbon monoxide, unburned hydrocarbon and oxides of nitrogen had
not been addressed prior to the ICAO regulations. The aviation industry at first resisted the
proposed regulations for reducing pollutants on the grounds of impracticality, but since then there
has been considerable progress and compliance with limits is obligatory. The first regulations,
the ones issued in 1981, are usually referred to as the ICAO regulations, but the later regulations
are known by the committee of ICAO which considers them, the Committee for Aviation
Environmental Protection (CAEP). After agreement is reached in CAEP the proposals have to be
agreed by ICAO and then ratified by the governments of all the member countries. The regulations
known as CAEP2 were issued in 1993 for compliance for new aircraft types from 1996 and
CAEP4 were issued in 1998 and are due to take effect in 2004.

 The ICAO regulations for CO, UHC, NO_x and smoke lay down procedures as well as
actual levels. (For smoke it is the maximum level to ensure invisibility.) For each species the
mass generated is summed for operation of the engine over a standard landing and take-off (LTO)
cycle: 42 seconds at 100% thrust, 2.2 minutes at 85% of maximum thrust (to simulate climb to
3000 feet), 4 minutes at 30% thrust (to simulate the approach) and 26 minutes at 7% thrust to
allow for taxi-ing and idle on the ground. This mass is then divided by maximum thrust in
kiloNewton at standard sea-level conditions.

 Generally it is NOx which is the hardest target to meet. The allowable amount of NO_x is
proportional to the pressure ratio to compensate for the additional difficulty in reducing this
pollutant when compressor outlet temperature increases; the formula for the original ICAO limit is
$40+2p_{03}/p_{02}$, so the effect of pressure ratio is highly significant.

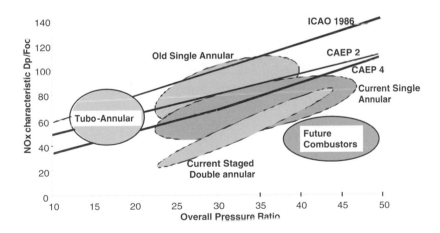

Figure 11.12. The ICAO regulations versus pressure ratio with the regions
of expected performance for different types of combustor.

The emission index used in the regulations (grams of pollutants per kilogram of fuel used)
effectively provides scaling for engine size. Figure 11.12 shows the original ICAO level for the

landing and take-off (LTO) cycle and the more stringent regulations: CAEP2, requiring a 20% reduction in NO_x from the original limit and CAEP4 which come into force in 2004. It is highly probable that the regulations will be tightened further, possibly including regulations for cruise.

Whereas NO_x and smoke are the main problems of high thrust conditions (when the fuel flow to the combustor is high and the temperatures are also high) the emission of CO and UHC, tend to be worst during the taxi and idle conditions. At least conceptually the removal of CO, UHC and smoke is straightforward: the combustion should be prolonged for as long as possible at high temperature in the presence of ample excess oxygen. This also has the effect of increasing the combustion efficiency, though it may not necessarily assist high-altitude re-light capability. The problem of reducing NO_x is much more subtle. NO_x is formed in chemical reactions which are much slower than those leading to the formation of CO_2 and H_2O, but the rate of formation of NO_x increases rapidly with temperature. Because of the *comparatively* slow rate of formation of NO_x, the amount created depends on both the temperature and the *residence time* at that temperature. Unfortunately the long residence time which would reduce CO, UHC and smoke would favour the formation of NO_x. Any subsequent breakdown of NO_x is much slower. Figure 11.13 shows a cross section through a modern combustor together with an inset plot showing schematically the production and consumption of NO_x and smoke.

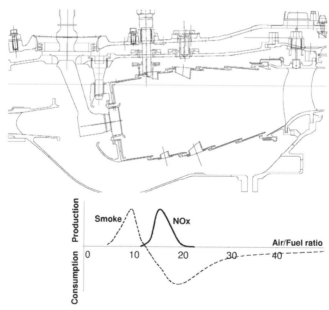

Figure 11.13 A cross-section through the combustor of a modern combustor with a sketch of the smoke and NOx formation.

The picture of the combustor in Fig. 11.13 illustrates how complicated the designs have become to achieve satisfactory emissions, with carefully placed chutes to direct the mixing air where it is most needed. The diagram shows that soot is formed where the mixture is rich (and oxygen correspondingly scarce), but since oxygen is scarce little NOx is formed at first; the NO_x forms rapidly when more air is added to burn the soot. What happens downstream is a balance between maintaining the temperature high enough in the presence of excess air to burn the soot and cooling the gas sufficiently to avoid forming high levels of NO_x. The standard approach to reducing NO_x is to minimise the residence time at high temperature as much as possible, having in mind the need to burn off soot, UHC and CO, and also the need to keep an acceptable level of combustion efficiency at high altitude and low fuel flow rate. The very hot gases, which locally contain pockets of stoichiomentric combustion, are therefore rapidly quenched by mixing with cool air to drop the temperature below that at which the NO_x formation is significant. The need to reduce residence time to avoid high levels of NO_x is a powerful driver to the reduction in size of the combustor in relation to the flow rates of air and fuel.

The strong dependence of NO_x formation on temperature means that there is a tendency for the level to increase as the temperature of air entering the combustor increases and as the temperature leaving the combustor increases. The requirement to keep NO_x low at high thrust (a short residence time at high temperature) and to keep CO, UHC low at low thrust (a long residence time) are fundamentally contradictory. For very low levels of emissions (or for acceptable levels at very high combustor inlet and outlet temperatures) the solution will probably lie with some form of *staged* combustor. For staged combustors different injectors and different regions of the combustor are used for low thrust and high thrust. For idle and low power there is a pilot stage and for high power there is a separate stage. The GE90 (on the Boeing 777) has a two-stage combustor and so do some versions of the CFM-56. At the time of writing, however, the lowest levels of NO_x for the Boeing 777 are from the aircraft with the Rolls-Royce Trent having the single annular combustor shown in Fig.11.13. The use of staged combustion introduces extra cost and complexity.

The progress that has been made in the case of NO_x is illustrated in Fig. 11.14, mainly for engines from the GE/SNECMA stable. The ordinate is the percentage of the limit laid down in the original ICAO regulation for the engines shown. Although the levels have fallen, the extent of the fall to date is not enormous. Moreover the ICAO levels include a term to raise the value in proportion to engine pressure ratio, so the drop in absolute level is less than indicated. What is certain, however, is that without the improvements in design to lower NO_x the level would have been very much higher. The graph also shows sketches of three types of combustor: the old can-annular (or tubo-annular) type, the annular combustor which is the main type in service now and the two-stage annular combustor.

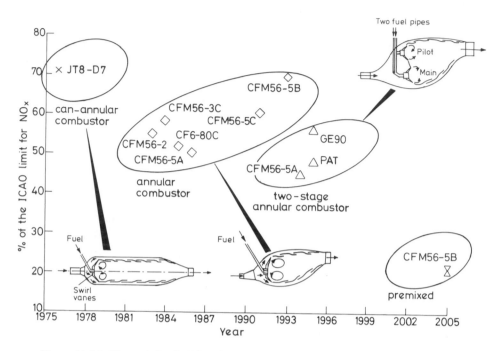

Figure 11.14. The variation in NO_x with date. (Based on figure supplied by SNECMA.)
Also shown are schematics of three types of combustor.

If even lower levels of NO_x are to be achieved, especially if it become necessary to limit production at cruise, then a premixed arrangement will be necessary; here the fuel will be mixed with the air and partially vaporised before entering the combustion region. A potential advantage of premixing is that it avoids the near-stoichiometric burning which takes place in most present-day combustors as fuel and air diffuse together and burn with high local temperatures. Although there are practical difficulties to be overcome before premixing becomes feasible for aircraft applications, it is now widespread for large gas turbines designed for land-based power generation.

The 'black box' approach

For the purpose of this book and the design based exercises the combustor will be treated as a 'black box' with a combustion efficiency of 100%. In using the 'black box' it should be recalled that the change in chemical composition and in temperature in the combustion process brings about a change in gas properties. Within the simplification adopted here γ is taken as 1.40 for air at entry and as 1.30 for the combustion products; there is a consequent increase in c_p. The change in c_p causes a complication in handling the combustor, which is addressed in Exercise 11.8. Because the gases leaving the combustor have a higher c_p than those entering, the energy input required to produce the rise in temperature is substantially greater than that which would be found using a constant value.

Exercise

11.8 Consider a calorimeter with inlet and outlet flows at 298 K for which the energy removed per kg of fuel is the lower calorific value *LCV*. (*LCV* is defined at 25°C and not 288 K.) Then consider the adiabatic combustor with inlet temperature T_{03} and outlet temperature T_{04}. If c_p and c_{pe} are the specific heat capacity of the gas at entry and exit, show that

$$\dot{m}_f LCV = (\dot{m}_{air} + \dot{m}_f) c_{pe} (T_{04} - 298) - \dot{m}_{air} c_p (T_{03} - 298), \qquad (11.7)$$

where $\dot{m}_f$ is the mass flow rate of fuel and $\dot{m}_{air}$ the mass flow of air entering the combustor. Taking *LCV* = 43 MJ/kg, c_p = 1005 and c_{pe} = 1244 J/kgK, find the mass flow of fuel needed per unit mass flow of air into the combustor, to create T_{04} = 1700 K when T_{03} = 917.5 K (from Exercise 4.1).

(**Ans:** 0.0272 kg/s)

Repeat the calculation taking $c_{pe} = c_p$ = 1005 J/kgK.   (**Ans:** 0.0186 kg/s)

Notes: 1) Calculations were also performed for the combustor using a method which treats c_p and γ as functions of temperature and integrates the change in enthalpy up to T_{03} = 917.5 K and to T_{04} = 1700 K, an approach which is essentially exact. Using this method the mass flow of fuel was found to be 0.0236 kg/s per kg of air.)

2) In practice a substantial amount of air is used to cool the nozzle guide vanes to the HP turbine and by the normal convention the turbine inlet temperature is the mixed-out temperature at the exit from these stators. This is the temperature used here T_{04}.

11.6 THE TURBINE

The pressure ratio, p_{04}/p_{05} of an HP turbine stage is plotted against the non-dimensional mass flow in Fig. 11.15.

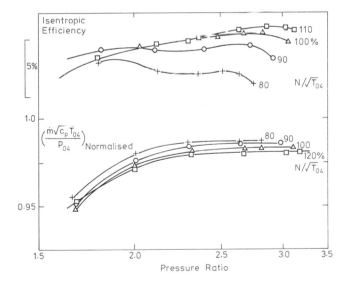

Figure 11.15. Characteristics of a modern HP turbine showing mass flow (non-dimensionalised by inlet conditions) and isentropic efficiency versus pressure ratio for lines of constant non-dimensional rotational speed.

In this case the non-dimensional mass flow is evaluated using the inlet stagnation pressure and temperature, p_{04} and T_{04}.   The results shown are for various non-dimensional speeds, $N/\sqrt{c_p T_{04}}$, but the performance is almost independent of the speed over the range shown.   In fact the turbine behaves to the upstream flow like a choked nozzle for all but the lowest speeds; because the choking flow is almost independent of speed this indicates that it is the nozzle row which gives most of the choking effect.   Although the turbine has a mass-flow/pressure ratio variation like a choked nozzle, most turbines are not actually choked, though the maximum average Mach numbers are close to unity.   The combination of several rows of blades, each nearly choked, simulates a truly choked row.   In fact the changes in efficiency are small enough to be neglected in the analyses which will be carried out here, in which the turbine is not far removed from its design condition. As a result the temperature ratio and non-dimensional power output will be assumed to be fixed only by the pressure ratio across the turbine.

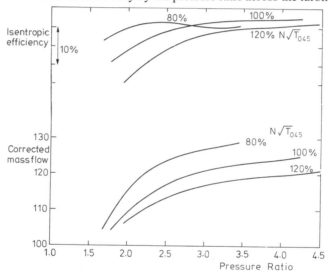

Figure 11.16. Characteristics of a modern LP turbine showing corrected mass flow and isentropic efficiency versus pressure ratio for lines of constant non-dimensional rotational speed.

Figure 11.16 shows the corresponding flow rate and efficiency curves for an LP turbine for a high bypass ratio engine. The LP turbine rotates comparatively slowly, because the fan which it drives cannot rotate very fast, and as a result the Mach numbers in such a turbine are lower than in the HP. To get the large work output the turning in the LP turbine blades is large and there are generally four or more stages. The effect of these is to give only a small variation in non-dimensional flow rate and efficiency with pressure ratio over a substantial range of rotational speed. Again, so far as mass flow rate is concerned, the LP turbine behaves very much like a choked nozzle and the dependence of efficiency on speed and pressure ratio is small enough to be neglected in the approximate analysis here.

Exercise
11.9 Replot the pressure ratio–mass flow characteristic of the turbine shown in Fig. 11.15 at 100% $N/\sqrt{T_{04}}$ in terms of the non-dimensional mass flow based on *outlet* conditions. Take the polytropic efficiency to be constant at 90%.

SUMMARY CHAPTER 11

For an ideal convergent nozzle there is a unique relation between non-dimensional mass flow and the ratio of inlet stagnation pressure to exit static pressure. For a ratio of inlet stagnation pressure to outlet static pressure exceeding 1.89 the nozzle is choked in the case of air at near ambient temperatures and the non-dimensional mass flow does not increase further. For a convergent–divergent nozzle the choking relation is also valid, but based on the area and static pressure at the throat in place of exit conditions.

For a fan or compressor the pressure ratio increases at a given rotational speed as the mass flow rate is reduced. This corresponds to pressure rise increasing as axial velocity falls and incidence increases. In non-dimensional form the pressure ratio and non-dimensional mass flow are strong functions of the non-dimensional rotational speed; pressure rise is roughly proportional to the square of rotational speed; at low speed the mass flow increases approximately linearly with rotational speed, but when choking becomes important the rate of increase is much smaller.

Although pressure rise and pressure ratio for a fan or compressor increase as the mass flow rate is reduced, there is a limit to this. The limit is marked on the performance map as the surge or stall line; attempts to operate above and to the left of this line result either in rotating stall (with a large drop in pressure rise) or surge (a violent pulsation of the entire flow). The working line of the fan or compressor, produced either by a nozzle, throttle or the other engine components, is approximately parallel to the surge line.

A serious problem of mismatching arises in compressors with a large design pressure ratio. This is most marked when the speed is reduced, leading to a tendency for the rear stages to choke and the front stages to stall. This is alleviated to some extent by having variable stagger stators in the front stages (arranged to become more nearly tangential as the rotational speed falls) and having bleed ports part way along the compressor which open at low speeds. Even with these the design of compressors for pressure ratios above about 20 is difficult.

Combustors are complicated components in which the designer has to simultaneously meet conflicting requirements. Not only do requirements for low pollution conflict with some operability aspects, but techniques for reducing carbon monoxide and unburned hydrocarbons can lead to an increase in oxides of nitrogen. There have been very remarkable improvements over the last few years to meet the regulations and to meet still stiffer rules will probably lead to widespread adoption of staged combustors with one stage for low thrust and another for high thrust operation. For the elementary treatment of engine matching it suffices to assume that the combustion efficiency is 100% and that there is a loss in stagnation pressure equal to 5% of the inlet stagnation pressure.

Turbines have pressure ratio versus non-dimensional mass flow characteristics very similar to those of a nozzle, and for most conditions the turbine behaves as if it were choked. The mass flow rate is barely affected by the rotational speed of the turbine. Turbine efficiency is dependent on non-dimensional rotational speed as well as pressure ratio, but much less than the efficiency of a compressor and to a sufficiently small extent that it may be neglected in the cycle analyses to be performed here. With this approximation the temperature ratio across the turbine is therefore fixed by the pressure ratio alone.

As an alternative to the isentropic efficiency the polytropic efficiency can be defined. This allows compressors or turbines of different overall pressure ratio to be compared without the bias introduced in the definition of isentropic efficiency. It also allows algebraic simplification in some cases.

CHAPTER 12

ENGINE MATCHING
OFF-DESIGN

12.0 INTRODUCTION

In Chapter 11 the performance of the main aerodynamic and thermodynamic components of the engine were considered. In earlier chapters the design point of a high bypass ratio engine had been specified and a design arrived at for this condition. At the design point all the component performances would ideally fit together and only the specification of their performance at this design condition would be required. Unfortunately engine components never exactly meet their aerodynamic design specification and we need to be able to assess what effect these discrepancies have. Furthermore engines do not only operate at one non-dimensional condition, but over a range of power settings and there is great concern that the performance of the engine should be satisfactory and safe at all off-design conditions. For the engines intended for subsonic civil transport the range of critical operating conditions is relatively small, but for engines intended for high-speed propulsion, performance may be critical at several widely separated operating points.

The treatment in this chapter is deliberately approximate and lends itself to very simple estimates of performance without the need for large computers or even for much detail about the component performance. The ideas which underpin the approach adopted are physically sound and the approximations are sufficiently good that the correct trends can be predicted; if greater precision is required the method for obtaining this, and the information needed about component performance, should be clear.

12.1 ASSUMPTIONS AND SIMPLIFICATIONS

In understanding the operation of an engine it should be appreciated that in normal operation the engine is affected only by the inlet air stagnation pressure and temperature and by the fuel flow. Of these inputs only the fuel flow may be treated as the control variable. Depending on the fuel flow are the thrust of the engine, mass flow of air, the rotational speeds of the shafts and the temperatures and pressures inside the machine; at a fixed condition the appropriate non-dimensional values of these quantities must be constant. The problem of determining these quantities can be set as a list of constraints and for this we assume that the engine is a multi-shaft machine.

1) The rotational speed of the compressor and turbine must be equal on each shaft.

2) The mass flow through the compressor and turbine must be equal (neglecting the mass flow removed in bleeds and the small mass flow of fuel).

3) The power output of the turbine must equal the power input into the compressor on the same shaft (neglecting the very small power losses in the bearings and windage and the substantial power off-take to supply electrical and hydraulic power to the aircraft).

4) The pressure rise in the compression process (including the intake) must equal the pressure drop in the expansion process, including the combustor, turbines and nozzle.

In general matching these constraints is a process which involves iteration, since the measured performance characteristics of the fan and compressors and of the turbines have to be used. In a calculation carried out for which detailed and accurate predictions are necessary, the pressure loss in the combustor, the bleed flows (for cooling the blades, de-icing the aircraft and cabin pressurisation) and the power off-take would all be included. For the present purpose, which is to show the trends, these complications can be neglected.

Two very important simplifications are possible as a result of the turbine operating characteristic. As was shown in Chapter 11 the dependence of non-dimensional mass flow on pressure ratio for the turbine is almost independent of rotational speed at pressure ratios likely to be encountered above idle conditions. Using the stagnation temperature and pressure at *inlet* to the turbine the form of this dependence is effectively identical to that for a choked nozzle. In other words, for each turbine in a multi-shaft engine

$$\frac{\dot{m}\ \sqrt{c_p T_{0in}}}{A_{in} p_{0in}} = \bar{m}_{in} = \text{constant}$$

where the area is an appropriate inlet area such as the throat area of the turbine nozzle guide vanes. Taking $\gamma = 1.30$ this constant is equal to 1.389, see section 11.2.

The second great simplification, which follows from the great tolerance of well designed turbine blades to incidence, is that the efficiency is little affected by the rotational speed over the range of speeds experienced above the idle condition. It therefore becomes possible to derive the ratio of stagnation temperature into and out of a turbine stage (and thence the temperature drop) from the corresponding pressure ratio using a constant value of polytropic efficiency; for example

$$\frac{T_{0out}}{T_{0in}} = \left(\frac{p_{0out}}{p_{0in}}\right)^{\frac{\eta_p(\gamma-1)}{\gamma}} .$$

The significance of the approximations applicable to the turbine, that it is effectively choked and that the efficiency is constant and independent of rotational speed, becomes apparent when there are two turbines in series, as in a two-shaft engine, or when a turbine operates upstream of a choked nozzle. The compressor performance is, as shown in Chapter 11, strongly dependent on rotational speed and it is the compressor characteristics which largely determine at what speeds the engine shafts rotate.

The engines of current interest are the modern two- and three-shaft engines. These engines, even if they are of the high-speed military type, usually now have a bypass stream. The multiple shafts and the bypass stream complicate the treatment so, as an introduction to the approach, it is worthwhile to look at a single-shaft turbojet engine, first designed more than 40 years ago, the Rolls-Royce Viper which is illustrated in Fig.5.1. For the conditions of interest the turbine is effectively choked and the propelling nozzle is choked. Because this simple configuration shows the important off-design effects it is treated in some detail, beginning with a graphical explanation of the behaviour.

12.2 A SINGLE SHAFT TURBOJET ENGINE

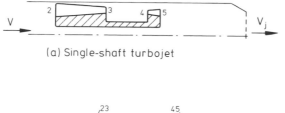

(a) Single-shaft turbojet

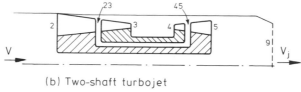

(b) Two-shaft turbojet

Figure 12.1. Schematic representation of turbojet engines.

A Viper turbojet engine is shown schematically in Fig. 12.1(a), using the standard numbering system. The final propelling nozzle and the turbine are treated as choked, which will be a good approximation near the design point. The information in Table 12.1 is for the engine at design point on a stationary sea-level test bed:

Table 12.1 Parameters for the Viper on a sea-level test bed
$T_{02} = T_a = 288$ K, $p_{02} = p_a = 101$ kPa

Gross thrust (static)	= 15167 N
Specific fuel consumption sfc	= 0.993 kg/h/kg
Air mass flow $\dot{m}_{air}$	= 23.81 kg/s
Fuel flow $\dot{m}_f$	= 0.4267 kg/s
Stagnation pressure ratio p_{03}/p_{02}	= 5.5

From the thrust and mass flow the jet velocity $V_j = F_G/\dot{m} = 637$ m/s.

The Viper compressor has eight stages and the turbine one. The polytropic efficiency of the compressor may be assumed to be about 0.90 and of the turbine about 0.85; the efficiency of the turbine is realistically put lower because, with only one stage, it is relatively highly loaded. (Later versions of the Viper engine, such as the one shown in Fig.5.1, had two turbine stages.)

Because the turbine output power must equal the compressor input power, with the same mass flow rate (neglecting the small bleed flow and the fuel addition)

$$c_p(T_{03} - T_{02}) = c_{pe}(T_{04} - T_{05}) \qquad (12.1)$$

where c_p and c_{pe} are average values of the specific heat appropriate for the compressor and turbine which are held constant. Through the compressor c_p and γ are taken to be equal to 1005 J/kgK and 1.40 respectively, whilst through the turbine 1244 J/kgK and 1.30 will be adopted.

The energy release during combustion is found, as in Exercise 11.7, to be

$$\dot{m}_f LCV = (\dot{m}_{air} + \dot{m}_f)c_{pe} (T_{04} - 298) - \dot{m}_{air} c_p (T_{03} - 298) . \qquad (12.2)$$

Exercise
12.1 Use the design pressure ratio across the Viper for operation on a sea-level test bed with standard atmospheric conditions to find the temperature ratio across the compressor and the stagnation temperature at compressor discharge, T_{03}. From the temperature rise across the compressor find the temperature drop across the turbine using suitable values of gas properties.
(Ans: 1.72, $T_{03} = 494.8$ K, $\Delta T_{0c} = 206.8$ K, $\Delta T_{0t} = 167.1$ K)
From the given air and fuel flow rates find the energy release in the combustion per kg of air flow. Take $LCV = 43$ MJ/kg. Hence find the turbine inlet temperature T_{04} and the stagnation temperature T_{05} and pressure p_{05} downstream of the turbine. Treating the nozzle as isentropic, find the jet velocity and compare with the value derived from thrust in the above table. **(Ans:** 771 kJ/kg, $T_{04} = 1063$ K, $T_{05} = 896$ K, $p_{05} = 233$ kPa, $V_j = 625$ m/s)

Note: a discrepancy of 12 m/s in jet velocity is fortuitously small having in mind the simplicity of the approach, the neglect of pressure loss in the combustor and the assumptions for efficiency.

The turbine pressure ratio and power output

The turbine and the nozzle are both effectively choked. As a result

$$\bar{m} = \frac{\dot{m}\sqrt{c_p T_{0in}}}{A p_{0in}} = 1.389 \qquad (12.3)$$

into each, or $\bar{m}_4 = \bar{m}_9 = 1.389$.

The propelling nozzle entry conditions also correspond to turbine exit. Assuming that the loss in stagnation pressure in the jet pipe is negligible, the stagnation pressure at the nozzle exit is equal to that at turbine exit, $p_{09} = p_{05}$, and likewise, $T_{09} = T_{05}$. It is now possible to divide the normalised mass flows to give

$$\frac{\bar{m}_4}{\bar{m}_9} = \frac{A_9}{A_4}\frac{p_{05}}{p_{04}}\frac{\sqrt{T_{04}}}{\sqrt{T_{05}}} = 1$$

or

$$\left(\frac{p_{05}}{p_{04}}\right)^2 = \left(\frac{A_4}{A_9}\right)^2 \times \frac{T_{05}}{T_{04}} \ . \tag{12.4}$$

This equation, relating the conditions upstream and downstream of the turbine, follows from the presence of two choked components, the turbine and the propelling nozzle, one after the other. (A similar relation will be found to hold between each of the turbines in a multi-shaft engine, as will be discussed below.)

The temperatures and pressures upstream and downstream of the turbine are also related by the polytropic efficiency, so that it is possible to write

$$\left(\frac{p_{05}}{p_{04}}\right)^{\frac{\eta_p(\gamma-1)}{\gamma}} = \frac{T_{05}}{T_{04}} \ . \tag{12.5}$$

Equations (12.4) and (12.5) each relate pressure ratio and temperature ratio across the turbine. It is a consequence of the form of these equations that they can only be satisfied by unique values of pressure ratio and temperature ratio. More surprisingly the magnitude of each ratio is determined by the ratio of areas at turbine inlet and at the nozzle, so eliminating T_{05}/T_{04} from (12.4) and (12.5) gives

$$\left(\frac{p_{05}}{p_{04}}\right)^2 = \left(\frac{A_4}{A_9}\right)^2 \times \left(\frac{p_{05}}{p_{04}}\right)^{\frac{\eta_p(\gamma-1)}{\gamma}} \ .$$

For $\gamma = 1.30$ and $\eta_p = 0.9$ it follows that $\eta_p(\gamma-1)/\gamma$ is equal to 0.208 , so the above equation can be written as

$$\frac{p_{05}}{p_{04}} = \left(\frac{A_4}{A_9}\right)^{1.11} \tag{12.6a}$$

leading to

$$\frac{T_{05}}{T_{04}} = \left(\frac{A_4}{A_9}\right)^{0.232} \ . \tag{12.6b}$$

Reducing the propelling nozzle area A_9 gives an *almost* proportional increase in p_{05}/p_{04}, that is it reduces the pressure ratio across the turbine. With the assumption of constant turbine efficiency the reduction in pressure ratio across the turbine corresponds to a reduction in power output. Increasing the area of nozzle guide vanes (i.e. the stator blades) A_4 has the same effect. Once the turbine and nozzle are choked the pressure ratio and temperature ratio across the turbine can only be changed by altering one of the areas and these are used in practice as the means of matching engines during their development.

If the temperature ratio T_{05}/T_{04} is a constant wholly determined by the area ratio, equation (12.6b), it is then easy to show that the temperature drop in the turbine is proportional to the turbine inlet temperature, and given by

$$T_{04} - T_{05} = T_{04}(1 - T_{05}/T_{04}) = k_H T_{04} \qquad (12.7)$$

where k_H is a constant fixed by the area ratios. In other words the turbine work per unit mass is proportional to the turbine inlet temperature and given by $c_{pe}k_H T_{04}$.

The power produced by the turbine must be equal to the power into the compressor. Neglecting the increase in mass flow through the turbine because of the fuel and the flow bled off the compressor for cooling and other purposes this balance in power can be expressed as

$$c_p(T_{03} - T_{02}) = c_{pe}(T_{04} - T_{05})$$

$$= c_{pe}k_H T_{04}. \qquad (12.8)$$

From equation (12.8) the temperature ratio in the compressor T_{03}/T_{02} is known once the inlet temperature T_{02} is specified. If the compressor efficiency can be taken as constant (which presumes that the engine can match at a condition in which the compressor is operating efficiently) this allows the pressure ratio to be found,

$$\frac{p_{03}}{p_{02}} = \left(1 + \frac{T_{03} - T_{02}}{T_{02}}\right)^{\eta_p \gamma/(\gamma-1)}$$

and

$$\frac{p_{03}}{p_{02}} = \left(1 + k_H \frac{c_{pe}}{c_p} \frac{T_{04}}{T_{02}}\right)^{\eta_p \gamma/(\gamma-1)} \qquad (12.9)$$

If the engine overall pressure ratio and turbine inlet temperature are known at one condition, such as the design point, the value of k_H can be determined. It is worth remarking that within the approximations adopted here, the pressure ratio of the compressor is wholly determined by the ratio of turbine inlet area to nozzle area and the turbine inlet temperature ratio T_{04}/T_{02}.

A graphical view of the turbine and nozzle in series

Figure 12.2 shows side by side the characteristics of the turbine and the final nozzle plotted as normalised mass flow versus pressure ratio. (The normalised mass flow $\dot{m}\sqrt{c_p T_0}/p_0$ is related to the non-dimensional mass flow $\bar{m}$, but for simplicity the area term has been removed. The area term has been dropped because it is independent of the flow but changes from one component to another.) For the turbine the pressure ratio is p_{04}/p_{05} and for the nozzle the pressure ratio is p_{05}/p_9, where p_9 is the static pressure at nozzle exit.

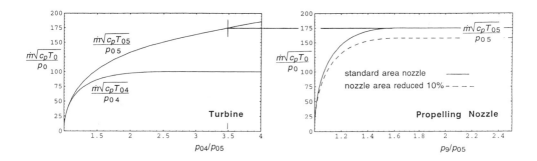

Figure 12.2. A graphical representation of the matching of a turbine with a downstream propelling nozzle. The horizontal line from the choked non-dimensional mass flow for the nozzle to the turbine fixes the turbine non-dimensional mass flow (in terms of variables at turbine exhaust) and the pressure ratio across the turbine.

The characteristic for the turbine in Fig. 12.2 is shown in two ways: the conventional way, where the stagnation pressure and temperature are the *inlet* values p_{04} and T_{04}, and also based on the *exit* conditions p_{05} and T_{05}. (Such characteristics would be obtained from tests carried out without a constant area choked nozzle downstream of the turbine.) The important feature of the turbine exit properties is that they form the inlet conditions to the nozzle. In terms of exit conditions the normalised mass flow $\dot{m}\sqrt{c_p T_{05}}/p_{05}$ does *not* become constant when the turbine chokes, but instead $\dot{m}\sqrt{c_p T_{05}}/p_{05}$ increases continuously because as the pressure ratio is increased the value of p_{05} falls. (In calculating the normalised mass flow in terms of exit conditions for Fig. 12.2 the polytropic efficiency of the turbine has been maintained constant at 0.85.)

Because the value of $\dot{m}\sqrt{c_p T_{05}}/p_{05}$ at turbine outlet is the value at propelling nozzle inlet, a horizontal line has been drawn on Fig. 12.2 to indicate this match. So long as the nozzle is choked there is only one value of $\dot{m}\sqrt{c_p T_{05}}/p_{05}$ at which the two components can be matched, and this therefore fixes the pressure ratio across the turbine. If the propelling nozzle area is reduced (the dotted curve in Fig. 12.2) there will be a proportional drop in normalised mass flow at choke and the pressure ratio across the turbine will also necessarily reduce. A reduction in nozzle area therefore leads to reduced power output from the turbine; reduced power from the turbine leads to less power to the compressor, which decreases the pressure rise in the compressor and in turn decreases the mass flow swallowed by the compressor. To swallow this reduced mass flow and produce the reduced pressure ratio requires a decrease in compressor rotational speed.

It can be seen from Fig. 12.2 that if the nozzle becomes unchoked, as it does at low rotational speeds and pressure ratios, nothing exceptional happens. The calculations are no longer so simple, however, because normalised mass flow can vary, and an iterative calculation is needed.

The same increase in complexity occurs when the dependence of turbine normalised mass flow or efficiency on speed is included in the calculation.

The case shown in Fig. 12.2 is the simplest possible. The dependence, however, is identical for two or more turbines in series; in this case the swallowing capacity, in terms of normalised mass flow, of the downstream turbine determines the pressure ratio in the upstream turbine, whilst for the most downstream turbine the pressure ratio is determined by the propelling nozzle.

Exercise
12.2 Use the design point conditions to determine k_H for the engine of Exercise 12.1. When the turbine inlet temperature is reduced to 900 K, find the temperature drop in the turbine and thence the temperature rise in the compressor. Assuming that the polytropic efficiencies of the compressor and turbine do not alter from the values assumed at design point, find the pressure ratio across the compressor and the stagnation pressure in the jet pipe. Treating the flow in the nozzle as isentropic, find the jet velocity.

(Ans: $k_H = 0.157$, 141.7 K, 175.1 K, pr $= 4.46$, 189 kPa, 502 m/s)

The compressor working line

Although the pressure ratio across the compressor is fixed by the ratio of turbine inlet area to propelling nozzle area, the mass flow of air through the engine is not yet determined. To get this it is necessary to again consider the choked condition at turbine inlet, equation (12.3),

$$\bar{m}_4 = \frac{\dot{m} \sqrt{c_{pe} T_{04}}}{A_4 p_{04}} = 1.389 \ ,$$

because the compressor will be forced to supply a pressure ratio and mass flow to satisfy this. We can assume that there is a negligible loss in pressure across the combustor, so $p_{03} = p_{04}$. Using equation (12.8), $T_{03} - T_{02} \propto T_{04}$ and the condition for mass flow can be rewritten as

$$\frac{\dot{m} \sqrt{c_p (T_{03} - T_{02})}}{p_{03}} = \text{constant},$$

which defines the working line, sometimes known as the operating line of the engine.

The compressor mass flow non-dimensionalised by conditions at *exit* is given by

$$\bar{m}_3 = \frac{\dot{m} \sqrt{c_p T_{03}}}{A_3 p_{03}} = \bar{m}_4 \frac{A_4}{A_3} \frac{\sqrt{c_p T_{03}}}{\sqrt{c_{pe} T_{04}}}$$

where the pressure drop in the combustor is neglected, $p_{04} = p_{03}$.

The more conventional non-dimensional mass flow for the compressor, based on *inlet* conditions follows from

$$\bar{m}_2 \quad = \frac{\dot{m}\ \sqrt{c_p T_{02}}}{A_2 p_{02}} \quad = \bar{m}_4 \frac{A_4}{A_3} \frac{\sqrt{c_p T_{03}}}{\sqrt{c_{pe} T_{04}}} \frac{p_{03}}{p_{02}} \frac{A_3}{A_2} \frac{\sqrt{T_{02}}}{\sqrt{T_{03}}}$$

$$= \bar{m}_4 \frac{A_4}{A_2} \frac{p_{03}}{p_{02}} \frac{\sqrt{c_p T_{02}}}{\sqrt{c_{pe} T_{04}}} . \tag{12.10}$$

Since both the non-dimensional mass flow $\bar{m}_2$ and the pressure ratio are given in terms of the turbine inlet temperature it is easy to plot the compressor working line on conventional axes.

Exercises

12.3 If for the engine of Exercise 12.1 the corrected mass flow at the design point is 23.8 kg/s, find the mass flow when the turbine inlet temperature is 900K. At this temperature find the gross thrust, the fuel mass flow and the specific fuel consumption.

(**Ans:** 20.9 kg/s, $F_G = 10.6$ kN, $\dot{m}_f = 0.288$ kg/s , $sfc = 0.958$ kg/h/kg)

12.4 Show that the minimum pressure p_{05} in the jet pipe to choke the nozzle of the Viper on a sea-level test bed is 185 kPa. Find the turbine inlet temperature at which this occurs, assuming (correctly) that the turbine remains choked. By using Fig.11.1 and considering equations (12.3) to (12.6), indicate how the working line will be altered by the unchoking of the final nozzle.

(**Ans**: $T_{04} = 884$ K)

Compatible values of compressor pressure ratio and non-dimensional mass flow have been calculated for the Viper engine from equations (12.9) and (12.10) and used to create the working line plotted on Fig. 12.3 for the Viper engine. The working line has been chosen so as to pass through the design point, so the value of mass flow shown in Fig. 12.3 is normalised by the value at the design point. Superimposed are values of the temperature T_{04} at each end of the working line, which is almost a straight line. The working line does *not* require the compressor rotational speed to be selected; rather, given the working line, the compressor 'chooses' the rotational speed which gives the necessary pressure ratio and mass flow. Alternatively the compressor may be imagined to select the speed which absorbs the turbine power output and passes the flow necessary to choke the turbine inlet guide vanes.

To predict the off-design rotational speed it is necessary to have the measured pressure ratio versus non-dimensional mass flow characteristics of the compressor.

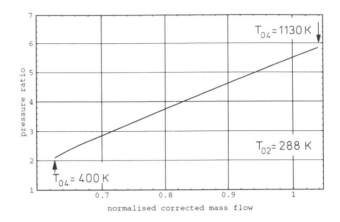

Figure 12.3. The pressure ratio versus non-dimensional mass flow
(working line) for the Viper.

It is a necessary presumption in the approach used here that the working line remains in a part of the compressor map such that it does not stall and that it is reasonable to assume that the efficiency remains constant. If the working line takes the compressor into areas of the map where the efficiency varies significantly then an iterative calculation is necessary, which requires measured or estimated values of the compressor performance. If the turbine or the nozzle cease to be choked, an iteration is also necessary. Codes used in industry would certainly iterate.

The results of the present simple calculation are compared with measurements for this engine compressor with measured engine working lines superimposed, Fig. 12.4. In Fig. 12.4(a) the working line was obtained with a normal sized propelling nozzle and the normal turbine nozzle guide vane area. The calculated and measured working lines move apart at low rotational speeds, largely because the efficiency of the compressor starts to fall. This reduction in compressor efficiency as speed is reduced along the working line becomes more pronounced as the design pressure ratio is increased, for reasons which are explained later in the chapter.

In Fig. 12.4(b) the working line was measured with the turbine nozzle guide vane area reduced by 20% and the propelling nozzle area reduced by 11%; compared to the nominal ratio, A_4/A_9 has been reduced to 0.90 times the nominal value, thereby affecting the turbine temperature ratio (and therefore the power output) according to equation (12.6). Superimposed on the measurements of Fig. 12.4(b) are the working lines from the above analysis. The agreement is reasonably good, particularly at the higher pressure ratios, showing the correct trend with mass flow for each working line and the correct effect of the change in areas. The discrepancy at the lower speeds and pressure ratios is almost certainly because in this region the compressor efficiency has fallen considerably and the nozzle may have unchoked.

Figure 12.4 allows the rotational speed to be estimated for the reduced operating temperatures: at $T_{04} = 1000$ K, for example, the pressure ratio was found in Exercise 12.2 to be 4.38, leading to a rotational speed of about 85% of the design value.

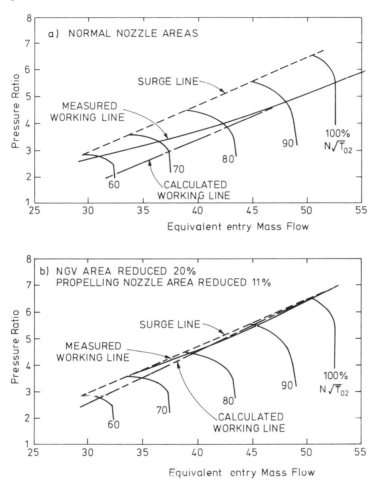

Figure 12.4. The predicted working line for the Viper superimposed on measured characteristics for the compressor. Case(a) with the normal NGV and propelling nozzle areas; case (b) with reduced areas.

Shown in Fig. 12.5 are the variations in mass flow and gross thrust with temperature ratio. Of particular note is the very rapid fall in the thrust with T_{04}/T_{02} .

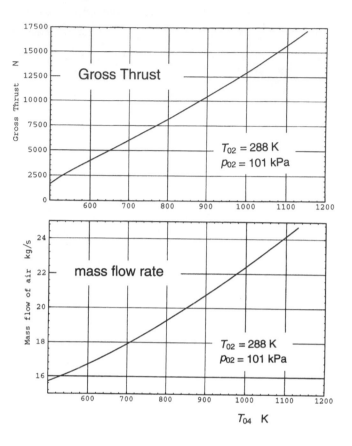

Figure 12.5. Predicted gross thrust and air mass flow rate through the Viper on a
sea-level test bed versus turbine inlet temperature.

SUMMARY OF SECTIONS 12.1 AND 12.2

The matching of conditions for the single spool turbojet requires that the rotational speed, mass flow and power are the same for the compressor and turbine, with the pressure changes compatible. In general this requires the characteristic maps of the compressor and turbine to be used in an iterative calculation.

Enormous simplification is possible when the special nature of the turbine characteristic is utilised in approximate calculations. The calculations are capable of indicating the correct trends. In terms of its upstream effect the turbine behaves as a choked nozzle; the efficiency of the turbine is only a weak function of rotational speed for a given pressure ratio. For a simple turbojet the propelling nozzle is choked for most conditions of interest, which makes the turbine effectively choked at inlet and outlet. Under this condition, and with the turbine efficiency constant, the turbine is forced to operate at a constant stagnation pressure ratio and stagnation temperature ratio. Turbine pressure ratio and temperature ratio are then fixed by the ratio of the turbine nozzle guide vane throat area and the propelling nozzle area. It follows immediately that the turbine enthalpy drop is proportional to the stagnation temperature at turbine inlet.

With the turbine work per unit mass flow, and therefore the compressor work, proportional to the turbine inlet temperature it is easy to find the pressure ratio across the compressor if the compressor efficiency is known as a function of turbine entry conditions. With the flow choked at combustor outlet (turbine inlet) the non-dimensional mass flow at compressor inlet can be found in terms of the turbine inlet condition and the compressor working line is then defined.

Although this section was devoted to a simple jet engine the approach is immediately transferable to gas turbines with a power turbine replacing the propelling nozzle: turboprops, land based or marine prime movers. If the power turbine is choked the analysis is then virtually identical.

Staying on the working line for the engine but reducing rotational speed leads to a drop in compressor efficiency. This causes a rise in the working line towards the surge or stall line. The tendency of compressors to surge or stall at low speed, see section 11.4, becomes greater as the design pressure ratio is increased.

12.3 A TWO-SHAFT TURBOJET ENGINE

The tendency of the working line to move the operating point towards the compressor surge line is a problem for all engines; the problem gets more acute as the design pressure ratio is increased. The most significant way of alleviating the effect is to use separate HP and LP compressors on concentric shafts which are able to rotate at different speeds. The compressors are able to select the speed at which they can meet the requirement for the non-dimensional mass flow and pressure ratio. Section 12.2 considered the simplest form of gas turbine and this treatment provides the approach here to a slightly more complicated engine with two shafts. On each shaft there is a compressor and turbine but, being a straight turbojet (there is no bypass), all of the flow passes through each component .

The engine is shown schematically in Fig. 12.1(b) with the numbering system for the stations shown. The power from the LP turbine must be equal to that absorbed by the LP compressor so again neglecting the mass flow rate of fuel and the cooling and bleed flows

$$c_p(T_{023} - T_{02}) = c_{pe}(T_{045} - T_{05}) \qquad (12.11a)$$

and likewise for the HP shaft

$$c_p(T_{03} - T_{023}) = c_{pe}(T_{04} - T_{045}). \qquad (12.11b)$$

It will be assumed that both the LP and HP turbines are effectively choked and that the final propelling nozzle is also choked. This gives rise to the following relation

$$\bar{m}_4 = \bar{m}_{45} = \bar{m}_9 = 1.389$$

i.e.
$$\frac{\dot{m}\sqrt{c_{pe}T_{04}}}{A_4 p_{04}} = \frac{\dot{m}\sqrt{c_{pe}T_{045}}}{A_{45} p_{045}} = \frac{\dot{m}\sqrt{c_{pe}T_{05}}}{A_9 p_{05}} = 1.389 \ . \qquad (12.12)$$

In the expression for $\bar{m}_9$ it has been assumed that there is negligible change in stagnation pressure or temperature between LP turbine outlet and the propelling nozzle exit, giving $p_{05} = p_{09}$ and $T_{05} = T_{09}$.

As before it is assumed that the turbine efficiencies are sufficiently insensitive to incidence that the polytropic efficiencies of both may be taken to be constant. Then across the HP turbine

$$\left(\frac{p_{04}}{p_{045}}\right)^{\frac{\eta_p(\gamma-1)}{\gamma}} = \frac{T_{04}}{T_{045}} \ . \qquad (12.13)$$

The choking relations, equation (12.12), gives rise to the relation between pressures and temperature for the HP turbine,

$$\frac{\bar{m}_4}{\bar{m}_{45}} = \frac{A_{45}}{A_4} \frac{p_{045}}{p_{04}} \frac{\sqrt{T_{04}}}{\sqrt{T_{045}}} = 1$$

or
$$\left(\frac{p_{045}}{p_{04}}\right)^2 = \left(\frac{A_4}{A_{45}}\right)^2 \times \frac{T_{045}}{T_{04}} \ . \qquad (12.14)$$

The reasoning here is just as it was for the turbine in the single-shaft engine considered in section 12.2. Equations (12.13) and (12.14) can be simultaneously satisfied only for particular values of stagnation pressure and stagnation temperature ratios; combining these two equations gives

$$\left(\frac{p_{045}}{p_{04}}\right)^2 = \left(\frac{A_4}{A_{45}}\right)^2 \times \left(\frac{p_{045}}{p_{04}}\right)^{\frac{\eta_p(\gamma-1)}{\gamma}} \ .$$

As in section 12.2, for $\gamma = 1.30$ and $\eta_p = 0.9$ the value of $\eta_p(\gamma-1)/\gamma$ is approximately equal to 0.208, so the above equation can be written

$$\frac{p_{045}}{p_{04}} = \left(\frac{A_4}{A_{45}}\right)^{1.11} \quad \text{and} \quad \frac{T_{045}}{T_{04}} = \left(\frac{A_4}{A_{45}}\right)^{0.232} \ . \qquad (12.15a)$$

The results are entirely equivalent for the LP turbine

$$\frac{p_{05}}{p_{045}} = \left(\frac{A_{45}}{A_9}\right)^{1.11} \quad \text{and} \quad \frac{T_{05}}{T_{045}} = \left(\frac{A_{45}}{A_9}\right)^{0.232} \ . \qquad (12.15b)$$

The pressure and temperature ratios of both turbines are therefore uniquely fixed by the ratio of areas, the pressure ratio being almost in linear proportion to the area ratio.

With T_{045}/T_{04} fixed across the HP turbine it follows that the HP turbine work per unit mass (equal to the power per unit mass flow rate) is given by

$$\dot{W}_{HP}/\dot{m} = c_{pe}(T_{04} - T_{045}) = c_{pe}T_{04}(1 - T_{045}/T_{04})$$

i.e.
$$T_{04}(1 - T_{045}/T_{04}) = k_{HP}T_{04} \qquad (12.16a)$$

where k_{HP} is a constant fixed by the relative areas of the two turbines. In an analogous way for the LP turbine

$$\dot{W}_{LP}/\dot{m} = c_{pe}(T_{045} - T_{05}) = c_{pe}T_{045}(1 - T_{05}/T_{045})$$

i.e.
$$T_{045}(1 - T_{05}/T_{045}) = \text{const} \times T_{045}.$$

Because, however, the ratio T_{045}/T_{04} is fixed by equation (12.16a), the LP turbine work is also proportional to the temperature entering the HP turbine, T_{04}, and is more conveniently written in terms of this and the constant k_{LP} as

$$T_{045}(1 - T_{05}/T_{045}) = k_{LP}T_{04}. \qquad (12.16b)$$

Using equations (12.11a) and (12.11b) the LP and HP compressor temperature rises can be shown to be equal to $k_{LP}T_{04}c_{pe}/c_p$ and $k_{HP}T_{04}c_{pe}/c_p$ respectively. The pressure ratio for the LP compressor is then given by

$$\frac{p_{023}}{p_{02}} = \left(1 + k_{LP}\frac{c_{pe}}{c_p}\frac{T_{04}}{T_{02}}\right)^{\gamma/(\gamma-1)} \qquad (12.17)$$

with a similar expression for the HP pressure ratio.

An example engine – the Olympus 593

There are not many examples of two-shaft turbojet engines. By the time the technology was sufficiently advanced to be able to build two-shaft engines it was also evident that for most applications a bypass stream was desirable[1]. For very high speed propulsion the turbojet is still a viable alternative and the Olympus 593 used in the Concorde is an example of such an engine. It is illustrated in Fig.5.2. Some information about its operation at cruise is as follows:

---

[1] It is sometimes supposed that the early jet engines were single-shaft turbojets because the designers knew no better. In fact, the delay in introducing multi-shaft bypass engines arose principally from the mechanical problems which had to be overcome with even the simplest turbojet. Dr. A.A. Griffiths, working for Rolls Royce in the 1940's, drew an engine scheme in 1941 with a bypass ratio of 8. a suggestion not far from what is now regarded as optimum for subsonic propulsion.

Altitude	51000 ft
Mach number	2.0
Overall pressure ratio in compressors p_{03}/p_{02}	11.3

The LP compressor and the HP compressor each have seven stages, the LP and HP turbines each have one stage. The engine is highly developed and polytropic efficiencies of 90% for all components may be pessimistic. At 51000 ft for the International Starndard Atmosphere (ISA) the ambient pressure is 11.0 kPa and the temperature is 216.65 K.

Exercises

12.5 Find the stagnation pressure and temperature of the air entering the Olympus 593 engine at cruise. Neglect stagnation pressure drop in intake.

(**Ans:** p_{02} = 86.0 kPa ; T_{02} = 390 K)

Assuming equal pressure ratios across the LP and HP compressors, find the stagnation pressure and temperature out of each. What are the temperature drops in each of the turbines? Neglect the mass flow rate of fuel.

(**Ans:** p_{023} = 289kPa, T_{023} = 573.0K, p_{03} = 971kPa, T_{03} = 841.9K, ΔT_{0HP} = 217.2K, ΔT_{0LP} = 147,8K)

If the temperature of the gas leaving the combustor is 1300 K, find k_{HP} and k_{LP} for the two turbines. Determine the temperature and pressure downstream of the LP turbine. (Ignore the pressure loss in the combustor. Note that at the cruise condition the afterburner is not used.) Find the jet velocity assuming that the nozzle is isentropic and fully expanded.

(**Ans:** k_{HP} = 0.167, k_{LP} = 0.114, T_{05} = 935 K, p_{05} = 199.0 kPa, V_j = 1065 m/s)

12.6 If the final propelling nozzle area is increased by 10% on the Olympus 593 whilst at cruise for the conditions given in Exercise 12.5, what is the effect on the engine if the inlet conditions and the turbine inlet temperature are held constant. Find the new k_{LP} and the jet velocity.

(**Ans:** k_{LP} = 0.130, p_{023}/p_{02} = 3.87, p_{03}/p_{023} = 3.21, p_{05}/p_{045} = 0.444, p_{05} = 197 kPa, V_j = 1051 m/s)

Note: on the Olympus 593 the variable nozzle is used to improve efficiency and reduce the noise during the climb. By opening the nozzle the LP power is increased and the LP shaft turns faster. This in turn increases the mass flow into the engine. The overall effect is to get the same amount of thrust with a higher mass flow and lower jet velocity than would have been the case with a constant nozzle area.

Dynamic scaling and non-similar conditions

In Chapter 8 the dynamic scaling and use of dimensional analysis was considered. It will be recalled that for this to be applied to an engine it was essential that the conditions inside the engine remained the same, that is to say all the pressure ratios, temperature ratios, $N/\sqrt{c_p T_0}$ for all the shafts, etc. were held constant. It was perhaps remarkable how much could be learned in this way, but it was not possible to make deductions when the engine conditions were significantly different.

Such circumstances are most important when the flight speed is high. For the cruise conditions adopted for a subsonic airliner the stagnation temperature entering the engine is about 259 K, not far removed from the sea-level temperature of the standard atmosphere 288 K. For a flight speed of Mach 2 the inlet stagnation temperature is very much higher, about 390 K; because of the constraint imposed by the maximum temperatures which the engine materials can withstand, at $M = 2.0$ the engine is forced to operate with a substantially lower pressure ratio across the compressor. An example is given below for the Olympus 593 at the cruise conditions of the Concorde.

Exercises

12.7   Taking the results from Exercise 12.5 for the Olympus, find the ratio of turbine inlet to compressor inlet temperature T_{04}/T_{02} at cruise. If this ratio were held constant, what would T_{04} be on a static sea-level test? During take off the turbine inlet temperature is allowed to increase to about 1450 K, what is T_{04}/T_{02}?   (**Ans:** $T_{04}/T_{02} = 3.33$, $T_{04} = 960$ K $T_{04}/T_{02} = 5.03$)
    With the propelling nozzle having its nominal area (before the increase in the last part of Exercise 12.6) find the overall pressure ratio, the pressure in the jet pipe and the jet velocity at the take-off condition with no afterburner. (Assume that the flow in the nozzle is isentropic.)   At this condition the engine passes 186 kg/s of air – what is the gross thrust?
                    (**Ans:** $p_{03}/p_{02} = 24.2$, $p_{05}/p_a = 4.95$, $V_j = 896$ m/s , $F_G = 167$ kN)

12.8   Find the overall pressure ratio at take off for $T_{04} = 1450$ K when the propelling nozzle area is increased by 10%.   What is the mass flow, jet velocity and gross thrust with this increased nozzle opening? Assume that losses in the engine are unchanged and the nozzle remains isentropic.
                    (**Ans**: $p_{03}/p_{02} = 27.1$, $\dot{m} = 208$ kg/s, $V_j = 885$ m/s, $F_G = 184$ kN)

The engine working lines

For the two-shaft engine, as for the single-shaft engine of section 12.3, the compressor power is effectively fixed by the turbine and nozzle area ratios and the turbine inlet temperature, as explained above. To find the mass flow requires the area ratio, pressure rise and temperature rise in the compressor in conjunction with the HP turbine nozzle guide vane area. The HP turbine nozzle row is assumed to be choked, so that

$$\bar{m}_4 \quad = \quad \frac{\dot{m}\ \sqrt{c_{pe}T_{04}}}{A_4 p_{04}} = 1.389 \ .$$

It is then easy to see that the non-dimensional mass flow into the LP compressor is given by

$$\bar{m}_2 \quad = \quad \frac{\dot{m}\ \sqrt{c_p T_{02}}}{A_2 p_{02}} = 1.389 \ \frac{\sqrt{c_p T_{02}}}{\sqrt{c_{pe}T_{04}}} \frac{A_4}{A_2} \frac{p_{03}}{p_{02}} \qquad (12.18a)$$

where the pressure into the turbine may be assumed equal to that leaving the HP compressor, $p_{03} = p_{04}$. The ratio of areas A_4/A_2 will have been chosen by the designer to give the required compressor mass flow at the design conditions and is fixed; the ratio of specific heats may also be regarded as constant but the pressure and temperature ratio vary with engine operating condition. The turbine inlet temperature is fixed by the temperature rise in the compressors and the fuel flow but the square root of temperature ratio varies very much less than the pressure ratio. As a result $\bar{m}_2$ falls steeply as the pressure ratio is reduced (in other words, as the fuel supply to the engine is reduced) and the rotational speed falls. The variation in $\bar{m}_2$ is mainly determined by the *overall* pressure ratio and is the same whether the engine is a single- or two-shaft engine. The variation of p_{023}/p_{02} with $\bar{m}_2$ is the working line of the LP compressor. The non-dimensional mass flow into the HP compressor is likewise given by

$$\bar{m}_{23} \ = 1.389 \ \frac{\sqrt{c_p T_{023}}}{\sqrt{c_{pe} T_{04}}} \frac{A_4}{A_{23}} \frac{p_{03}}{p_{023}} \tag{12.18b}$$

and it is primarily the pressure ratio across the HP compressor which determines its variation. Since the pressure ratio across the HP is less than the overall pressure ratio, the excursion in non-dimensional mass flow is correspondingly smaller. The variation of p_{03}/p_{023} with $\bar{m}_{23}$ defines the working line of the HP compressor.

Exercises

12.9 For the Olympus 593 find the mass flow, the gross and net thrust at the cruise condition with the nominal nozzle area. Use the information given in Exercise 12.7.

(**Ans:** $\dot{m}$ = 78.0 kg/s, F_G = 83.1 kN, F_N = 37.1 kN)

12.10 It was shown in connection with the single-shaft turbojet that the compressor working line is defined by $\dot{m}\sqrt{(\Delta T_0)}/p_{03}$ = constant. Confirm that this applies to the HP compressor of a two-shaft engine and show that the working line of the LP compressor can be described by

$$\frac{\dot{m} \sqrt{(\Delta T_{0LP})}}{p_{023}} \frac{\sqrt{(\Delta T_{0HP})}}{\sqrt{(\Delta T_{0LP})}} \frac{p_{023}}{p_{03}} = \text{constant}.$$

The need for multiple shafts

The fall in $\bar{m}_2$ as engine speed is reduced, with the tendency of the operating point to move towards the compressor surge line, is a problem for all engines and one which increases as the design pressure ratio is increased. The primary means of alleviating the effect is to adopt separate HP and LP compressors on separate shafts which are able to rotate at different speeds. The compressors are able to select the speed at which they can meet the requirement for the non-dimensional mass flow and pressure ratio. We can draw a working line on the pressure ratio versus mass flow map for each compressor to determine what rotational speed is required.

The use of multiple shafts has an additional benefit for the design point operation. The front stage of a compressor must not operate with a blade speed which is too high in relation to local sonic velocity lest the efficiency falls sharply. The temperature rises through the compressor, so that the blade speed for the later stages could be significantly higher without high losses. With a compressor on a single shaft there is a severe limit on how much the blade speed can be increased (because this can only be done by increasing the mean radius) but by splitting the compressor into two or more parts it is possible to let the HP compressor rotate substantially faster.

Figure 12.6 shows the working lines calculated using equations (12.17) and (12.18) for the HP and LP compressors of the Olympus 593. The calculations have been carried out with T_{04}/T_{02} in the range 2 to 5. 2, the latter being the value at take off. In working out these the k_{HP} and k_{LP} have been taken from Exercise 12.3. Note that the pressure ratio at take off is substantially greater than the cruise value of 11.3 overall, a result of the higher values of T_{04}/T_{02} being allowed at take off. The variation in non-dimensional mass flow into the LP is the same as it would be for the single-shaft version, since this is determined by the overall pressure ratio which is the same in each case, but the variation into the HP is very much smaller. The compressor maps are not available for this engine, but a common property can be used to assess the changes in the shaft speeds needed to accommodate the changes in flow rate. For compressors with modest peak pressure ratios, such as that shown in Fig.11.4, the non-dimensional mass flow along a working line falls rapidly with rotational speed: about a 10% reduction in flow for a 10% reduction in speed is shown here. (For a machine giving a high pressure ratio, such as that shown in Fig.11.5, the mass flow falls very much faster as the speed is reduced.)

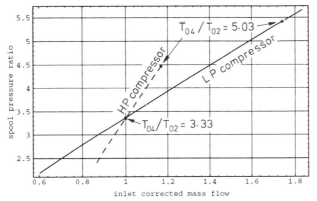

Figure 12.6. The working lines for the Olympus engine compressors. The LP and HP pressure ratios are assumed to be equal at cruise (T_{04}/T_{02} = 3.33), at which condition the non-dimensional mass flow into each is normalised to unity.

Because the working line for the LP compressor shows a much greater variation in mass flow than the corresponding HP compressor, Fig. 12.6, the excursions in LP speed needed to match these variations are very much greater.

This is common to all multi-shaft engines, with the innermost shaft changing speed much less than the outer ones: for a three-shaft engine the non-dimensional rotational speed of the HP may vary by less than 5% over much of the engine operating range[1] .

Exercise

12.11 Sketch the working lines in Fig. 12.6 on the compressor map, Fig.11.5. If the non-dimensional shaft speeds are 100% at the cruise condition, what are they at take off?

(**Ans:** $N_H/\sqrt{T_{023}} \approx 110\%$, $N_L/\sqrt{T_{02}}$ too large to be estimated on this figure.)

Note: The compressor used to produce the map in Fig.11.4 is very different from those in the Olympus engine. This comparison only serves to show how large the speed excursions need to be on the LP shaft compared with those on the HP.

SUMMARY OF SECTION 12.3

The behaviour of multi-shaft engines can be determined easily, if approximately, by using the particular flow features of the turbine and nozzle in a way introduced earlier for a single-shaft engine: the final propelling nozzle is choked and for each shaft the turbine behaves as if it were choked and operates with almost constant efficiency over a wide range of speeds and pressure ratios. Combining this means that the ratios of stagnation temperature and pressure are constant across each turbine, leading to the work per unit mass flow from each turbine being proportional to T_{04}, the temperature at entry to the HP turbine.   Since the power from each turbine is the power into the compressor on the same shaft, the temperature rise in each compressor is also proportional to T_{04}.   Assuming constant polytropic efficiency, the pressure ratios in the compressors can be found.

The mass flow is fixed by the choking of the HP turbine and the final nozzle. If the area of the final nozzle is increased the pressure is lowered in the jet pipe. This has the effect of increasing the power output of the LP turbine, which raises the speed of the LP shaft, in turn raising the mass flow swallowed by the LP compressor.   Similar effects can be predicted for changes in the area of the turbine nozzle guide vanes.

As the engine is throttled back and the pressure ratio and non-dimensional mass flow fall, the front stages of a compressor move towards stall. Having two shafts allows the speed of each shaft to vary to meet the needs of the compressor, thereby reducing the mismatch of the front stages of each compressor. The changes experienced by the LP compressor are much greater than the HP and the changes in rotational speed are correspondingly larger.

---

[1] An analogous effect is found with turbocharged diesel engines. The reciprocating part of the engine is equivalent to the core of the jet engine, while the turbocharger is equivalent to the LP part of a jet engine. As the speed of the reciprocating part is altered the turbocharger alters proportionately much more.

12.4 A TWO-SHAFT HIGH BYPASS TURBOFAN ENGINE

The groundwork has been laid for the consideration of a high bypass ratio engine of the type used in the design exercise in the earlier chapters. So far as the combustor, turbine and core nozzle are concerned, the treatment is identical to the two-shaft turbojet. With the compressor the flow treatment is complicated somewhat because only a small fraction of the flow through the fan enters the core compressor.

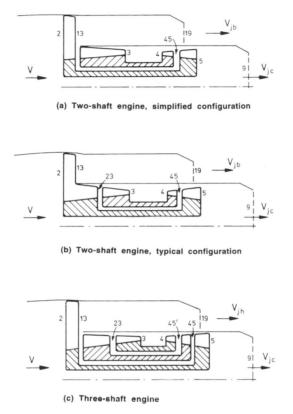

(a) Two-shaft engine, simplified configuration

(b) Two-shaft engine, typical configuration

(c) Three-shaft engine

Figure 12.7. Schematic layouts and engine station numbering schemes for
high bypass ratio engines.

It has been shown that there are practical difficulties in using compressors with large pressure ratios on one shaft and the pressure ratio on one shaft is normally kept to no more than 20. The cause of the difficulty is that the non-dimensional mass flow at exit from the compressor is nearly constant and the reduction in overall pressure ratio with reduction in rotational speed leads to a drop in mass flow at inlet to the compressor which can be large in comparison with the reduction in rotational speed. The engines designed in the early chapters had an overall pressure ratio of 40 at cruise, of which 1.6 was produced in the fan root and 25 in the core. The simplified design based on this had the fan alone on the LP shaft and all the remaining core compression on the HP shaft, see Fig. 12.7(a). Looking at the designs of two-shaft engines adopted by Pratt and Whitney, Fig.5.4(b), and General Electric, Fig.5.4(c), it can be seen that some of the compression of the core flow takes place on the LP shaft in booster stages – for the reason given at the start of this paragraph. It is only sensible to take note of this and a more realistic layout for a two-shaft engine is shown schematically in Fig. 12.7(b).

Just as for the turbojet considered in section 12.3, the power balance for the HP shaft is, assuming equal mass flow rate in the HP compressor and HP turbine, given by

$$c_{pe}(T_{04}-T_{045}) = c_p(T_{03} - T_{023}) . \tag{12.19}$$

Again, as for the turbojet, the pressure and temperature ratios across the turbines are fixed by the turbine NGV areas and the propelling nozzle area, this time the nozzle for the core, all of which are assumed choked. As a result, for the HP turbine the work per unit mass is

$$c_{pe}(T_{04}-T_{045}) = k_{HP}c_{pe}T_{04} \tag{12.20a}$$

and for the LP $$\qquad c_{pe}(T_{045}-T_{05})= k_{LP}c_{pe}T_{04}. \tag{12.20b}$$

The mass flow rate through the core $\bar{m}_c$ is found in a way exactly analogous to that of the turbojet. For the core the constraint of the choked HP turbine NGV gives, with $\gamma = 1.30$,

$$\bar{m}_{c4} = \frac{\dot{m}_c\sqrt{c_{pe}T_{04}}}{A_4 p_{04}} = 1.389. \tag{12.21}$$

The bypass flow is always choked at the cruise conditions and very nearly choked at take-off conditions, so for this flow, with $\gamma = 1.40$,

$$\frac{\dot{m}_b \sqrt{c_p T_{013}}}{A_{19}p_{013}} = 1.281 . \tag{12.22}$$

The mass flow conditions, equations (12.21) and (12.22), can be simplified to a form convenient for calculation as

$$\dot{m}_c = k_c \, p_{03}/\sqrt{T_{04}}$$

for the core flow and

$$\dot{m}_b = k_b p_{013}/\sqrt{T_{013}} \approx k_b (T_{013})^{2.65} \tag{12.23}$$

for the bypass stream.

The non-dimensional mass for the core at fan inlet is, following from equation 12.21,

$$\bar{m}_{c2} = \frac{\dot{m}_c \sqrt{c_p T_{02}}}{A_2 p_{02}} = 1.389 \frac{\sqrt{c_p T_{02}}}{\sqrt{c_{pe} T_{04}}} \frac{A_4}{A_2} \frac{p_{03}}{p_{02}} . \qquad (12.24)$$

The major difference between the bypass engine and the turbojet comes when the power balance for the LP shaft is considered. When the mass of fuel and the air removed in bleeds from the compressor are neglected the equation is,

$$\dot{m}_c c_{pe}(T_{045} - T_{05}) = c_p\{\dot{m}_b(T_{013} - T_{02}) + \dot{m}_c (T_{023} - T_{02}) \}$$

where T_{013} is the temperature downstream of the fan in the bypass and T_{023} is the temperature downstream of the booster stages.

The equation for the power balance can be changed to

$$\frac{c_{pe}}{c_p} k_{LP} T_{04}/T_{02} = \frac{\dot{m}_b}{\dot{m}_c}(T_{013}/T_{02} - 1) + (T_{023}/T_{02} - 1) \qquad (12.25)$$

where the ratio of mass flow through the bypass to that through the core, $\dot{m}_b/\dot{m}_c = bpr$ is the bypass ratio. The difficulty arises because the bypass ratio does not remain constant as the engine operating condition is altered: whilst T_{04}/T_{02} is an input variable, equation (12.25) contains 3 unknowns, bpr, T_{013}/T_{02} and T_{023}/T_{02}. To make the problem tractable with the minimum of empirical input we make the simple but plausible assumption that the temperature rise for the core flow across the fan and booster stages $T_{023} - T_{02}$ is proportional to the temperature rise of the bypass flow across the fan $T_{013} - T_{02}$. In other words, comparing the temperature rises on-design and off-design the ratios are constant,

$$\frac{T_{023} - T_{02}}{T_{013} - T_{02}} = \frac{\{T_{023} - T_{02}\}_{design}}{\{T_{013} - T_{02}\}_{design}} = k_t . \qquad (12.26)$$

It is assumed that the efficiency of the compressors will remain constant (which is plausible provided the excursion from the design point is not too large) and that the compressor will be able to adopt a speed at which the simultaneous requirements of the working line for pressure ratio and mass flow are simultaneously satisfied. Because the bypass ratio alters with the engine operating point the fraction of the air through the fan which enters the core also changes. The flow pattern inside the fan is therefore also changed and a precise calculation would require detailed knowledge of the way the fan behaves as a function of speed, overall mass flow and bypass ratio.

The power balance for the LP shaft, equation (12.25), then simplifies to

$$\frac{c_{pe}}{c_p} k_{LP} T_{04}/T_{02} = bpr\{\frac{T_{013}}{T_{02}} - 1\} + k_t\{\frac{T_{013}}{T_{02}} - 1\}$$

$$= \{bpr + k_t\}\{\frac{T_{013}}{T_{02}} - 1\}. \tag{12.27}$$

Equations (12.23) and (12.27) are easily solved iteratively to give the temperature ratio across the fan T_{013}/T_{02} for a given turbine inlet temperature ratio, T_{04}/T_{02}. All of the constants (k_{LP}, k_{HP}, k_c, k_b and k_t) can be found from the design performance specification of the engine. With the temperature ratio across the fan known all the other variables are easily calculated.

The application of this model will be illustrated for the example of bypass ratio 6 evaluated in Exercise 7.1. At the design point (start of cruise at 31000 ft and $M = 0.85$) the turbine inlet temperature $T_{04} = 1450$ K and the inlet stagnation temperature $T_{02} = 259.5$K, giving the ratio $T_{04}/T_{02} = 5.588$. Taking the design overall pressure ratio to be 40, Exercise 7.1 shows that for a bypass ratio of 6 the fan pressure ratio is 1.81. Whereas in the design exercise the pressure ratio across the fan as it affects the core was specified to be only 1.6, we shall take $p_{023}/p_{02} = 2.50$ across the fan and boosters into the core at the design point. With this assumption the pressure ratio across the HP compressor is $p_{03}/p_{023} = 40/2.5 = 16.0$ at the design point. We shall also assume *polytropic* efficiencies for all components equal to 0.90 – this is not strictly compatible with the assumptions in the design leading to Exercise 7.1 where *isentropic* efficiencies were used. However, the discrepancy will be relatively small and the convenience great. (By taking the more realistic values of $\gamma = 1.30$ and $c_{pe} = 1244$ J/kgK for the flow through the turbine in this example we have also introduced an inconsistency with example 7.1 where $\gamma = 1.40$ and $c_p = 1005$ J/kgK were used throughout the engine.)

Exercises
12.12   Verify that in equation (12.27) all the terms can be shown to depend on T_{04}/T_{02}; in other words that this is the sole independent variable which defines the non-dimensional operating point of the engine.

12.13   For the two-shaft high bypass engine laid out in Exercises 7.1 and 7.2 find the temperature rise of the core flow in the compressors on each shaft at start of cruise, 31000 ft and $M = 0.85$. The overall pressure ratio in the engine is 40 at this condition and the pressure ratio for fan and booster stages is 2.5. For the case with bypass ratio of 6 take the pressure ratio of the fan at design to be 1.81 for the bypass stream.
(**Ans:** $T_{013} - T_{02} = 53.8$ K, $T_{023} - T_{02} = 87.6$ K , $T_{03} - T_{023} = 490.1$ K)
Find the values of k_{LP} and k_{HP} if the turbine inlet temperature $T_{04} = 1450$ K, that is when $T_{04}/T_{02} = 5.589$. (**Ans:** for $bpr = 6.0$; $k_{LP} = 0.229$, $k_{HP} = 0.273$)

For ease of calculation take the *polytropic* efficiencies to be equal to 90% for all compressor and turbine components. Take γ equal to 1.4 for unburned air and 1.30 for the products of combustion.

Figure 12.8 shows the working lines for the fan and core compressor calculated from the iterative solution of equations (12.23) and (12.27). The non-dimensional mass flow has again been normalised by the value at design condition for which $T_{04}/T_{02} = 5.588$; the corresponding pressure ratio is 1.81 for the fan in the bypass, 2.5 for the fan and boosters in the core and 16 for the HP compressor. Superimposed on the working lines are points showing values of the turbine inlet temperature ratio T_{04}/T_{02}, going from 6.0 at the upper end down to 4.0. It will be seen from Table 5.1 that the top-of-climb condition is $T_{04}/T_{02} = 6.07$, when the engine is at its highest non-dimensional condition, whereas take off is somewhat lower at about 5.9 for a standard day and $T_{04} = 1700$ K. The working lines do not in themselves allow the relative speeds of the LP and HP shafts to be obtained, these must be found when the performance map of the compressor is available.

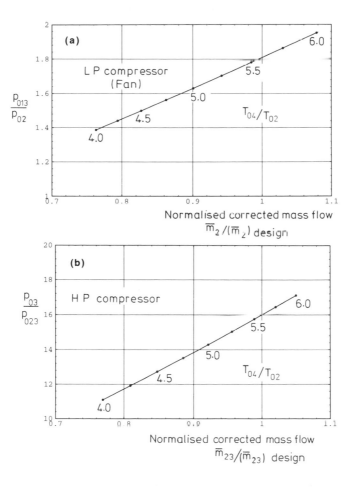

Figure 12.8. The working lines for the LP compressor (fan) and HP compressor for
a two-shaft engine with a bypass ratio of 6.0.

Exercises

12.14 For the two-shaft high bypass engine discussed in the previous section, show that the design point pressure ratio of the core compressor, 16.0, intersects the 95% speed line for the E^3 compressor shown in Fig.11.6 close to the marked working line. Take the 95% line to be the design speed line for the HP compressor of the new engine, occurring when $T_{04}/T_{02} = 5.588$. Take combinations of pressure ratio and normalised mass flow from Fig. 12.8 for $T_{04}/T_{02} = 4.0$ and 6.0 and plot them on Fig.11.5. Draw a straight line between these points to indicate the approximate working line of the new engine – does the simple model give a working line of an engine close to that measured by GE?

Use the speed lines to estimate the HP speed when $T_{04}/T_{02} = 4.0$ as a percentage of that speed at the engine design point.

(**Ans:** For $T_{04}/T_{02} = 4.0$, $N_{HP}/\sqrt{T_{02}} \approx 95\%$ of value at design point)

12.15 A measured fan characteristic is shown in Fig.11.4. At the design point for the new engine the fan pressure ratio is 1.81. Show that if the fan is sized so that this pressure ratio is achieved close to the conditions for maximum efficiency, the normalised mass flow is about 1.005 and the 'design corrected speed' is about 101%. Take the fan pressure ratio and normalised mass flow for $T_{04}/T_{02} = 4.0$ and 6.0, shown in Fig. 12.8, and superimpose these on the measured fan characteristic shown in Fig.11.3. Draw a straight line to indicate an approximate working line. Estimate the LP shaft speed corresponding to $T_{04}/T_{02} = 4$.

(**Ans:** For $T_{04}/T_{02} = 4.0$, $N_{LP}/\sqrt{T_{02}} \approx 77\%$ of value at design point for the engine)

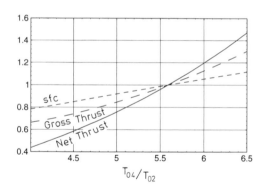

Figure 12.9. Variation in gross thrust, net thrust and sfc with T_{04}/T_{02} for the two-shaft engine with a bypass ratio of 6.0. (Results normalised by values for design point at $M = 0.85$ at 31000 ft with $T_{04} = 1450$ K, $T_{04}/T_{02} = 5.588$).

The calculation for the high bypass engine can be extended to calculate many of the overall performance parameters. Figure 12.9 shows the variation in net thrust, gross thrust and specific fuel consumption with T_{04}/T_{02} over the range 4.0 to 6.5, all for start of cruise, $M = 0.85$ at 31000 ft. All the values are normalised by the condition at the design point, for which $T_{04}/T_{02} = 5.588$. The highest temperature ratio $T_{04}/T_{02} = 6.5$ would correspond to a turbine inlet temperature of 1687 K at cruise conditions and 1870 K on a sea-level test bed – not far above what may soon be acceptable. The net thrust varies more in proportional terms than the gross thrust because the constant 'ram drag' is subtracted from the latter. The steep variation in net thrust with turbine inlet temperature means that when T_{04}/T_{02} is increased from the cruise value of 5.588 to about 6.07 at top-of-climb (see Table 5.1) the net thrust is increased by about 24% above the design point value; reducing T_{04}/T_{02} down to 4.0 reduces the net thrust to about 44% of the design point value.

In section 8.4 the issue of engine-out at take off was addressed and at this condition gross thrust is more relevant because the forward speed is relatively low. Allowing an increase in temperature to 1850 K to cope with this situation at sea-level and low forward speed (so $T_{02} =$ 288 K) would raise T_{04}/T_{02} to about 6.4, giving about 30% increase in gross thrust. This is still not enough to give the margin required with a two-engine plane having engines of bypass ratio equal to 6 and sized just large enough to give the required thrust at start of cruise at 31000 ft. As noted in Chapter 8 the two-engine aircraft is likely to have engines sized for cruise at a higher altitude to give adequate thrust during take off with one engine.

As T_{04}/T_{02} is decreased there is a fall in jet velocity and an increase in bypass ratio. The fall in V_j leads to an increase in propulsive efficiency. Together these effects more than offset the drop in thermodynamic efficiency associated with the fall in pressure ratio and maximum temperature so the net effect of reducing T_{04}/T_{02} is a reduction in sfc evident in Fig. 12.9.

Figure 12.10(a) shows the strong monotonic increase in overall pressure ratio inside the engine, p_{03}/p_{02}, with turbine inlet temperature, T_{04}/T_{02}. The corresponding increase in fan pressure ratio is even larger. As T_{04}/T_{02} is increased bypass ratio falls, which can be understood when the different constraints on mass flow through the fan and the core are understood. Because the bypass nozzle is choked the non-dimensional mass
flow downstream of the fan in the bypass stream must be constant. The non-dimensional mass flow through the core is constant at HP turbine entry and this imposes the mass flow constraint on the non-dimensional mass flow through the core at fan inlet $\bar{m}_{c2}$. As T_{04}/T_{02} is reduced the overall pressure ratio falls steeply, and so too does $\bar{m}_{c2}$. The constant non-dimensional mass flow in the bypass and the reduced non-dimensional mass flow into the core as T_{04}/T_{02} is decreased lead to an increase in bypass ratio.

The pressure inside the core jet pipe, p_{05}, is shown in Fig. 12.10(b) non-dimensionalised in one case by the ambient pressure p_a for cruise at $M = 0.85$ and in the other by the inlet stagnation pressure p_{02}. The ratio p_{05}/p_a is appropriate to the cruise Mach number of 0.85 and gives the stagnation to static pressure ratio across the propelling nozzle of the core. The ratio p_{05}/p_{02} is appropriate for operation on a stationary test bed (or approximately for an aircraft at low speeds) when the inlet stagnation pressure is equal to the ambient pressure. Figure 12.10(b) also shows the minimum pressure ratio at which the nozzle will choke, 1.83 for exhaust gases with $\gamma = 1.30$. For cruise conditions the propelling nozzle remains choked down to T_{04}/T_{02} ≈ 5.1, whereas when stationary the nozzle is unchoked below $T_{04}/T_{02} \approx 6.2$. As noted in section 12.2, the variation in non-dimensional mass flow with pressure ratio is small down to pressure ratios across a nozzle of about 1.4, see Fig.11.1; the analysis of this chapter will therefore not be seriously in error as soon as the nozzle unchokes and for the cruise condition should be good down to $T_{04}/T_{02} \approx 4.5$. Significantly this figure shows the difficulty of simulating correct engine conditions on a stationary engine and indicates why facilities to simulate forward speed, or flying test beds, remain invaluable. The effect of unchoking the core propelling nozzle is to drop the pressure in the jet pipe for the same mass flow, and this increases the power from the LP turbine, thereby increasing the fan pressure rise and flow rate. To take this into account requires an iterative solution which incorporates the variation in non-dimensional mass flow shown in Fig.11.2.

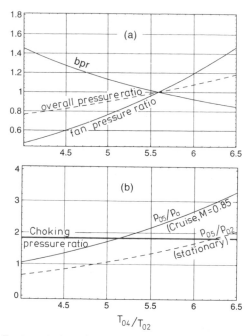

Figure 12.10. For two-shaft engine with bypass ratio of 6.0 at design point
($M = 0.85$ at 31000 ft with $T_{04} = 1450$ K, $T_{04}/T_{02} = 5.588$).
Case (a) Variation with T_{04}/T_{02} of bpr, overall pressure ratio and fan pressure ratio values normalised by design point values. Case (b) Variation with T_{04}/T_{02} of jet pipe stagnation pressure referred to ambient static pressure p_a and to engine inlet stagnation pressure p_{02}.

The bypass nozzle will be choked at cruise to substantially lower values of T_{04}/T_{02}; even at $T_{04}/T_{02} = 4$ the fan pressure ratio is 1.39 (Fig. 12.8) and a forward speed of $M = 0.85$ gives a ram pressure ratio of 1.60 leading to a pressure ratio across the bypass nozzle $p_{013}/p_a = 2.23$, well above the choking value for air, 1.89. With the engine stationary, however, the nozzle will be unchoked for T_{04}/T_{02} less than about 5.7.

Exercise

12.16   The mass flow at cruise is given in Exercise 7.2 with T_{04} = 1450K.   If the turbine inlet temperature were increased by 100 K, use Fig. 12.8 to find the new value of fan pressure ratio p_{013}/p_{02}. (Use of Fig. 12.8 is recommended to remove the need for iterative calculations.) Hence find the new mass flow through the core and bypass (remember to include the 'ram' pressure rise in the intake). Assuming that the acceleration of the flow in the propelling nozzles is reversible, find the jet velocity for the core and bypass and then find the gross and net thrust

(Ans: $\dot{m}_C$ = 85.2 kg/s, $\dot{m}_b$ = 471 kg/s, V_{jc} = 625 m/s, V_{jb} = 424 m/s, F_G = 253 kN, F_N = 110 kN)

Note: The increase in net thrust in Exercise 12.16 seems large relative to the value in Exercise 7.1 scaled up using the curve in Fig. 12.9.   This is because the parameters used in the calculation in Exercise 12.16 have been altered to make them more realistic, most significantly the use of different cp and γ for the combustion products. With these parameters the engine at its nominal condition, T_{04}/T_{02} = 5.588, gives F_N= 92 kN, compared with F_N =75 kN in Exercise 7.1. The difference shows up most clearly in the higher jet velocity of the core.

In section 12.3, in the context of the two-shaft turbojet, the dependence of turbine pressure ratio and temperature ratio, and thence power output, on area ratio was shown. Because each turbine was choked at inlet and because the polytropic efficiency for each was effectively constant, it is possible to write for the HP turbine

$$\left(\frac{p_{045}}{p_{04}} \right)^2 = \left(\frac{A_4}{A_{45}} \right)^2 \times \left(\frac{p_{045}}{p_{04}} \right)^{\frac{\eta_p(\gamma-1)}{\gamma}}$$

and a similar expression for the LP turbine. With γ = 1.30 and η_p = 0.9, $\eta_p(\gamma-1)/\gamma$ is equal to 0.208 and simple expressions can be found, as set out in equation (12.15a). The selection of the throat areas of the HP and LP turbine nozzle guide vanes and the area of core propelling nozzle, are what determines the power input to the shafts. By setting the ratios of these areas an engine is matched in a practical sense.

Exercise

12.17 For the two-shaft high bypass engine considered above (in Exercises 12.13 to 12.16) consider the effect of a 5% reduction in area of the HP nozzle guide vane, the other areas remaining unchanged. The turbine inlet temperature remains at 1450 K. Find the new values of k_{LP} and k_{HP} and indicate the direction of change in the LP and HP compressors.

(Ans: k_{HP} = 0.282, k_{LP} = 0.226)

SUMMARY OF SECTION 12.4

The treatment here has been for the civil transport engine which has formed the basis of the design in the early chapters. The engine was assumed to have two shafts and a simple assumption was introduced to accommodate the pressure ratio of the booster stages on the LP shaft; this simplification is not expected to alter the trends substantially. An underlying assumption, that the core propelling nozzle is choked, was found to be violated at turbine inlet temperatures which were surprisingly high in relation to the design values; this is not a major problem at cruise, when the Mach number is 0.85, but it does have a major impact for operation on a test bed.

   The ideas carry across into the consideration of modern military engines which are mainly turbofans of the two-shaft type with typical design values of bypass ratio in the range 0.3 up to 1.0. Whereas the non-dimensional operating point for civil engines does not normally alter very much, for military engines it varies greatly and an understanding of the matching 'off-design' is essential.

   In Chapter 8 dynamic scaling of the engine was considered when the non-dimensional operating point was held constant. One way of specifying this constancy was that T_{04}/T_{02} was constant. In the present chapter T_{04}/T_{02} was found to be the independent variable in the equations which determine the operating condition and the effect of changes in the ratio were considered. In other words the treatment here is compatible with the stipulation in Chapter 8 that if T_{04}/T_{02} is constant the engine operating point is fixed.

12.5 A THREE-SHAFT HIGH BYPASS TURBOFAN ENGINE

The treatment of the three-shaft is analogous to that of the two-shaft engine, though there is some increase in complexity, mainly in nomenclature. The way two- and three-shaft engines behave off-design is rather different, and this is discussed towards the end of the section.

   With this configuration, illustrated schematically in Fig. 12.7(c), the LP spool consists only of the fan on one end and the LP turbine on the other. We will assume again that the stagnation pressure and temperature are uniform in the radial direction downstream of the fan; conditions at this station will be denoted by 13 from hub to casing, while those out of the IP compressor and into the HP are 23. For the turbine the numbering system is, as before, 4 into the HP turbine, 45 into the LP turbine and 5 out of the LP turbine. The conditions entering the IP turbine are denoted by 45'. It is assumed that at the design point the fan pressure ratio of 1.81 extends to the hub, so the combined pressure ratio of the core at this condition is 40/1.81 = 22.1.

If the pressure ratios of the IP and HP compressors are taken to be equal it follows that at the design point $p_{023}/p_{013} = p_{03}/p_{023} = 4.70$.

The power balance for the LP and HP shafts is essentially identical to those for the two-shaft engine, equation (12.19) and (12.25). Similarly equations (12.23) for the mass flow through the bypass and the core are the same. The extra conditions are for the IP shaft and these, derived in the same way as those for the two-shaft engine, are

$$c_p(T_{023} - T_{013}) = c_{pe}(T_{045'} - T_{045}) = k_{IP}T_{04} \qquad (12.28)$$

and
$$\bar{m}_{c13} = k_c\, p_{03}/p_{013}/\sqrt{T_{04}/T_{013}}. \qquad (12.29)$$

As for the two-shaft engine, the equation for the power balance in the LP shaft, (12.28), must be solved numerically to find T_{013}/T_{02}, the temperature ratio across the fan.

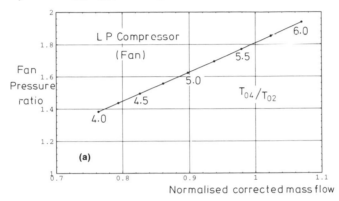

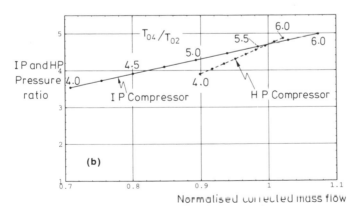

Figure 12.11 The working lines for the LP compressor (fan), IP compressor and
HP compressor for a three-shaft engine with a bypass ratio of 6.0.

In the off-design treatment of the two-shaft engine in section 12.4 two types of results were presented. One was the overall performance, such as the variation in thrust, sfc, bypass ratio and overall pressure ratio. These results will be identical for the three-shaft engine with the same parameters such as design pressure ratio and constant polytropic efficiency. The difference occurs in the pressure ratio–mass flow variations for the compressors. In place of Fig. 12.8 for the two-shaft engine is Fig. 12.11 for the three-shaft engine, with the temperature ratio T_{04}/T_{02} varied from 4.0 to 6.0. In Fig. 12.11(b) both the IP and HP working lines are shown together, with the mass flow into each normalised by the value at the design point. It is immediately apparent that the range of operating conditions is very much greater for the IP than the HP. It may also be noticed that the excursion in non-dimensional mass flow for the IP is very similar to that for the core compressor of the two-shaft engine in Fig. 12.8 and would be identical if the booster stages were removed but the overall pressure ratio maintained. This is a direct consequence of the overall core pressure ratio being involved in the specification of the core non-dimensional mass flow.

   The fan working line shown in Fig. 12.11(a) is essentially identical with that for the two-shaft engine.

Exercise
12.18 From the IP and HP working lines in Fig. 12.11(a) determine the pressure ratios and normalised mass flows for $T_{04}/T_{02} = 4.0$ and 6.0. Plot these on the compressor map in Fig.11.5 and connect them with straight lines as approximate working lines.
Put the design point $p_{023}/p_{013} = p_{03}/p_{023} = 4.70$, for the IP and HP compressor on the 100% $N//\sqrt{T_{02}}$ line, which fixes the normalised mass flow at design. Estimate the reduction in speed for each compressor when T_{04}/T_{02} is reduced from the design point value of 5.588 down to 4.0. Do any potential difficulties arise?   (**Ans:** $N_{HP}/\sqrt{T_{023}} = 93.5\%$, $N_{IP}/\sqrt{T_{013}} = 86\%$)

The reduced non-dimensional mass flow at low values of T_{04}/T_{02} is a consequence of the overall pressure ratio and is the same for two- and three-shaft engines. The way the compressors are forced to behave is rather different. The working line for the IP compressor tends to be rather shallow, so that at low non-dimensional mass flow rates the pressure ratio is comparatively high, and there is a risk of stall or surge. The strongest effect in reducing the rotational speed is the magnitude of the mass flow, so the IP compressor slows down to accommodate this, with the working line 'moving up' the constant speed line to give the necessary pressure ratio.

   For the two-shaft engine the working line is steeper, so that the fall in overall pressure rise is greater as a proportion of the total. The reduction in speed to allow the working line to be satisfied by the compressor characteristic is smaller than was the case of the IP compressor in the three-shaft machine. The problem for the two-shaft engine is the high incidence in the front stages at low values of T_{04}/T_{02} consequent on the large reduction in non-dimensional mass flow at inlet and relatively small reduction in rotational speed. Although the whole compressor may be well removed from the stall line the front stages may be approaching stall; the remedy is variable stagger stator blades for the front stages.

SUMMARY CHAPTER 12

The behaviour of multi-shaft engines can be determined easily, if approximately, by using the particular flow features of the turbine and nozzle. For each shaft the turbine behaves as if it were choked and the final propelling nozzle is choked over most of the important operating range. In addition the efficiency of each turbine is almost constant over a wide range of speeds and pressure ratios. Combining choking and constant efficiency means that the ratios of stagnation temperature and pressure are constant across each turbine, leading to the work per unit mass from each turbine being proportional to T_{04}, the temperature at entry to the HP turbine. Since the power from each turbine is the power into the compressor on the same shaft, the temperature rise in each compressor is also proportional to T_{04}. Assuming constant polytropic efficiency, the pressure ratios in the compressors can be found.

The mass flow through the core in a bypass engine is fixed by the choking of the HP turbine nozzle guide vanes and the non-dimensional mass flow is therefore constant at this location. The condition at turbine entry determines the non-dimensional mass flow at the front of the HP compressor (or, in the case of three-shaft engine, the IP compressor). The non-dimensional mass flow in the core is proportional to $(p_{03}/p_{02}) \sqrt{(T_{02}/T_{04})}$; since the pressure ratio varies much more rapidly than the square root of temperature ratio so the non-dimensional mass flow falls rapidly as the temperature into the turbine falls. This leads to high incidence and possible stall of the front stages of the compressor and it is to reduce the extent of this that the multi-shaft configuration is used.

The present analysis defines the working lines for the compressor and engine. It does not in itself fix the rotational speed; nor does it determine whether the engine can operate at the condition chosen. When the working line is superimposed on the compressor map of pressure ratio versus non-dimensional mass flow the shaft rotational speed at each turbine inlet temperature can be obtained. If the working line extends above and to the left of the surge line operation is not possible without a redesign of the compressor; if the working line takes the compressor into a region where the efficiency is very low a recalculation is necessary.

The system described in this chapter is designed to allow *simple* calculations which will illustrate the way in which engines operate off-design. To make this simple not only are the component efficiencies held constant and the turbine and final nozzle assumed choked, but some effects are neglected. Such neglected effects are the cooling and other bleeds and the power off-takes, but it is not conceptually difficult to include these. To include the variation in compressor or turbine efficiency simply requires an iteration based on the measured (or estimated) characteristics of the components. When the nozzles cease to be choked the simple constancy of non-dimensional mass flow is no longer valid and the measured relation between mass flow and pressure ratio of the turbine and final nozzle must be used. The way that this can be performed is illustrated in Cohen, Rogers and Saravanamuttoo (1996). For most of the operating range of jet engines where performance is critical, that is omitting starting and idling conditions, the simple approximations give appropriate indications of the trends.

Part 3

Design of Engines for a

New Fighter Aircraft

13.0 INTRODUCTION

This part of the book begins the consideration of the engine requirements of a new fighter aircraft. In parallel with the treatment in earlier chapters for the engines of the new large civil aircraft, the approach chosen is to address the design of engines for a possible new aircraft so that the text and exercises can be numerically based with realistic values. The specifications for the New Fighter Aircraft used here have a marked similarity to those available for the new Eurofighter.

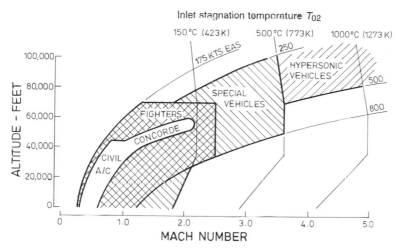

Figure 13.1. Aircraft operating envelopes, altitude versus Mach number, with contours of inlet stagnation temperature. (From Denning and Mitchell, 1989.)

The topic of the present chapter is the nature of the combat missions and the type of aircraft involved. Figure 13.1 shows the different regions in which aircraft operate in terms of altitude versus Mach number, with the lines of constant inlet stagnation temperature overlaid. We are concerned here with what are referred to in the figure as fighters, a major class of combat aircraft. Figure 13.1 shows the various boundaries for normal operation. Even high-speed planes do not normally fly at more than $M = 1.2$ at sea level because in the high density air the structural loads on the aircraft and the physiological effects on the crew become too large. At high altitude high-speed aircraft do not normally exceed $M \approx 2.3$, largely because the very high stagnation temperatures preclude the use of aluminium alloys without cooling. The boundary to the left in Fig. 13.1 corresponds to the aircraft having insufficient speed to create the necessary lift; this boundary can be expressed in terms of the equivalent airspeed, that is the airspeed at sea level having the same value of $\frac{1}{2}\rho V^2$. The stalling boundary flattens at very high altitude to give the operational ceiling. Some special aircraft operate outside these boundaries but they will not be considered here.

For civil transport aircraft at subsonic conditions the lift-drag ratio is around 16 for an aircraft of the 747 generation, rising to 20 for newer ones. For high-speed aircraft the lift-drag ratio is lower and a value of between 6 and 7 seems realistic. This shifts the optimum type of engine from a relatively heavy one with low fuel consumption for civil transport to one which is lighter but with higher fuel consumption. In any case the requirements in a fighter for acceleration, turning and high speed demand a much higher thrust-weight ratio of the engine than is to be expected for transport aircraft. Engine weight is very important because an increase in the weight of one component, for example the engine, requires changes in other parts of the aircraft, most obviously wing area and structural strength of the airframe, and this adds additional weight. A rule of thumb is that growth in the aircraft will take place such that a 1kg increase in the weight of any component leads to between 4 and 5 kg increase in overall aircraft weight for a subsonic aircraft and between 6 and 10 kg increase for an aircraft capable of supersonic flight. In other words, 1kg of extra weight in each engine of a two-engine aircraft able to fly at supersonic speeds would increase the aircraft weight by between 12 and 20 kg.

13.1 THE TYPES AND NEEDS OF COMBAT AIRCRAFT

The civil transport has a comparatively easily defined task - to carry the largest payload between airports at the lowest cost. Although the precise optimum depends on the distance between the airports and the fraction of the maximum payload which is likely to be used, the duty is relatively straightforward. In contrast the task of a combat aircraft is normally complicated with the performance classified under three headings: field requirements (for take off and landing), mission requirements (where the performance, in particular fuel consumption, over the whole mission must be adequate) and point requirements (to enable acceleration, turning, etc., to take place). Indeed, if an aircraft is recognisably deficient in one area, an opponent can immediately seek out this vulnerability.

Different aircraft are designed for different duties, but pressures to reduce cost often lead to large overlaps. Sometimes there is a decision to design an aircraft with several disparate missions from the start. A clear example of this is the European Multi-Role-Combat-Aircraft, later called the Tornado, which had quite different requirements for the different countries taking part in the programme. Although there were variations between the different classes of Tornado, the same configuration (and, in particular, the same engines) had to be able to operate as a high-altitude interceptor or as a low-altitude ground attack aircraft. The most interesting aircraft with multiple roles at the moment is the American Joint Strike Fighter, recently renamed the F-35, to be built by Lockheed-Martin. This aircraft is intended to enter service in 2007, initially with the US Air Force in a conventional form, then with the US Navy in an aircraft carrier based form and then later to the US Marines and British Royal Navy in a short or vertical take and landing configuration. This aircraft is therefore designed to cover a wide range of different tasks.

Some of the classifications of combat aircraft and their attributes are summarised below; they can be divided into two broad classes. In the first (air superiority and interception) the aircraft is involved in combat with other aircraft, in the second (battlefield interdiction and close air support) the idea is to use the aircraft to attack ground targets. The situation becomes more confused when aircraft designed with one clear primary mission are adapted for a totally different one. An example of this might be the F-15, which was designed primarily as a high-speed, air-superiority fighter, but is also equipped to carry bombs in a secondary attack role.

Air superiority

This is the role of the traditional fighter aircraft. Manoeuvrability is the key performance issue and speed now seems to be held to be less crucial – it is believed that most fighting would take place in the subsonic range between $M \approx 0.7$ and 0.9. At the highest speeds that many aircraft are capable of, around $M = 2$, the fuel consumption is so large that the duration of combat is necessarily very short; in addition the turning circle is so high that conventional fighting may be impossible. Examples of air-superiority aircraft are the US Air Force F-15 and F-16, the US Navy F-14 and F-18, the Russian Mg-29 and Su-27, the French Mirage 2000. The new[1] F-22, the French Rafaele and the Eurofighter will be of this class.

Manoeuvrability is the ability of the aircraft to be able to accelerate and turn very quickly. To be able to turn quickly means that there must be a relatively large wing area for the mass of aircraft. The area of wing is usually expressed in terms of the wing loading, the load per unit area of wing when the lift is equal to the normal maximum weight of the aircraft. The normal weight of the aircraft corresponds to 1g loading, in other words, weight $=$ $g \times$ mass. In many manoeuvres the load on the wings is much greater than this, perhaps going as high as 9g, (referred to as a *load factor* of 9) which is about the upper level which the human body can withstand. A low wing loading corresponds to large wings for the normal weight of the aircraft; large wings allow a very large lift force to be created before the wings stall, which allows rapid turns with a comparatively small radius of curvature. Large wings, however, create more drag and can, by their inertia, make the aircraft less responsive in roll. They can also make the ride exceedingly bumpy in low-altitude, high-speed flight.

---

[1]The term 'new' takes on a rather special meaning here. For example, the need for the what has become the F-22 was formally identified in 1981. In 1986 two companies were selected by the USAF to build prototypes and in 1991 the Air Force selected the Lockheed F-22 with the P&W F119 engine, awarding a \$9.55 Billion contract for engineering and manufacturing development. The aircraft is currently expected to enter service with USAF in 2005. Eurofighter has been developed over the same period at a rate which is not significantly faster.

In specifying the optimum aircraft it is normal to calculate the performance over idealised typical missions. Such a mission is shown in Fig. 13.2 for an air-superiority aircraft where a significant period of loitering is involved; loitering involves remaining in an area for relatively long periods using as little fuel as possible. It should be noted that whereas the whole mission lasts about two hours, the combat portion is only one minute. Fuel consumption during the cruise and loiter period can therefore be far more influential to the total aircraft weight at take off, and therefore its size and cost, than the fuel used during the brief period of combat.

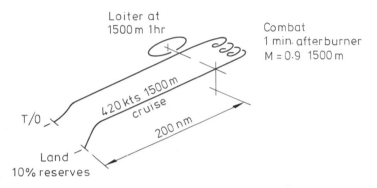

Figure 13.2. A simple, idealised mission for an air-superiority aircraft.
(Derived from Garwood, Round and Hodges, 1995.)

Both rapid turns and high acceleration in the direction of travel require a high level of thrust and a high value of the ratio of engine thrust to aircraft weight. During cruise and loiter phases of a mission the thrust will be much less than the maximum used during combat and the principal requirement then is low fuel consumption. Some of this variation in thrust is achieved by the use of the afterburner, sometimes also called reheat, which is common to most fighter aircraft. In afterburning operation fuel is burned in the jet pipe, downstream of the turbine but upstream of the nozzle, to give a boost in thrust of the order of 50%. When the afterburner is not being used the operation is referred to as 'dry'.

The balance which determines the design of the aircraft and its engines depends on geo-political factors as well as technical issues. An important issue relates specifically to fuel consumption whilst not engaged in combat. During the cold war the United States needed to be able to ferry its combat aircraft over the Atlantic to meet a potential threat in Europe and even with in-flight refuelling the fuel consumption at the optimum condition for range was very critical. The Soviet Union, on the other hand, did not have the requirement to ferry its planes long distances over water and could give less attention to the range and duration of flight.

Interception

This type of aircraft is part of an air defence system, with the aircraft despatched to intercept intruders detected by radar. When attack by bombers was perceived as a major strategic threat special interceptor aircraft were built, but the need for these special aircraft has receded somewhat and modern air-superiority fighters also usually serve as interceptors.

For the specialised interceptor long range is not normally required, but what is needed is high climb rate and speed, with manoeuvrability and fuel consumption being less important. Many of the aircraft specifically of this type are old, for example the British Lightning. The Russian MiG-25, with a maximum speed of Mach 3.3 at high altitude, is a good, if extreme, example of this type. Aircraft such as the American F-15, with its high maximum operational speed, and the F.3 version of the European Tornado are also able to operate as interceptors. The new American F-22 also has interception as one of its design missions.

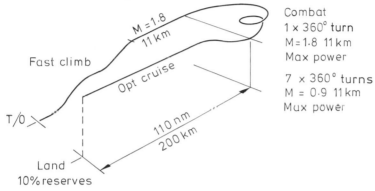

Figure 13.3. A simple, idealised mission for an interceptor aircraft.
(Derived from Garwood, Round and Hodges, 1995.)

Figure 13.3 shows a typified intercept mission with a supersonic dash at $M = 1.8$ over a considerable distance to the combat zone. There are advantages, including reduced fuel consumption, if this supersonic cruise can be carried out without use of the afterburner, a condition referred to as 'supercruise'. Here the interception would be carried out beyond visual range (BVR) using missiles. Some tight turns are envisaged as part of avoidance manoeuvres after the release of air-to-air missiles, and operation without the afterburner is highly desirable here too since heat-seeking missiles can more easily lock into the afterburner plume.

Interdiction

The proposed role here is to operate 50 km or more behind the battle front and attack roads, bridges, railways, radar installations, etc. The requirement for the aircraft and engine is transonic operation at low altitude, medium-to-long range and good manoeuvrability. In terms of the engine this means that fuel consumption is vitally important. The American F-111 and the F/A-18 and the European Tornado are intended for this role and are capable of flight at high speeds. The Russian MiG-27 is similar. An example of a more specialised aircraft of the type is the American A7.

Close air support

This category can be sub-divided, but the role is to operate within 50 km of the battle front to attack ground targets. The general characteristics needed are the ability to cruise at low level, to carry a large payload and to be able to loiter with low fuel consumption. The aircraft must be highly manoeuvrable, but range can be relatively short.   Some examples are the American A10, the British/American Harrier AV8 and the Russian Yak-38.

13.2   THE REQUIREMENTS FOR A NEW FIGHTER AIRCRAFT

To make the treatment more concrete a hypothetical aircraft is considered for which the design decisions can be made. So far as the engines are concerned the most demanding role is for the air-superiority aircraft and the subject of the design is therefore a New Fighter Aircraft. In fact it is because there are so many different performance requirements for an air-superiority aircraft that considerable adaptability is needed.

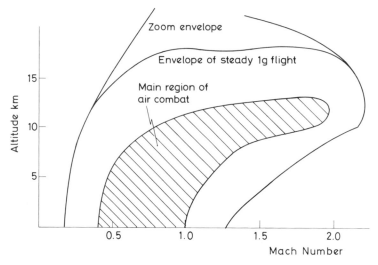

Figure 13.4. The flight envelope for a typical combat aircraft showing the main region envisaged for combat. (The zoom envelope represents the extreme altitude which can be attained taking full advantage of kinetic energy at lower altitude.)

The entire flight envelope for an air-superiority and interception aircraft are sketched in Fig. 13.4 with the main region for air combat shown shaded. Of particular note is that most combat is expected to take place at comparatively low speed, mainly below about M = 1.5, and mostly below an altitude of about 11km (36089 feet). This altitude is the tropopause in the International Standard Atmosphere, above which the temperature is assumed to remain constant at 216.7 K. (It may be noted that whereas for civil aviation feet are the normal units of altitude, in military aviation metres are widely used.)   The temperature at altitude is approximately defined by the standard atmosphere, but at sea level the range can be enormous: perhaps 230 K (ISA-58°C) in the Arctic and 320 K (ISA+32°C) in some hot regions of the world. For the present, as in earlier sections of the book, the temperature at sea level will be taken to be the standard value of 288 K. The stagnation temperature into the engine T_{02} has a powerful effect on the behaviour and performance. Temperature T_{02} is not only determined by the ambient static, assumed here to be given by the standard atmosphere, but also by the flight Mach number.

Some of the possible important operational points for combat aircraft are set out below:

1   Low level combat and escape manoeuvres at Mach numbers between about 0.5 and 0.9 and altitudes below about 4 km.   (This combination of speed and altitude requires very tight and rapid turns.)

2   Acceleration near sea level from around $M \approx 0.3$ to 1.2.

3   Medium-altitude combat around 6 km at $M \approx 0.7$ to 1.2, with acceleration being important in this altitude range from $M \approx 0.5$ to 0.9.

4   High-altitude combat in the range 9 km to 11 km at $M \approx 0.9$ to 1.6.   Acceleration is important from $M \approx 0.9$ to 1.6.

5   Supercruise (without afterburner), M at least 1.5, altitude 11 km above .

6   Supersonic steady level flight at $M \approx 2.0$ or higher, altitude 11 km or above.

Not every design of aircraft would be able to meet all the requirements, and judgement would be required to determine which were most important. Small changes to the same basic design might allow different versions to excel in some, but not all, of the possible missions. Only by trying a large number of different combinations of airframe and engine could anything like an optimum be said to have been chosen. The exercises in this book will concentrate on the two extremes for temperature of the standard atmosphere which occur at sea level and at the tropopause – both are vitally important to a fighter aircraft and encompass many of the critical operating points. At the tropopause, operation will be considered at three Mach numbers (0.9, 1.5 and 2.0) whilst at sea level the static condition and $M = 0.9$ will be used.

It is only peripheral to the discussion here, which primarily concerns propulsion, but an additional requirement now exists for most combat aircraft – some measure of stealth. There are two aspects, infra-red and radar reflection, both of which can be used by missiles to seek their target. To minimise infra-red visibility it is desirable to avoid the afterburner and to have the final nozzle shielded from the ground by the wing or tail plane.   The worst sources of radar reflections for older aircraft are the engine inlet and exhaust. Frequently now the exhaust nozzle is rectangular and is shielded from below by the wing or tail. The intake is often curved and vanes may be used to absorb the incident radar wave. Surfaces of the intake, and the aircraft itself, may be covered in radar absorbing material. Aircraft such as the F-117 and the F-22 contain many stealth features.

13.3   THE PROPOSED NEW FIGHTER AIRCRAFT

The hypothetical New Fighter Aircraft (NFA) provides an example, analogous to the New Large Aircraft for civil application considered in the Chapters 1 to 10.   Just as the New Large Aircraft was presented as a step forward from existing large civil aircraft, any new fighter aircraft would also take advantage of previous experience.   In the present case it is convenient to use as precursors the American F-15 and F-16, both very successful aircraft designed and developed in the early 1970s.   The F-15 has two engines while the F-16 has only one; current thinking seems to prefer the two-engine layout, because of the added security in case of engine failure, and two are selected for the NFA.   Other parameters necessary to the design choices are given in Table 13.1.   Cost increases with two engines and the F-35 (formerly known as the JSF) has only one.

Table 13.1.   Major parameters for two existing fighter aircraft and a
possible New Fighter Aircraft (NFA)

		F-15	F-16	NFA
Number of engines		2	1	2
Date of first flight		1972	1976	
Operating weight empty	tonne	12.2	7.1	
Max. gross take-off weight	tonne	20.2	10.8	18.0
Nominal max. weapon load	tonne	10.7	5.8	
Nominal max. internal fuel	tonne	8.0	3.2	
Engine weight / max. gross T-O weight		0.135	0.127	0.10
Max. dry thrust /max. gross T-O weight		0.66	0.62	0.66
Max. thrust with ab /max. gross T-O weight		1.07	1.00	1.0
Wing loading (max. take-off weight) N/m^2		3600	3400	3500
Maximum Mach number, sea level		1.2	1.2	1.2
Maximum Mach number (approx. 40000 feet)		2.5	2.05	2.0
Service ceiling		50000 feet	60000 feet	50000 feet
Service radius, no external fuel tanks			314 nm	200 nm
Ferry range, with external fuel tanks		2880 nm	2415 nm	2000+ nm
Maximum acceleration by wing lift			9 g	+9g, −3g

ab refers to use of the afterburner and dry refers to operation without the afterburner.
The afterburner will be described in Chapter 14.
Gaps in Table 13.1 correspond to information being unavailable (for the existing aircraft)
or not yet decided (for the New Fighter Aircraft).

Table 13.1 gives the maximum take-off weight of the New Fighter Aircraft to be 18 tonne but the more extreme manoeuvres would not be carried out at this weight.   For the sake of simplicity a single value of weight, 15 tonne, will be used for all the calculations (other than take off which will be at 18 tonne) allowing for some fuel to have been burned in take off, climb and loiter.

It is also helpful to relate some of the statistics to those of a current civil aircraft and as an example Table 13.2 compares the F-16 and the Boeing 747-400. The fraction of the weight which is fuel is similar. For the fighter the maximum thrust–weight ratio at take off is about four times that of the civil aircraft, whereas the fraction of the total weight due to the engines is about three times greater for the fighter. The greater fraction of weight due to the engines puts particular pressure to increase the engine thrust-to-weight ratio for the fighter. Current engines, such as that in the F-16, have a thrust–weight ratio of about 8 with the afterburner in use; engines soon to be introduced will have a ratio of about 10 and in the future ratios in the range 12 to 15 seem likely.

Table 13.2. Comparison of salient parameters for a current large civil aircraft and current fighter aircraft

	747-400	F-16
Take-off thrust/ max. take-off weight	0.25	0.66 'dry'
		1.00 a/b
Total engine weight /max. take-off weight	0.045	0.13
Max. fuel weight/max. take-off weight	0.43	0.40
Maximum wing loading (N/m^2)	7600	3400

Exercise

13.1 For the New Fighter Aircraft use the data to find the wing area.   (**Ans:** 50.5 m^2)
During combat the aircraft mass may be assumed to be 15 tonne (3 tonne less than the maximum take-off mass). Find the wing loading for this condition.   (**Ans:** 2917 N/m^2)

13.2 The Pratt & Whitney F100-PW-100 engine (powering the F-15 and F-16 aircraft) has an approximate maximum thrust–weight ratio on a sea-level test bed of 7.9 with the afterburner lit and 4.7 'dry'. The specific fuel consumption is about 2.6 kg/h/kg with the afterburner and about 0.69 when dry. Find the time required in each operating condition to burn a mass of fuel equal to the engine mass.
          (**Ans:** with afterburner, 2.9 minutes; 'dry' 18.5 minutes)

It is perhaps relevant here to refer to one additional aspect of combat aircraft, the unmanned (sometimes called uninhabited) aircraft. At one extreme is the cruise missile, with a small gas turbine, and at the other an aircraft which might be expected to engage conventional fighters in aerial combat. In between there is a wide range of reconnaissance vehicles, from the very small up to the size of small manned aircraft. In terms of the aerodynamics and thermodynamics of the propulsion the task is essentially the same as that for manned aircraft. In terms of other aspects of the engine there may be major differences. Naturally, if the aircraft is to be destroyed in the mission (as for a cruise missile) a short life (of a few hours) suffices and low cost is important. If the aircraft is loaded with expensive reconnaissance equipment the aircraft and engine would expect to be used many times and long life would be important. Some of the other requirements may be more surprising. For example an unmanned fighter would not need to be used for practice by the pilot sitting at remote control station – the practice could presumably be provided by a simulator. Whereas for manned combat aircraft in times of peace practice is the major use of the aircraft, this would not be the case for an unmanned aircraft. The unmanned aircraft engine must therefore be capable of being stored for several years without deterioration and then to start reliably and quickly.

SUMMARY CHAPTER 13

The fighter aircraft is the topic of Chapters 13–18 and this chapter has described some features of this class of aircraft, comparing it with some other types of combat aircraft. The fighter is required to be manoeuvrable and it will be seen in Chapter 14 that to be able to make rapid turns requires a high level of thrust if altitude or speed are not to be lost. Fighter aircraft therefore must have a high thrust-weight ratio and the same is true of the engines themselves.

Although fighter aircraft must have a higher thrust-weight ratio than civil aircraft, which indicates a lower bypass ratio, this is not the feature which makes the aerothermal design of the engines so interesting. Because fighter aircraft are normally expected to be able to fly fast, the inlet temperature can rise well above the temperature experienced for a sea-level take off. This, it will be shown, has a radical effect on the behaviour of the engine and therefore on its design.

CHAPTER 14   LIFT, DRAG AND THE EFFECTS OF MANOEUVRING

14.0 INTRODUCTION

A fighter aircraft is required to be agile, which requires it to turn sharply, to accelerate rapidly and usually to travel fast. It is no surprise that accelerating rapidly or travelling fast require large amounts of thrust from the engine. What may be more of a surprise is that rapid changes in direction require high levels of engine thrust. The reason is that the drag of the aircraft rises approximately with the square of the lift coefficient and making rapid turns demands high lift from the wings. An aircraft normally banks in order to turn so that the resultant of the gravitation acceleration and the centripetal acceleration is normal to the plane of the wings, Fig.14.1, and the

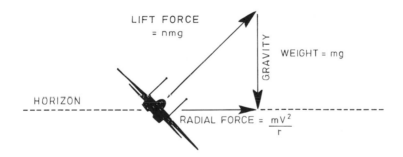

Figure 14.1. An aircraft making a horizontal banked turn.

force they produce is exactly balanced by the wing lift. It is normal to express the increase in acceleration in terms of the *load factor*, denoted by *n*: a load factor of unity corresponds to an acceleration g perpendicular to the wing, when the lift is the normal weight of the aircraft, whereas a load factor of 5 corresponds to an acceleration of 5g and the lift is five times the weight. For a modern fighter aircraft structures are designed to withstand the approximate limit on acceleration set by the human pilot and load factors can be as high as 9.

For the civil airliner the turns are normally so gentle that the lift on the wings is little more than the weight of the aircraft, and the size of the engine is normally fixed by requirements at the top of the climb. As will be shown below for the fighter the size of the engine is more likely to be set by the maximum rate of turn required during some parts of the flight envelope.

The treatment of aircraft aerodynamics is necessarily very superficial here, but a much more complete treatment is given by Anderson (1989).

14.1 LIFT AND ACCELERATION

The lift force L is related to the dynamic pressure $\frac{1}{2}\rho V^2$ by the lift coefficient

$$C_L = \frac{L}{\frac{1}{2}\rho A V^2} \qquad (14.1)$$

where A is the wing area. For steady level flight the lift force is equal to the weight of the aircraft mg and at this condition mg/A defines the *wing loading*. Note from Table 13.2 how the wing loading for the F16 is less than half the value for the 747-400. The low wing loading, meaning large wings in relation to the aircraft weight, is needed to give high manoeuvrability. Although the shape of the wings affects the maximum lift force which can be produced for a given flight speed and air density, the most important parameter in determining maximum lift is the area of the wings. The wing area is set by the aircraft weight and the selected wing loading – reducing the wing loading therefore increases the maximum lift which a wing can produce. For the New Fighter Aircraft the wing loading is taken to be 3500 N/m².

For take off and landing the lift coefficient is often raised by the use of flaps, slats or blowing, but for higher speed flight this is less common. The upper limit on lift coefficient depends on Mach number, but is likely to vary little between different designs of fighter aircraft. For the present purpose a maximum value of lift coefficient of about 1.00 is realistic at $M = 0.7$, falling to about 0.4 at $M = 1.5$; for simplicity a linear dependence on Mach number will be assumed between these values.

The maximum acceleration or load factor that can be achieved is determined by the peak value of lift and this means that the maximum turning rate also depends on lift coefficient. This is known as the maximum *attained* turning rate and will normally cause the flight speed and/or altitude to decrease, for reasons described below. The engine performance has only a small effect on maximum attained turn rate, which is mainly fixed by the wing area and the dynamic pressure.

Exercise

14.1 The New Fighter Aircraft flies with a mass of 15 tonne. The wing loading is given in Exercise 13.1.

a)For a constant-altitude 9g banked turn at sea level (ambient temperature 288 K) and a Mach number of 0.9, find the angle of bank and the radius of curvature. What is the lift coefficient ?
(**Ans:** bank = 83.6° ; radius = 1.07 km; C_L = 0.457)

b*)Find the radius of curvature and lift coefficient for 3g turns at Mach numbers of 0.9, 1.5 and 2.0 at the tropopause (11 km altitude; $T = 216.65$K, $\rho = 0.365$ kg/m3).
(**Ans:** radius = 2.54, 7.06, 12.55km; C_L = 0.680, 0.245, 0.134)

c)If the lift coefficient is not to exceed 0.85, find the highest turning acceleration possible flying at $M = 0.9$ at the tropopause. If the maximum lift coefficient is 1.00, find the lowest Mach number at which a 9g turn is possible close to sea level . **(Ans:** 3.76 g; $M_{min} = 0.609$)

14.2 Show that for constant Mach number flight the dynamic pressure $1/2 \rho V^2$ is proportional to the local ambient pressure. Show that the ambient pressure p at height h_T above the tropopause is given by
$$p/p_T = \exp\{-gh_T/RT\}$$
where $p_T = 22.7$ kN/m^2 is the ambient pressure at the tropopause and h_T is in metres.

If the maximum lift coefficient at $M = 0.9$ is equal to 0.85 and the wing loading is 2917N/m^2, find the maximum altitude for steady level flight at that Mach number (if sufficient thrust were available).

(Ans: 19.4 km)

14.2   DRAG AND LIFT

For the civil transport aircraft treated in Chapter 2 consideration was only given to the case when the lift was equal to the weight of the aircraft. In other words the case treated was when the load factor was taken to be equal to unity and any effect of turning was neglected. At this condition the drag was obtained from a fixed lift-drag ratio. For combat aircraft, however, the load factor is frequently sufficiently high that the effect of lift on drag must be included.

The drag coefficient C_D is defined by

$$C_D = \frac{D}{\frac{1}{2}\rho A V^2} \tag{14.2}$$

which can be written, with some simplification, as

$$C_D = C_{D0}(M) + k(M)C_L^2, \tag{14.3}$$

where C_L is the lift coefficient and C_{D0} is the drag coefficient at zero lift. Both C_{D0} and the parameter k are dependent on flight Mach number as well as the aircraft geometry. (For combat aircraft the minimum drag corresponds fairly closely with the drag at zero lift, which is not the case for transport aircraft.) C_{D0} is altered by the configuration of the aircraft, and is much increased by weapons or extra fuel tanks attached to the outside of the aircraft. For combat the extra fuel tanks would normally be dropped, even if partly full. The exact values for C_{D0} and k for specific aircraft are normally secret, but Figure 14.2 shows approximate[1] values which serve to demonstrate how drag is affected by aircraft Mach number and lift coefficient. These results are displayed as *drag polars* in Fig.14.3(a) and for comparison some measured results for the F 16 are shown in Fig.14.3(b).

---

[1] Figure 14.2 was originally compiled by Fred Jonas of the US Air Force Academy and incorporated in the book by Mattingley, Heiser and Daley (1987). These plots are used in the solution of exercises in Chapter 14, but it is hard when reading numbers from such diagrams to obtain better accuracy than two significant figures. The answers to the exercises are nevertheless given to three significant figures to assist consistency, to show changes and to help in checking.

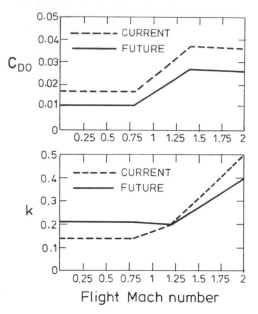

Figure 14.2. Drag coefficient for zero lift and lift dependent drag factor versus Mach number for fighter aircraft. (Derived from Mattingley, Heiser and Daley, 1987.)

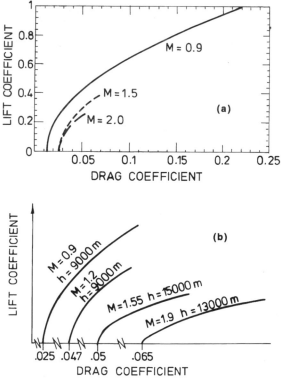

Figure 14.3. Drag polars, showing lift plotted versus drag for different Mach numbers
(a) derived from the curves in Fig.14.2;
(b) measured results for F-16 (from Buckner and Webb, 1974).

At subsonic flight speeds the lift-dependent drag, $k(M)C_L^2$, is usually referred to as the induced drag and is produced by trailing vortices behind the wing. These are revealed in Fig.14.4. In this figure a Tornado is making a steeply banked turn, producing high lift on the wings, and this creates strong trailing vortices. At the core of the vortices the pressure and temperature are relatively low and water vapour in the air condenses to reveal their presence. The strength of the vortices, and therefore the velocity associated them, is linearly proportional to the lift coefficient. The induced velocity requires the wing incidence relative to the direction of motion to be increased in order to maintain the lift force, inclining the wing backwards. The lift vector therefore has a component acting backwards parallel to the direction of travel. Since the backward inclination of the lift due to the induced velocity is proportional to the lift coefficient, the induced drag is proportional to the square of the lift coefficient. At supersonic flight speeds the lift-dependent drag is principally wave drag, but again it is proportional to C_L^2.

Figure 14.4. A photograph of a Tornado in a high-g turn with the trailing vortices made
visible by condensation of water vapour. (Reproduced by permission of
British Aerospace Defence Ltd, Military Aircraft Division)

Because the drag rises steeply with lift coefficient high turn rates require high levels of engine thrust if speed and altitude are to be maintained. The maximum *sustained* turning rate is the highest rate which can be achieved at maximum thrust without loss of speed or altitude; it is a function of flight Mach number and altitude. Whereas the maximum *attained* turn rate depends mainly on the wing area and maximum lift coefficient, the maximum *sustained* rate is engine thrust dependent.

We are neglecting a complication when an aircraft operates at high incidence. The jet is then directed at a substantial angle to the direction of flight and the thrust is no longer in the direction of the drag, but there is a substantial component in the direction of the lift. Although this would be important for specification of a real design, the omission does not affect the conclusions reached here.

Exercises

14.3 If the wing loading in steady flight is 2917 N/m^2 use the future-aircraft curves for C_{D0} and k in Fig.14.2 to estimate the following drag coefficients:

    a) 11 km altitude and M = 0.9, 1.5 and 2.0 in straight and level (i.e. 1 g or n =1) steady flight;

    b) sea level at M= 0.9 straight and level (i.e. 1 g) and also for 5g and 9g turns;

    c) 11 km at M= 0.9 and 1.5 for 3g turns;

                  (**Ans:** a) 0.0246, 0.0289, 0.0268; b) 0.0145, 0.0275, 0.0577; c) 0.110, 0.0435)

14.4 Use the results of Exercise 14.3 to estimate the thrust needed from *each* of the two engines of the New Fighter Aircraft (with a mass of 15 tonne) for the following conditions:

    a*) 11 km altitude and M = 0.9, 1.5 and 2.0 in straight and level (i.e. 1 g) steady flight;

    b*) sea level at M= 0.9 straight and level (i.e. 1 g) and also *sustained* 5g and 9g turns;

    c*) 11 km at M=0.9 and 1.5 for *sustained* 3g turns;

                  (**Ans:** a) 8.0kN, 26.0kN, 44.3kN; b) 21.0kN, 39.8kN, 83.5kN; c) 35.8kN, 39.3kN)

14.5 Show that in steady level flight the required thrust from the engines is a minimum if the net thrust is directed downwards at an angle θ to the horizontal, where $\tan\theta = dD/dL$.

    For the New Fighter Aircraft flying steady and level (n=1) at the tropopause at M = 1.5 find the lift-drag ratio, L/D using Exercise 14.3a.   (**Ans:** L/D= 2.83)

    Find the value of θ for minimun net thrust and the magnitude of this thrust from each engine for cases a) and b) below. Then, if the jet velocity relative to the aircraft is 1100 m/s, determine the angle β at which the jet should be directed downward for minimum engine thrust:

    a) assuming that L/D is constant ;   (**Ans:** θ = 19.4°, F_N= 24.5 kN, β=11.4°)

    b) using equation 14.3 to determine drag.   (**Ans:** θ=2.6°, F_N= 25.6 kN, β=1.6°)

14.3 ENERGY AND SPECIFIC EXCESS POWER

The *energy state* is the combined potential and kinetic energy given by

$$E \quad = \quad m(gh + V^2/2),$$

with the *specific energy* $E_S \quad = \quad gh + V^2/2.$ (14.4)

At constant energy state it is possible to trade kinetic energy for potential energy, so possessing a high energy state gives a combat aircraft great initiative. Neglecting any difference between engine thrust and aircraft drag (i.e. maintaining constant energy state) an aircraft travelling at M = 0.9 at sea-level (V = 306 m/s) can 'zoom' to an altitude of 4777 m, at which point it would have no forward speed. More significantly, an aircraft at an altitude of, say, 15 km at M = 0.9, for which V = 266 m/s, could zoom to 18.6 km, beyond the altitude it could fly continuously. Manoeuvres which exceed the sustainable turn rate lead to a reduction in energy state whereas when thrust exceeds drag the energy state is increasing.

    The ability to accelerate or climb depends on the excess thrust over and above that required to balance the drag in steady level flight. The measure of this is the excess thrust divided by the aircraft weight mg, which can be written

$$(F_N - D)/mg \, ,$$

where F_N is the maximum net engine thrust and D is the aircraft drag. (Both engine thrust and drag are functions of altitude and speed.) More conventionally the above quantity is multiplied by the flight speed V to give *specific excess power*, *SEP*

$$SEP = V (F_N - D)/mg \, , \qquad\qquad (14.5)$$

with units of velocity. During manoeuvres such as turning the drag rises, as discussed in section 14.3, so for the same engine thrust the *SEP* falls. The available maximum rate of climb at constant speed is equal to *SEP*, whereas the available rate of acceleration in level flight (i.e. increase in forward speed) is equal to the specific excess thrust (*SEP /V*)*g*.

The specific excess power is related to the specific energy by

$$SEP = \frac{\mathrm{d}(E_S/g)}{\mathrm{d}t} \, , \qquad\qquad (14.6)$$

and *SEP* = 0 corresponds to flight at constant speed and constant altitude with the engine producing maximum thrust. Figure 14.5 shows a schematic plot of the *SEP* = 0 curve for a representative aircraft at the condition when the lift on the wings is equal to the aircraft weight, the 1g case. The boundary of the *SEP* = 0 curve is set by the aircraft aerodynamics and weight and by the thrust of the engine. For Fig.14.5 the engine thrust would be at its maximum available at every altitude and Mach number. Because the aircraft drag rises around $M = 1.0$ the *SEP* = 0 curve dips in this region, implying a lower steady ceiling around sonic velocity. The aircraft may be forced to operate somewhat inside the boundary set by *SEP* = 0 for reasons indicated on the diagram.

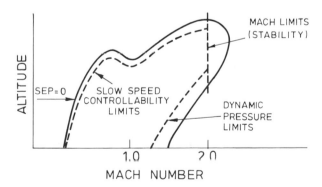

Figure 14.5. A schematic plot of altitude versus Mach number showing the line *SEP* = 0 for a combat aircraft. Possible operating limits for steady flight are superimposed. (Derived from Shaw, 1988.)

Curves of *SEP* = constant are shown in Fig.14.6 for a fighter aircraft at three levels of load factor. The figures show contours plotted on axes of Mach number and altitude; for the plot at a load factor of 1, which corresponds to straight and level flight, curves of energy state are superimposed. The curve *SEP* = 0 corresponds in each case to the limit of steady operation at maximum engine thrust and moving inside this contour the value of *SEP* rises. As the load factor is increased the size of the region for which *SEP* ≥ 0 decreases. For turning, when the load factor is larger than unity, the boundary *SEP* = 0 gives the limit of *sustained* turn rate. The highest values of *SEP* occur in all cases at low altitude and at Mach numbers in the high subsonic range. At this condition the high density air gives low values of lift coefficient (and therefore low induced drag) and also high net thrust from the engine.

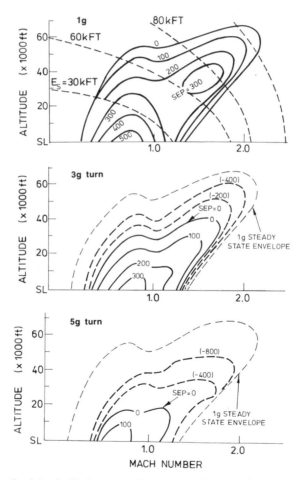

Figure 14.6. Schematic plots of altitude versus Mach number for a combat aircraft at load factors of 1,3 and 5. Shown are lines of *SEP* = constant (units ft/s),drawn solid when positive, broken when negative. The *n* = 1 case also shows lines of constant energy state (units ft).
(Derived from Shaw, 1988.)

To be slightly more concrete, the curves in Fig.14.6 show that for a 3g turn at 30000 ft altitude (9.1 km) and $M = 0.9$ the value of *SEP* is about zero. In other words 3g gives the maximum *sustained* turn rate at this combination of altitude and Mach number so that at this condition the turn could be accomplished at constant energy state. The corresponding plot for 1g flight at this combination of altitude and M shows that $SEP \approx 150$ ft/s, so with the engines giving their maximum thrust the aircraft would be increasing its energy state, either climbing at 150 ft/s (47.5 m/s) or accelerating in the direction of flight. The forward speed for $M = 0.9$ at this altitude is 273 m/s and the acceleration in that direction is given by $(SEP/V)g = (47.5/273)g$, or $0.17g$. For a 5g turn, however, the corresponding plot in Fig.14.6 shows with this combination of altitude and Mach number that $SEP \approx -400$ ft/s (-122 m/s) and the aircraft would either be losing altitude at this rate or decelerating at the rate of $(122/273)g = 0.45g$.

Figure 14.7 shows altitude–Mach number diagrams pertaining to two aircraft, the F-5 and a later derivative, the F-20. The wider steady, sustained operating envelope of the F-20 is very clear, but still more obvious is the smaller operating range in terms of speed or altitude at a load factor of 4g compared with 1g. The primary battle zone is at altitudes from sea level to the tropopause and for Mach numbers below unity. Because of the low density at high altitude, turning becomes difficult, whilst at high supersonic Mach numbers the turning radius becomes so large that combat is difficult.

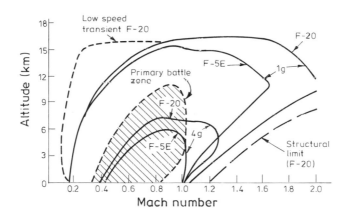

Figure 14.7 Altitude–Mach number diagrams for two aircraft: the F-5E and the later derivative the F-20. (From L'Aeronautique et L'Astronautique, 1983–5.)

14.4 OPERATION AT LOW THRUST AND DRAG

Up to now most of the discussion has centred on the manoeuvres which require large lift and create large drag, therefore requiring high levels of engine thrust if they are to be sustained. High thrust almost invariably involves high fuel consumption and this can shorten the mission.

For many combat missions there are phases of cruise to and from the combat zone and periods of loitering or patrolling. For both the cruise and the loiter it may be possible to fly at a condition approaching that for minimum fuel consumption. In this simplified treatment we will assume that the specific fuel consumption (mass flow rate of fuel/net thrust) is constant in the subsonic range of conditions likely. Under this assumption fuel consumption is therefore proportional to drag.

Condition of minimum fuel consumption at constant speed and altitude

This condition corresponds to that for maximum flight duration and occurs for minimum drag. The drag is given by

$$D = qAC_D = qA(C_{D0} + kC_L^2) \qquad (14.7)$$

where q is the dynamic pressure $1/2\rho V^2$ while C_{D0} and k are defined in section 14.2. The weight of the aircraft mg is fixed, neglecting here the variation as fuel is burned or weapons released, so the lift coefficient C_L can be replaced by mg/qA, where A is the wing area, and mg/A is the wing loading. The drag is therefore given by

$$D = qAC_{D0} + \frac{k}{q} A(mg/A)^2 \qquad (14.8)$$

and the minimum will occur when

$$q = (mg/A) \sqrt{\frac{k}{C_{D0}}} . \qquad (14.9)$$

Minimum fuel consumption during cruise

The condition for this is identical to that required for maximum range, considered in Chapter 2 for the civil transport. The requirement is that VL/D should be a maximum. This can be found by differentiating VL/D after again replacing the lift coefficient by the wing loading divided by the dynamic pressure, see Exercise 14.6.

Exercises
14.6 The New Fighter Aircraft may be required to loiter at a Mach number of 0.6. Find the dynamic pressure and thence the altitude which will give the minimum drag . Hence determine the thrust required from each engine. Take the aircraft mass to be 15 tonne.   (**Ans:** 12.75 kPa, 5.6 km, 7.1 kN)

14.7 Show that the minimum fuel consumption to cover a given distance at given altitude will occur when the dynamic pressure is given by

$$q = (mg/A) \sqrt{\frac{3k}{C_{D0}}} . \qquad (14.10)$$

Hence find the optimum altitude and thrust necessary for cruise of the New Fighter Aircraft with a mass of 15 tonne at $M = 0.8$.   (**Ans:** 5.7 km, 8.1 kN)

14.5 VECTORED THRUST

There are some benefits in being able to vary the pitch of the propelling nozzle and thereby alter the direction of the jet and the gross thrust. (The net thrust also varies in direction, but the direction of ram drag is fixed by the forward motion of the aircraft, of course.) It is well known that the F-22 and all versions of the F-35 will have vectored thrust. Some Russian combat aircraft also have vectored thrust and the Harrier has used it for many years.

The benefits claimed for vectored thrust vary. It can be used to lower landing speed, shorten take-off distance and reduce cruise drag. Following the theme of this chapter, it is immediately apparent that deflecting the jet is a way of producing large thrust normal to the direction of flight when the flight speed is low but the air density high. In other words it can be used to give a large normal force when the wings are not capable of giving large lift, and can therefore be used to increase manoeuvrability at low speed. The normal force is equal to the *gross* thrust times the sine of the angle of deflection and at high speed quite small deflections of the jet away from the direction of travel produce a large normal force. This is because at high speed the ram drag is a large fraction of the gross thrust so the gross thrust is generally large in relation to the net thrust.

The disadvantages of vectored thrust are the weight, complexity and cost of the variable nozzle. On top of this is the issue of reliability, especially if the vectored thrust is used in place of control surfaces on the tail.

SUMMARY CHAPTER 14

Turning manoeuvres are an essential part of combat and load factor is used to express the increase in lift needed for the turn. The drag increases in proportion to the square of the lift coefficient.

The maximum *attained* turn rate is fixed by the wing area and the dynamic pressure $1/2\rho V^2$ but the drag at this condition is likely to exceed the net thrust from the engines. In this case there is a reduction in flight speed or altitude during the turn; equivalently, a reduction in energy state. The maximum turning rate which can be maintained without loss in speed or height (when the engine thrust equals the drag) is the *sustained* turning rate and at this condition the specific excess power is zero ($SEP = 0$).

Most combat is expected to take place between sea level and the tropopause and at Mach numbers below unity. A requirement also exists for cruising at higher Mach numbers. The specification of the aircraft demands that at take off there is a thrust–weight ratio of 1.0 with the afterburner and 0.66 'dry', but the maximum thrusts required of the engines at a sample of other conditions were found in Exercise 14.4. The maximum required thrusts are summarised in Table 14.1 for use in later chapters.

Any one of the conditions which require high thrust could be imagined to prescribe *the* design point for the engine, but any engine will be expected to operate satisfactorily at many of these points. The steady level flight condition for $M = 1.5$ does not call for very high thrust, but there the practical interest is whether the engine can produce sufficient thrust 'dry' (i.e. without using the afterburner) to achieve 'supercruise'.

In Chapter 15 characteristics of the engines for combat aircraft will be explored. In Chapter 16 different conditions will be taken as design points and the performance of the corresponding engines calculated and compared.

In Exercise 14.4 it is shown that the highest thrust for a 9g turn at sea level was over 84 kN per engine, whilst from Exercise 14.5 the thrust required for loiter could be as low as about 7 kN; the engine needs to be able to accommodate these large variations. In Chapter 17 the performance at off-design conditions will be examined, both for high thrust conditions critical for the mission and also for low thrust, such as cruise or loiter, when low fuel consumption becomes a primary concern.

Table 14.1   Required net thrust (kN) from each engine for the
New Fighter Aircraft for total aircraft mass of 15 tonne

			Mach number		
			0.9	1.5	2.0
Tropopause	Steady level flight	1g	8.0	26.0	44.3
	Banked turn	3g	35.8	39.3	
Sea level	Steady level flight	1g	21.0		
	Banked turn	5g	39.8		
		9g	83.5		

Take-off thrust required at sea level with total aircraft mass 18 tonne:

'dry'   thrust–weight ratio = 0.66 $F_N = 58.3$ kN

afterburning   thrust–weight ratio = 1.00 $F_N = 88.3$ kN

(At tropopause $T_a = 216.7$ K, $p_a = 22.7$ kPa; at sea level $T_a = 288.2$ K, $p_a = 101$ kPa)

CHAPTER 15

ENGINES FOR
COMBAT AIRCRAFT

15.0 INTRODUCTION

Figure 15.1 shows a cross-section through a modern engine for a fighter aircraft and the large differences between it and the modern engines used to propel subsonic transport aircraft, Fig. 5.4, are immediately apparent. Above all the large fan which dominates the civil engine, needed to provide a high bypass ratio, is missing. Engines used for combat aircraft typically have bypass ratios between zero (when the engine is known as a turbojet) and about unity; most are now in the range from 0.3 up to about 0.7 at the design point, though the bypass ratio does change substantially at off-design conditions.

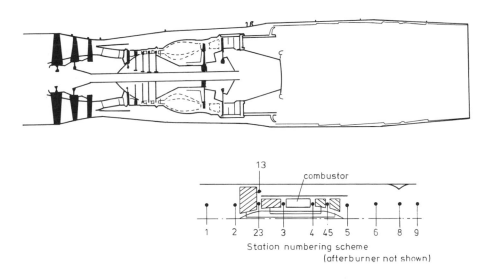

Figure 15.1. A cross-section through a modern combat engine with a schematic
indicating the numbering system adopted.

This chapter seeks to explain why fighter engines are the way they are. It begins with some discussion of specific thrust, since this is a better way of categorising engines than bypass ratio; fighter engines have higher specific thrust than civil transport engines. Then the components of the engine are described, pointing out special features of components common to the civil engine, and giving a general treatment of the special features: the mixing of the core and bypass stream, the high-speed intake, the afterburner and the variable nozzle. A brief treatment of the thermodynamic aspects of high-speed propulsion leads into the constraints on the performance of engines for combat aircraft and the rating of engines.

In earlier chapters the cooling air supplied to the turbine was neglected in calculating the cycles, so too was the mass flow rate of fuel in the gas through the turbine. These are now included, together with more realistic properties for the gas through the turbine, where it is assumed that $\gamma = 1.30$ and $c_p = 1244$ J/kgK.

15.1 SPECIFIC THRUST

The energy released in combustion at high pressure is converted into mechanical energy relatively efficiently and this has to appear as jet kinetic energy of the jet streams. The low bypass ratio engine therefore has a very much higher jet velocity than the high bypass ratio civil engine. The high velocity leads to high thrust per unit mass flow of air, that is to high specific thrust, but also to low propulsive efficiency and to a large amount of noise. (The type of engine required for supersonic transport aircraft will also need to be of the low bypass type and the problem of noise is particularly serious there.) It is more normal to characterise combat aircraft engines by their specific thrust than by their bypass ratio.

The net thrust from an engine is given by

$$F_N = V_j \, (\dot{m}_a + \dot{m}_f) - V \dot{m}_a \tag{15.1}$$

where $\dot{m}_a$ is the mass flow of air entering the engine, $\dot{m}_f$ is the mass flow of fuel, V_j is the jet velocity and V is the flight speed. The specific thrust is the net thrust per unit mass of air flow, and is given by

$$F_N / \dot{m}_a = V_j \, (\dot{m}_a + \dot{m}_f) / \dot{m}_a - V$$

$$\tag{15.2}$$

with units of velocity (m/s). When the mass flow of fuel is neglected the specific thrust reduces to

$$F_N / \dot{m}_a = V_j - V. \tag{15.3}$$

The industry typically expresses specific thrust in the units of pounds thrust per unit mass flow in pounds per second; the numerical value is identical in pounds or in kilograms. (Numerically the magnitude in m/s is 9.81 times the value in kg/kg/s.) Specific thrust for a given engine varies with altitude and Mach number, so the value quoted always refers to a particular flight condition. The size of an engine is primarily determined by the mass flow rate of air it swallows, so the combination of aircraft thrust requirement and engine specific thrust essentially determines the engine size.

Figure 15.2 shows the regions where aircraft operate and the different range of specific thrust involved. Whereas civil turbofan engines operate with values of specific thrust below 200 m/s, combat aircraft have values between about 500 and 1000 m/s. For propulsion at Mach numbers above about 2.5 a conventional engine becomes an encumbrance and the normal solution is to dispense with it and use a ramjet; a more innovative possibility is to use the fuel to pre-cool the air before it enters the compressor, but this has not been proven to be feasible.

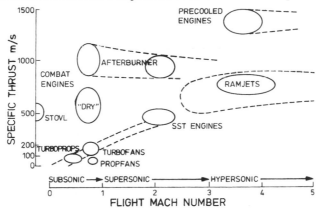

Figure 15.2. The variation in specific thrust with design flight Mach number.
(SST refers to supersonic transport. From Denning and Mitchell, 1989)

The reason for selecting high specific thrust will be explored further in this chapter and the next, but certain general reasons can be given. There is intense pressure to reduce weight, mainly to increase acceleration and manoeuvring ability. As the size of an engine is increased to pass a greater mass flow of air the engine weight rises; as a result, high specific-thrust engines, which give a large increase in velocity to a small mass flow of air, normally weigh less for the same thrust.

The net thrust is less than the gross thrust because of the ram drag, $\dot{m}_a V$. The proportional reduction in thrust with forward speed is therefore less when the jet velocity is very high, in other words when the specific thrust is high. The reduction in thrust with forward speed is normally referred to as thrust lapse and lapse is less when the specific thrust is high.

High speed propulsion gives special reasons for wanting compact engines which pass relatively small amounts of air. At supersonic speeds it is important to shape the aircraft to produce low drag, and this would be impossible with low specific thrust (i.e. high bypass ratio, and therefore large frontal area) engines. In addition there is a loss in stagnation pressure in the inlet to an engine at supersonic flight speeds. For a low specific-thrust engine, which passes a large mass flow but gives it only a small increment in kinetic energy, the stagnation pressure of the jet is much nearer to the inlet stagnation pressure than for a high specific-thrust engine. This means that for the low specific-thrust engine the loss in stagnation pressure across the shocks of the inlet is proportionately larger and low specific-thrust engines are undesirable at supersonic speeds.

The bypass air is used to cool the outside of the engine and it makes certain variations in the nozzle easier, so current practice favours a bypass ratio of at least 0.3. As the bypass ratio increases the boost in thrust with afterburning is increased. This means that the difference in thrust between the engine operating 'dry' to achieve low fuel consumption at cruise, for example, and afterburning, to achieve maximum thrust, is also increased. The choice of bypass ratio and specific thrust at the design point will be discussed in the next chapter.

15.2 THE LAYOUT OF HIGH SPECIFIC-THRUST ENGINES

The diagram of an engine in Fig. 15.1 reveals many of the important features of a combat engine. It can be seen that the LP compressor (colloquially referred to as the fan) is driven by the single-stage LP turbine, whilst the core compressor is driven by the single-stage HP turbine.

One of the most distinctive features of the majority of combat engines is the mixing of the core and the bypass streams in the jet pipe upstream of the final propelling nozzle. In this chapter the gas properties of the mixed stream are calculated by an approximate method based on the the relative mass of core gas (with $c_{pe} = 1244$ J/kgK for the gas which has gone through the combustor) and bypass air (with $c_p = 1005$ J/kgK). The specific heat of the mixed out gas c_{pm} is given by

$$(1+bpr)c_{pm} = c_{pe} + bpr\ c_p\ .$$

The value of γ_m can be obtained from $\gamma_m = c_{pm}/(c_{pm} - R)$ where $R = 287$ J/kgK. When the afterburner is lit it is assumed that $\gamma = 1.30$ and $c_p = 1244$ J/kgK for the exhaust gas. Because the mixing of the core and bypass is so important for the performance of the engine, this is addressed first.

Mixing of core and bypass streams

All combat aircraft, with the exception of the Harrier, have engines which mix the core and bypass streams before they enter the final nozzle. There is a relatively long jet pipe, comparable in length to the rest of the engine, to allow mixing to take place and it is normally assumed in simple treatments that the flow is fully mixed out to uniform before the final nozzle.

The condition where mixing takes place is most accurately modelled by assuming equal *static* pressure for each stream. The remainder of the cycle analysis, however, uses *stagnation* pressure, so it is comparatively complicated to use the static value. (In addition, to use static pressures would require the flow areas to be specified so that the Mach numbers, and thence the ratio of stagnation to static pressures, could be found.) In fact the Mach numbers out of the core and in the bypass duct are normally relatively low, so equating static and stagnation pressures for this purpose does not introduce a major error. Furthermore, if the Mach numbers are well below unity the loss in stagnation pressure in the mixing process itself will be much less than the absolute value of the stagnation pressure and we will neglect this loss here.

Because the core and the bypass are assumed to have equal stagnation pressure when they mix the pressure rise across the fan must be just sufficient for its outlet flow to have equal stagnation pressure to that leaving the LP turbine. This puts an extra constraint on the matching of the fan and core streams.

The two streams are mixed because there is an increase in thrust by doing so. A simple explanation for the gain in thrust by mixing can be made by considering two streams of equal stagnation pressure, each having a constant mass flow. For a given stagnation pressure upstream of the final nozzle the jet velocity is proportional to $\sqrt{(c_p T_0)}$ and the gross thrust is proportional to the jet velocity. Imagine that in place of a constant pressure mixing the two streams are kept separate but there is a constant pressure heat transfer from the hot core stream to the relatively cool bypass stream until both have the same temperature. If the bypass ratio were unity, the temperature change of each stream would be equal (assuming for simplicity that c_p is equal for both). The *proportional* increase in stagnation temperature, however, for the cooler bypass stream would inevitably be greater than the *proportional* drop in temperature of the warmer core stream. This gives a larger increase in V_j for the bypass than the corresponding fall in V_j for the core and consequently a net gain in thrust. The true situation is more complicated than this and mixing inevitably leads to a loss in stagnation pressure. Heating a moving gas also lowers stagnation pressure, but for low Mach numbers this loss in stagnation pressure is small.

The LP compressor or fan

For the engine of Fig. 15.1 the fan has three stages and the pressure ratio in the jet pipe is therefore much higher than would be normal in a civil engine; a figure in excess of 4 on a sea-level test bed might be regarded as typical. The flow from the fan is divided at the splitter with about half going to the core and half to the bypass duct. The bypass flow mixes with the core flow downstream of the LP turbine and the passage of the mixed stream through the final nozzle therefore determines the back pressure behind the fan.

All the stages of the combat aircraft engine are heavily loaded in the aerodynamic sense, in other words $\Delta h_0/U^2$ is high, typically near the upper level at which satisfactory performance is possible. The high loading leads to substantially reduced efficiency at maximum speed conditions and there is often a substantial rise in efficiency when the engine is 'throttled back' to a lower thrust setting and the Mach numbers in the blades are decreased.

The pressure and temperature leaving the fan will not in general be equal for the streams entering the bypass and the core but as a simplification which will suffice for the present purpose we will take $T_{023} = T_{013}$ and $p_{023} = p_{013}$, see numbering scheme in Fig. 15.1.

The core compressor

The engine in Fig. 15.1 has a five stage compressor, suggesting a pressure ratio at design across this component of about 7, giving an overall pressure ratio for the engine on a test bed of about 30. The optimum design will normally lead to efficiencies which are lower than those associated with a civil engine because of the higher value of $\Delta h_0/U^2$.

The combustor

The combustor is similar in its duty to those in a civil engine though the exit temperatures are higher than service temperatures in a civil engine. At some operating points the mass flow of fuel is higher in relation to the mass flow of air (i.e. the equivalence ratio is higher) than for the civil engine because the compression ratio is lower so the compressor delivery temperature is lower too. As discussed in Chapter 11 it is plausible to assume a pressure loss of 5% of the stagnation pressure at compressor outlet in the combustor.

The turbines

The turbines are likely to be more heavily loaded than would be normal in a civil engine and as a result the assumption that they are choked is more generally valid than was the case for a civil engine. Efficiencies may be expected to be a little lower than those for a civil engine. The entry temperature to the HP turbine can be as high as 1850 K for maximum thrust operation; this temperature is the mixed-out temperature at outlet from the HP nozzles and includes the effect of the cooling air provided to the nozzles. A consequence of the high turbine inlet temperatures is that the expected operating life of the blades in military engines is less than those on civil aircraft.

The afterburner

It has already been noted that the maximum thrust required may be more than ten times the minimum thrust required whilst the aircraft is loitering. One of the ways by which this large variation is accommodated is the use of the afterburner, sometimes known as reheat or thrust augmentation. The ability to switch the afterburner on and off gives a special flexibility to this type of engine.

Most combat aircraft engines have an afterburner to allow the exhaust to be raised in temperature. The jet pipe contains a system of fuel injectors and gutters to allow fuel to be burned downstream of the turbine. When the afterburner is not being used (so the engine is operating 'dry') the exhaust temperature is typically around 1000K. When lit the afterburner gives an increase in temperature to about 2200K with an increase in thrust up to about 50% and perhaps a three-fold increase in specific fuel consumption. The fuel flow with the afterburner lit may therefore be five or more times that with the engine 'dry', see Exercise 13.2.

If an engine is called upon to produce high thrust for only short periods of time the afterburner can be the most efficient overall solution because the specific thrust can be high and the engine consequently small and light. If the high thrust is needed for a substantial fraction of the flight the afterburner may be inappropriate since the weight of fuel consumed would be so large.

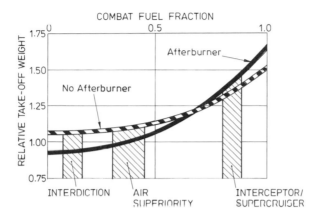

Figure 15.3. The effect of aircraft design role on the decision whether to have an afterburner.
Relative take-off weight is the measure of quality (lighter is better) and combat fuel
consumption is the fraction of fuel burned during the high thrust part of a typical mission.
(Denning and Mitchell, 1989.)

This is illustrated in Fig. 15.3, produced by Rolls-Royce, where two possible styles of engine are compared for a range of duties: one with an afterburner and the other without. The duty is characterised by the fraction of the total fuel which is used with the engine at combat rating (i.e. giving the maximum thrust). Three possible missions are marked on the figure, interdiction, air superiority and interception. From Chapter 13 it will be recalled that an air-superiority aircraft (the typical fighter) spends a significant amount of time getting to the combat zone or loitering, whereas an interceptor spends a large part of its mission travelling as fast as possible to intercept the enemy. The relative merit in the figure is established by the relative take-off weight: basically the lighter the aircraft capable of carrying out the mission the lower the cost is likely to be. It can be seen from Figure 15.3 that the interceptor is likely to be slightly lighter (and therefore better) if it achieves its necessary thrust without use of an afterburner, whereas the other two aircraft will be substantially lighter if an afterburner is used.

   For military engines enough fuel may be added to the afterburner to use virtually all the oxygen and the temperature approaches the stoichiometric value. A value of 2200 K is a realistic value to use here. The amount of fuel can be less than this to give a smaller increase in thrust and fuel consumption. (In the case of the Concorde, for example, the afterburner is used for take off and to pass through $M = 1.0$ and the temperature rise is much more modest, the stagnation temperature in the jet rising to about 1450 K and 1300 K for take off and passing through the sonic speed respectively.)

The propelling nozzle

The conventional operation of the afterburner attempts to keep the jet pipe pressure unchanged when the afterburner is lit; with the jet pipe pressure unaltered the operating point of the engine is also unchanged and the engine is unaware of whether the afterburner is on or off. To cope with the rise in temperature in the jet pipe when the afterburner is in use the propelling nozzle must be made of variable area. The required change in propelling nozzle throat area can then be found, since $\bar{m}_8 = \dfrac{\dot{m} \sqrt{c_p T_{08}}}{A_8 p_{08}}$ is constant at the throat of the choked nozzle to maintain the jet pipe pressure constant. If the gas properties (c_p and γ) were unaffected by the afterburning the nozzle area would need to rise in proportion to the square root of the temperature. (A word of caution is appropriate here. The temperatures from the afterburner are sufficiently high that dissociation is significant and the specific heat varies with temperature and pressure. In the industry it is therefore found to be better to work with enthalpy rather than c_p and T.)

For a reversible nozzle the jet velocity is given by

$$V_j = \sqrt{2 c_{pm} T_{08} \{ 1 - (p_a/p_{08})^{(\gamma-1)/\gamma} \}} \ . \tag{15.4}$$

The gas properties of the exhaust, the specific heat c_{pm} and the ratio of specific heats γ, depend on the amount of bypass air (and on the temperature T_{08} but we neglect this aspect). For operation without the afterburner T_{08} varies comparatively little, and the main function of the afterburner is to increase this, typically by a factor of about two. The other important term is the pressure ratio across the nozzle p_{08}/p_a. The pressure across the fan is in the range 2 to 5, but for high-speed flight the inlet stagnation pressure p_{01} is much higher than the ambient pressure p_a so the ratio p_{08}/p_a may get as high as 16. For the nozzle to be reasonably efficient at $p_{08}/p_a \approx 16$ it is important that it be of the convergent–divergent form, see Fig.11.3.

The behaviour of a convergent–divergent nozzle is given in Fig.6.2 (ignore for the present that this figure is for $\gamma = 1.40$ whereas the exhaust gases have a lower value). For a pressure ratio of 16 the exit Mach number would be about 2.46 and $\bar{m}_9$ approximately 0.4 times the throat value. In other words the area at nozzle outlet would be 2.5 times the throat area, which might be larger than the cross-sectional area of the rest of the engine. Such a large nozzle exit might be impractical and the area increase downstream of the throat may be limited to a smaller value than this. The expansion is then not fully reversible, but the loss of gross thrust can be kept reasonably small. Figure 15.4 shows the variation in gross thrust with pressure ratio for three different nozzles as a function of pressure ratio p_{08}/p_a calculated for $\gamma = 1.30$, a value representative of exhaust gases. The nozzles are a convergent nozzle, a fully reversible convergent–divergent nozzle and a nozzle which is truncated where the exit area is 1.6 times the value at the throat. (An exit area 1.6 times the throat area was selected here because it has been used for the F-16 fighter.) At a pressure ratio of 16 a loss in gross thrust of about 1.5% would be incurred as a result of the truncated divergent section, but the thrust is about 10% higher than that from a simple convergent nozzle. It can also be seen from Fig. 15.4 that up to a pressure ratio of about 5 across the nozzle there is little penalty in having the simpler and lighter convergent nozzle.

An ideal nozzle would vary the throat area A_8 and the exit area A_9 independently to maintain the correct area ratio corresponding to the pressure ratio imposed. The required throat area is determined by the mass flow, stagnation pressure and stagnation temperature from the engine so as to keep $\bar{m}_8$ constant. In some engines the two areas are made to vary independently, but in others there is a fixed schedule of A_9/A_8 as a function of A_8 so that a single set of actuators varies both areas at once. Having only a single set of actuators saves cost and weight, but with some loss in thrust and increase in sfc over parts of the operating envelope.

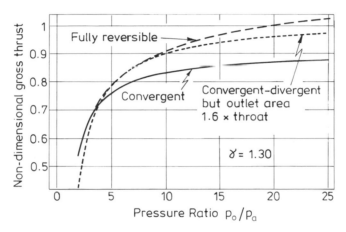

Figure 15.4. Non-dimensional gross thrust from nozzles, $F_G/(\dot{m}\sqrt{(c_p T_0)})$, versus the ratio of upstream stagnation pressure to dowmstream static pressure.

The requirement for stealth means that new aircraft frequently have nozzles of special shape. These are frequently rectangular and can be above the wing or tail plane to provide shielding from the ground. Aerodynamically the shape is likely to be less good than a simple round nozzle, but the principle of operation is unchanged.

The inlet for high-speed flight

By convention the inlet is the responsibility of the airframe manufacturer, but in terms of understanding the engine behaviour it needs to be included in any assessment of engine configuration. For flight at subsonic conditions a simple 'pitot' intake, similar in principle to those fitted to subsonic civil aircraft, with suitably rounded and shaped leading edges to accommodate incidence effects can produce very small loss. Such an intake can also be satisfactory at low supersonic conditions, but at supersonic flight speeds some loss is unavoidable. With the simple pitot intake at supersonic speeds the loss is created by a normal shock wave ahead of the intake. For Mach numbers approaching 2.0 the loss from a normal shock is relatively large and aircraft for which a critical role involves such high Mach numbers tend to have intakes which decelerate the inlet flow in a series of oblique shocks. The stagnation pressure loss decreases as the deceleration is spread through a greater number of shocks.

The American F-15 and F-16 aircraft were both designed at about the same time, but the F-15 is intended to be capable of operation up to $M = 2.3$, whereas the F-16 is not expected to have much role beyond $M = 1.6$, though it can reach $M = 2.0$. Because of their different roles the two aircraft have very different intakes. The F-16 is a pitot type, whereas the F-15 has variable ramps to produce deceleration in three oblique shocks followed by a weak normal shock, Fig. 15.5. The measured losses in the F-15 and F-16 intakes are shown in Fig. 15.6 as a function of flight Mach number; also shown are the losses from a normal shock and from an empirically based US military standard MIL-E-5007/8 used in the industry for design studies. As expected, the loss for the F-16 pitot intake is very close to the normal shock loss, but more surprisingly the F-15 follows MIL-E-5007/8 closely.

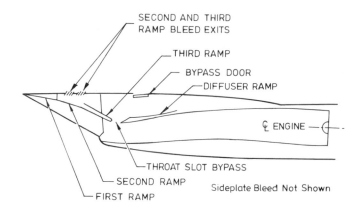

Figure 15.5. The engine intake for the F-15 fighter.

With p_{01} and p_{02} denoting the stagnation pressure at entry to and exit from the inlet, the empirical expressions for MIL-E-5007/8 are given by

$$p_{02}/p_{01} = 1.0 \qquad\qquad \text{for } M \le 1.0$$

and $\qquad\qquad p_{02}/p_{01} = 1.0 - 0.075(M-1)^{1.35} \qquad \text{for } M > 1.0 .$

$$(15.5)$$

The F-15 intake has comparatively sharp leading edges, so its loss at low subsonic flight speeds are higher. The three ramps have to be adjusted to accommodate variations in Mach number and engine condition and to operate these and to open the bypass door it is necessary to have three actuators, with associated control inputs, in each intake. It becomes easy to appreciate that such an intake entails substantial extra weight and cost compared with a simple fixed geometry pitot type which can only be justified when high flight speeds are sufficiently important to the main design mission.   The position is further complicated by the wish to include aspects of stealth technology, providing a curved path along the intake (to stop line of sight to the engine inlet face) and to include vanes coated in radar-absorbent material in the intake.

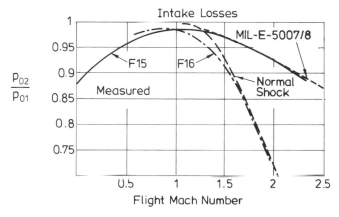

Figure 15.6. Intake pressure ratio showing measured results for the F-15 and F-16, the normal shock loss and the empirical relation from MIL-E-5007/8.

Exercises

15.1 **a)** An aircraft flies at $M = 0.9$ at the tropopause ($T_a = 216.65$ K) and the fan pressure ratio is 4.5. Find the temperature downstream of the fan T_{013} if the compressor polytropic efficiency is 0.85.

If the bypass ratio is 0.67, find the value of c_{pm} for the mixed flow through the nozzle using a simple mass weighting of the two streams and taking $\gamma = 1.30$ for the flow leaving the turbine. Thence find γ_m for the mixed flow. What is the pressure ratio p_{08}/p_a across the nozzle?

(**Ans:** $T_{023} = 417.4$ K; 1148 J/kg/K, 1.333 , 7.61)

b) Consider two alternative configurations of the above turbofan engines.   In one case the core and bypass mix with negligible loss in stagnation pressure before entering the propelling nozzle;   in the other case each flow passes through a separate nozzle.   The engines are identical upstream of where the streams can mix and in both cases the stagnation pressure *at nozzle entry* is equal to that out of the fan. In both cases the nozzles are reversible. If the temperature downstream of the LP turbine is $T_{05} = 1200$ K, find the mixed-out temperature T_{06}. Find the gross thrust per unit mass flow of air for the mixed and unmixed flows.

(**Ans:** $T_{06} = 925$ K,   unmixed $F_G/\dot{m} = 877$ m/s; mixed $F_G/\dot{m} = 919$m/s)

15.2 For the mixed engine of Exercise 15.1, find the proportional increase in throat area needed when the afterburner raises the temperature to 2200 K.   The jet pipe pressure (and all the properties of the engine upstream) remains unchanged by the afterburner. Assume that γ is equal to 1.30 for the jet with afterburning. Assume the nozzle is isentropic and neglect the mass flow rate of fuel.   From Chapter 6 note that at the throat of a choked nozzle

$$\overline{\dot{m}_8} = \frac{\dot{m}\sqrt{c_p T_{08}}}{A_8\, p_{08}} = \frac{\gamma}{\sqrt{\gamma-1}}\left(\frac{2}{\gamma+1}\right)^{(\gamma+1)/\{2(\gamma-1)\}}.$$

What is the increase in gross thrust F_G produced by the afterburner if the nozzle remains isentropic? For a flight Mach number of 0.9 at the tropopause, what is the increase in *net* thrust?

(**Ans:** $\Delta A_8 = 56\%$, $\Delta F_G = 56\%$, $\Delta F_N = 78.2\%$)

15.3 Figure 15.7 shows a possible intake for supersonic flight with inlet Mach number 2.0 and exit Mach number 0.89. The ramp angles are marked and the shocks are shown with broken lines.

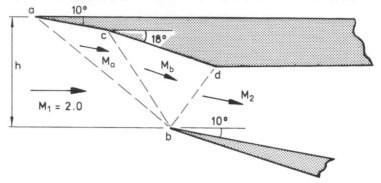

Figure 15.7. A simplified intake for an aircraft designed to operate at $M = 2.0$.

Find the stagnation pressure ratio p_{02}/p_{01} and compare this with the pressure ratio across a normal shock at $M_1 = 2.0$ ($p_{02}/p_{01} = 0.721$) and with the ratio for MIL-E-5007/8.

(**Ans:** three oblique shocks 0.959; MIL-E-5007/8 0.925)

If the height of the intake between points marked a and b (measured normal to the inlet flow direction) is h, what are the lengths of the ramps ac and cd?   (**Ans:** ac = 0.634h; cd = 1.060h)

Note: To solve this exercise easily requires tables or charts for oblique shock behaviour. To facilitate this some relevant data is given below, where subscripts u and d refer to upstream and downstream of the shock respectively. The shock angle is measured from the upstream flow direction.

M_u	Flow deflection	M_d	Shock wave angle	p_d/p_u	T_d/T_u	
2.0	10°	1.64	39.3°	1.71	1.17	
1.64	8°	1.36	46.6°	1.49	1.12	
1.36	8°	0.89	71.0°	1.76	1.18	(Strong branch)

15.3 THE THERMODYNAMIC CYCLE OF COMBAT AIRCRAFT ENGINES

As will be discussed in Chapter 16 the operating condition of a combat engine varies greatly between the different operating points. The most significant change comes from the variation in inlet stagnation temperature and pressure as the flight Mach number increases. This is summarised for some key operating conditions in Table 15.1.

It will be recalled that for a gas turbine the ratio T_{04}/T_{02}, the ratio of turbine inlet temperature to compressor inlet temperature, is very significant. Since the turbine inlet temperature cannot be increased beyond a limit fixed by the material and the cooling technology, the maximum permissible value of T_{04}/T_{02} falls as T_{02} goes up as the flight speed increases. The effect of this is to reduce the engine operating point as speed increases, with consequent reductions in non-dimensional rotational speeds, non-dimensional mass flows and pressure ratios. For the present studies the maximum turbine inlet temperature, $T_{04} = 1850$ K, which is in line with current advanced performance, will be used for the maximum power conditions when T_{04} is what limits the upper performance.

Table 15.1   Representative conditions for consideration of combat requirements.

M	Sea level		Tropopause	
	T_{01}	p_{01}/p_a	T_{01}	p_{01}/p_a
0.0	288.2	1.0	216.7	1.0
0.9	344.8	1.69	251.7	1.69
1.2	371.3	2.24	279.0	2.42
1.5			314.1	3.67
2.0			390.0	7.82

The tropopause is at 11 km altitude, $T_a = 216.7$ K, $p_a = 22.7$ kPa.

The ratio of inlet stagnation pressure to ambient pressure p_{01}/p_a rises with Mach number even more rapidly than the temperature ratio. (For supersonic flight there is in general a loss in stagnation pressure in the inlet, so $p_{02} < p_{01}$, but the loss is likely to be less than about 10% within the normal operating range.)   Because the pressure ratio produced in the inlet is so substantial at high flight Mach numbers, the pressure ratio required from the turbomachinery inside the engine is correspondingly lower; for $M = 2.0$ a fan pressure ratio of only 2 gives an overall pressure ratio of almost 16 across the propelling nozzle, and as Fig. 15.4 shows, there is little gain in thrust from increases in pressure ratio beyond this value.

Exercises
15.4 For flight Mach numbers of 0.9, 1.5 and 2.0 at the tropopause find for the air entering the inlet the stagnation temperature T_{01} and stagnation pressure p_{01}.

(**Ans:** $T_{01} = 251.7$ K, 314.1 K, 390.0 K; $p_{01} = 38.3$ kPa, 83.3 kPa, 177.6 kPa)

For the same conditions find the stagnation pressure entering the engine p_{02} using the inlet loss given by MIL-E-5007/8.   (Note that $T_{01} = T_{02}$)   (**Ans:** $p_{02} = 38.3$ kPa, 80.8 kPa, 164.3 kPa)

15.5 For a turbofan engine with an afterburner producing a constant exit temperature T_{0ab}, show that the jet velocity is a function of the fan pressure ratio and flight Mach number only.   (Neglect any losses.)

If the fan pressure ratio is 4.5, the flight Mach number is 0.9 at the tropopause and $T_{0ab} = 2200$ K, find the jet velocity, the fuel flow per kg of air, the specific thrust and the sfc.   Take $\gamma \approx 1.30$ for combustion products.   (**Ans:** $V_j = 1431$ m/s, $\dot{m}_f/\dot{m}_a = 0.0594$, $F_N/\dot{m}_a = 1250$ m/s , sfc = 1.68 kg/h/kg)

The thermodynamic behaviour of the engines may be better understood by looking at temperature entropy diagrams. These are shown in Fig. 15.8 for two flight speeds at the tropopause, $M = 0.9$ and 2.0, with the engine operating 'dry' and with the afterburner in each case. The turbine inlet temperature is equal to 1850 K in all the cases and the afterburner temperature raises the jet stagnation temperature to 2200 K. For simplicity, in Fig. 15.8 (but not elsewhere) the gas properties are those associated with $\gamma = 1.40$ for the burned and unburned air. The chosen fan pressure ratio and overall pressure ratio in Fig. 15.8 is lower for the engine at $M = 2.0$, in line with what will be found to be necessary.

Considering the 'dry' cases first, the pressure first rises in the intake, from $p_1 = p_a$ to p_{02}. The pressure is then raised in the fan to $p_{013} = p_{023}$ (it being assumed here that the pressure is uniform in the radial direction at outlet from the fan) and this sets the stagnation pressure for the flow in the jet pipe, $p_{08} = p_{013}$. In words, the stagnation pressure of the flow entering the propelling nozzle is therefore set by the fan exit pressure. The pressure of the core flow is further raised in the HP compressor to p_{03} and then raised in temperature, at constant pressure, to the turbine inlet temperature, T_{04}.

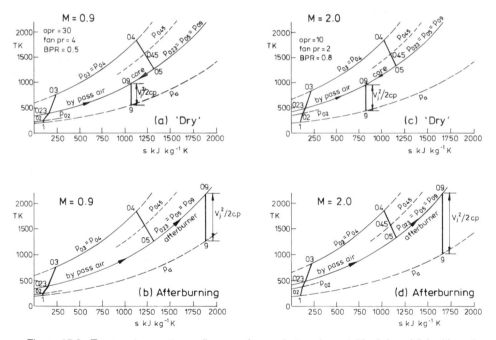

Figure 15.8. Temperature–entropy diagrams for combat engines at $M = 0.9$ and 2.0 with and without afterburning. (For these diagrams c_p taken as equal for burned and unburned gas.)

The pressure at the HP turbine outlet, p_{045}, is such that the HP turbine enthalpy drop is equal to the enthalpy rise in the HP compressor. The subsequent temperature drop in the LP turbine is used to drive the fan. The bypass air at temperature T_{013} and the core air at T_{05} are mixed at constant pressure in the jet pipe to give the exhaust temperature T_{09}. The comparable plots with the afterburner take the mixed flow and raise the temperature to 2200 K.

The overall pressure ratios and bypass ratios selected are similar to those which will be found to be optimum with the conditions prescribed; at the higher Mach number a much lower overall pressure ratio and fan pressure ratio are appropriate. The jet kinetic energy is higher for the M = 2.0 case because the pressure ratio in the jet pipe is higher, notwithstanding the lower fan pressure ratio for this case. The higher jet velocity for flight at M = 2.0 is even more apparent when the afterburner is in use.

The thermal efficiency is markedly lower, and the specific fuel consumption higher, when the afterburner is used, mainly because the pressure in the jet pipe relative to the ambient pressure p_{08}/p_a at which the equivalent heat input takes place is relatively low, much lower than the pressure p_{03}/p_a for the heat input in the main combustor. When the flight speed is high p_{08}/p_a is higher too and the penalty for using the afterburner is smaller; this benefit increases rapidly with Mach number so that beyond about M = 2.5 the optimum propulsion device is the ram jet which relies entirely on the inlet compression to obtain the pressure rise.

Exercise

15.6 Neglecting loss in the intake find the flight Mach number at which the inlet stagnation pressure is equal to 16 times the ambient pressure. For this Mach number at the tropopause what is the inlet stagnation temperature?   (**Ans:** M = 2.46, 478.4 K)

A ram-jet travels at the tropopause at the Mach number determined above. Fuel is burned to raise the stagnation temperature to 2200 K. If the combustion produces no loss in stagnation pressure and the expansion in the nozzle is reversible, find the jet velocity, the mass of fuel burned per kg of air, the specific thrust (net thrust per unit mass flow of air) and the specific fuel consumption. Take γ =1.40 for the air and γ=1.30 for the products of combustion, and for the fuel LCV = 43 10^6 J/kg.
              (**Ans:** V_j = 1608 m/s, $\dot{m}_f/\dot{m}_a$ = 0.0538, $F_N/\dot{m}_a$ = 969 m/s, sfc = 1.96 kg/h/kg)

Now consider the ram-jet with an intake loss at the above Mach number according to MIL-E-5007/8, and a further loss in stagnation pressure in the combustion process of 5%. Recalculate the jet velocity, the specific thrust and the sfc. The stagnation temperature of the jet is unchanged at 2200 K, so the fuel flow rate per unit mass flow of air is also unchanged.
              (**Ans:** V_j = 1569 m/s $F_N/\dot{m}_a$ = 928 m/s, sfc = 2.05 kg/h/kg)

Note: At this Mach number the ram-jet is an attractive engine, with high specific thrust and specific fuel consumption which is similar to an afterburning gas turbine engine. The complexity, weight and cost are low. The attractiveness of the ram-jet increases as the Mach number goes up, with the compressor and turbine of a gas turbine becoming an encumbrance. One of the problems is accelerating the vehicle to a high Mach number to get the ram-jet working; this can be achieved by using rockets or by launching the vehicle from an aircraft at high speed.

15.4 SOME CONSTRAINTS ON COMBAT AIRCRAFT ENGINES

The most familiar limit on engine operation is the turbine inlet temperature T_{04}, which is fixed by the material properties of the turbine blades, the amount of cooling air used, how effective the cooling is and the expected life of the blades. For the current work an upper level of 1850 K will be assumed which would be permitted during take off and for combat manoeuvres.

The non-dimensional operating point of the engine is set by the ratio of turbine inlet temperature to compressor inlet temperature T_{04}/T_{02}, where $T_{02} = T_{01}$ is the engine inlet stagnation temperature given by

$$T_{02} = T_a [1 + \tfrac{1}{2}(\gamma - 1) M^2] ,$$

T_a being the ambient air temperature and M the flight Mach number.

Another limit on engine performance is the compressor delivery temperature T_{03}. The limit here comes from the material properties of the compressor disc and blades at the rear of the compressor, and a value of 875 K will be taken to represent the current maximum allowed. (This temperature is an appropriate upper limit for titanium alloys, but with nickel-based alloy about 100 K higher temperature could be tolerated. Use of nickel-based alloys would, however, increase the weight.) The temperature of the gas at the rear of the compressor is determined by the inlet temperature to the engine (and therefore by the atmospheric temperature and flight speed) and by the overall pressure ratio and efficiency of the compressor, since

$$T_{03} = T_{02} (p_{03}/p_{02})^{(\gamma - 1)/\gamma \eta} .$$

If T_{04}/T_{02} is held constant, and the nozzle remains choked with a constant throat area, the engine stays at a constant non-dimensional condition and the pressure ratios throughout the engine also stay constant. Under this condition the compressor delivery temperature is proportional to T_{02}, the engine inlet temperature.

There is a third non-dimensional parameter to be considered, which is related to the Mach number of flow in the blades, and is characterised by $N/\sqrt{(c_p T_{02})}$, where N is the rotational speed of one of the shafts[1]. If T_{04}/T_{02} is constant, the engine is at a fixed non-dimensional condition and $N/\sqrt{(c_p T_{02})}$ will also be constant. (Where there is more than one shaft there are as many rotational speeds as there are shafts; for constant engine condition there is a fixed ratio between the speeds of the shafts.) The turbomachinery aerodynamic performance, particularly of the compressor, is sensitive to $N/\sqrt{(c_p T_{02})}$; if this parameter becomes too high the efficiency falls precipitously (see sections 11.3 and 11.4) and there is a risk of the self induced aero-elastic vibration known as flutter. The LP compressor experiences much larger changes in $N_L/\sqrt{(c_p T_{02})}$ as T_{04}/T_{02} is varied than the corresponding changes in $N_H/\sqrt{(c_p T_{023})}$ for the HP compressor. (Note that for fixed T_{04}/T_{02} the ratios T_{023}/T_{02} and N_H/N_L are both constant.)

The maximum value of N is primarily limited by the mechanical stresses in the discs holding the rotor blades. If N were held at its limit a higher than acceptable value of $N/\sqrt{(c_p T_{02})}$ could occur if the inlet temperature were lower than the value used in design; this

---

[1]The parameter $N/\sqrt{(c_p T_{02})}$ is derived from a non-dimensional term $ND/\sqrt{(c_p T_{02})}$, which is proportional to the relative Mach number into the first set of rotor blades. It is normally understood that T_0 is the stagnation temperature *into* the component or engine.

could occur, for example, with an engine designed for sea-level static conditions and flown at low Mach number at high altitude. By holding T_{04}/T_{02} at is its sea-level design value when T_{02} is below the design value, this over-speeding is prevented. When the inlet temperature T_{02} is higher than the value used in the design it is unlikely that the design value of $N/\sqrt{(c_p T_{02})}$ will be exceeded because this would require T_{04} to increase too; what normally happens as T_{02} rises is that T_{04}/T_{02} and $N/\sqrt{(c_p T_{02})}$ are both reduced.

The effects of the constraints imposed on T_{04}, T_{03} and $N/\sqrt{(c_p T_0)}$ are illustrated in Fig. 15.9 for an engine with an overall pressure ratio of 30 at design. The abscissa shows the inlet stagnation temperature and the ordinate shows the turbine inlet temperature and the compressor delivery temperature. The point marked **A** is designated the design point and has been selected here to be at $T_{02} = 288$ K, the sea-level temperature for the standard atmosphere; at this condition $T_{04} = 1850$ K is selected, the maximum allowed. At **A** $N/\sqrt{(c_p T_0)}$ and all the pressure ratios have their maximum values; this is true too along the line to the left of **A** because along this line the engine is at the same non-dimensional condition. Constant T_{04}/T_{02} is achieved along the line to the left of **A** by reducing the fuel supply to lower T_{04} and thereby maintain the temperature ratio constant. Operation along the line to the left of **A** could be during take off on a cold day ($T_a < 288$ K) or could be a high altitude and low speed.

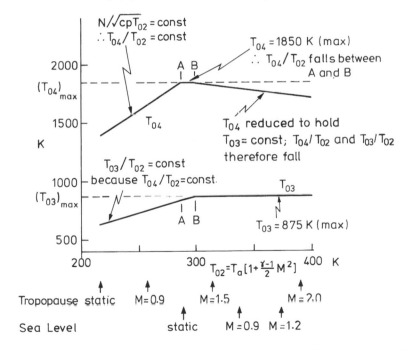

Figure 15.9. Turbine inlet temperature T_{04} and compressor delivery temperature T_{03} versus engine inlet temperature T_{02}. Engine overall pressure ratio $p_{03}/p_{02} = 30$ at the design point when $T_{02} = 288$ K. Compressor polytropic efficiency 0.90.

To the right of A the same T_{04}/T_{02} cannot be maintained without exceeding the limit on T_{04}. Therefore moving to the right of A the engine is at progressively lower non-dimensional operating points and all the pressure ratios and $N/\sqrt{(c_pT_0)}$ will decrease.

To the left of **A** the compressor delivery temperature varies in proportion to T_{02} but between **A** and **B** it increases more slowly with T_{02} since the pressure ratio is decreasing. At point **B** the compressor delivery temperature has reached its upper limit. If T_{02} is to be increased further the overall pressure ratio must be decreased to reduce T_{03}/T_{02}; reducing the pressure ratio is achieved by lowering the turbine delivery temperature below the maximum.

The range of engine inlet temperatures which cause the engine output to be limited by the turbine inlet temperature, the region from **A** to **B**, is quite small. Much larger are the regions where it is non-dimensional rotational speed (to the left of **A**) or compressor delivery temperature (to the right of **B**) which restricts thrust. **A** is at the value of T_{02} where maximum T_{04} and maximum $N/\sqrt{(c_pT_0)}$ both occur. The designer is able to select the value of T_{02} where **A** occurs, thereby fixing the design point of the engine. (Equivalently, point **A** is at the value of T_{02} where maximum T_{04} and maximum T_{04}/T_{02} coincide.) The position of **B** is less arbitrary because it follows directly from the choice of overall pressure ratio and efficiency.

It would seem sensible to put **A** where the highest value of T_{04}/T_{02} will be achieved, that is where T_{02} is low – but there is a snag. If the engine is going to have to operate efficiently at higher values of T_{02} it is going to be operating off-design and the further to the left is point **A** the further off-design some high speed conditions will be. Selection of point **A** is therefore an important design choice which reflects the most important duties expected of the engine.

Exercise

15.7 Find the overall pressure ratio at which the points A and B in Fig. 15.9 coincide. Take polytropic efficiency for the compression system to be 0.9.   **(Ans: 33.1)**

For an aircraft flying at the tropopause, $T_a = 216.65$, determine the Mach number at which T_{02} is equal to 288.15 K, the ambient temperature at sea level?   **(Ans: 1.285)**

15.5 ENGINE RATING

Military engines now use electronic control systems to ensure that operation does not exceed one of the many restrictions. (FADEC – full authority digital electronic control is the abbreviation for the control system used on recent engines.) Some of the selection remains, of course, in the hands of the pilot, but nevertheless there are restrictions on the time for which certain engine operating conditions may be maintained. The terms frequently used to refer to critical engine operating points are:

Combat Rating

This is operation with the afterburner lit (if one is installed) and the engine itself operating at its maximum allowed condition (which may correspond to maximum turbine inlet temperature, maximum compressor delivery temperature or maximum $N/\sqrt{(c_pT_0)}$). Operation is normally limited to a short period, say 2.5 minutes.

Maximum Dry Rating

This gives the maximum performance without the afterburner. In the simplest mode of operating the afterburner the engine condition upstream of the turbine outlet would be the same as that in Combat Rating.

Intermediate or Military Rating

In this condition operation is allowed for much longer, about 30 minutes. The afterburner is not normally lit for this condition.

Maximum Continuous Rating

This is the condition at which the engine may be operated without restriction on operating time.

SUMMARY CHAPTER 15

Just as there is no single aircraft type to fulfil all the various roles in an optimum way, so too with the engine the final choice must be a compromise. The process of selection for the engine, and for the aircraft, will involve numerous simulations to establish how well each of the many combinations accomplishes the various missions attempted. A good combination will be able to accomplish most, but it may be that a small number will need some compromise, such as additional fuel supplies from tanker aircraft. In general it can be said that the optimum engine will have high specific thrust without use of the afterburner (i.e. 'dry') if the combat phases of a mission use a large fraction of the total fuel; such a mission would be of the intercept type. Conversely an aircraft with a primary mission for air superiority (the classic fighter) will use less than half its fuel in the combat phase and fuel consumption during cruise and loiter will be more important; for such an aircraft a relatively small engine giving adequate 'dry' thrust for many phases of the mission, but using the afterburner for the short time of combat, will be best.

The requirement to have high thrust–weight ratio directs the engine towards a low bypass ratio with a high jet velocity; as a way of categorising engines the specific thrust is rather better than the bypass ratio. With the exception of the very specialised engine for the Harrier, all combat engines mix the core and bypass streams before the final propelling nozzle, obtaining thereby a small increase in thrust. In most cases an afterburner is installed between the mixing plane and the final nozzle. If there is an afterburner it is essential that the final nozzle has a variable throat area; for supersonic flight it is common, though not universal, to have a convergent–divergent nozzle, though the extent of the divergent portion is usually less than that to give isentropic expansion down to ambient pressure.

For aircraft with a major mission which involves flight at Mach numbers approaching 2.0 a variable inlet is common so as to obtain small shock losses from a succession of (up to 3) oblique shocks. For aircraft having their primary role at a lower speed the optimum is likely to involve a simpler and lighter intake without variable ramps or vents.

A crucial decision in the design and specification of an engine is the choice of design point at which the maximum pressure ratio and non-dimensional speeds occur simultaneously with the highest allowed turbine inlet temperature. If this occurs at a relatively high inlet temperature it implies that the maximum thrust capability of the engine is not being achieved at low flight speeds and high altitudes; if it occurs at too low an inlet temperature the engine may be a long way off-design, with much reduced pressure ratio, at high flight speeds. Over only a small range of inlet temperature is the engine limited by the maximum allowable turbine inlet temperature. The subsequent chapters will take the required thrusts and operating points from Chapter 14 and use these with engines of the type discussed in this chapter.

CHAPTER 16

DESIGN POINT FOR A COMBAT ENGINE

16.0 INTRODUCTION

In this chapter we will consider three separate engine designs corresponding to distinct operating conditions. For convenience here the three design points are at the tropopause (altitude 11 km; standard atmosphere temperature 216.65 K and pressure 22.7 kPa) for Mach numbers of 0.9, 1.5 and 2.0. The thrusts required for these conditions were determined in Exercise 14.4. At each condition a separate engine is designed – this is quite different from designing the engine for one condition and then considering its operation at different conditions, which is the topic of Chapter 17.

For this exercise all design points will correspond to the engine being required to produce maximum thrust, even though the ultimate suitability of an engine for its mission may depend on performance, particularly fuel consumption, at conditions for which the thrust is very much less than maximum. The designs will first be for engines without an afterburner (operation 'dry') and then with an afterburner; the afterburner will be assumed to raise the temperature of the exhaust without altering the operating condition of the remainder of the engine so the stagnation pressure entering the nozzle is unchanged.

The engines considered will all be of the mixed turbofan type – such an engine was shown in Fig.15.1 with a sketch showing the station numbering system adopted. Note that the numbering shows station 13 downstream of the fan in the bypass and station 23 downstream of the fan for the core flow; in the present simplified treatment it will be assumed that $p_{023} = p_{013}$ and that $T_{023} = T_{013}$. There are small losses associated with mixing, but these will be neglected here. As a result if the fan pressure ratio is fixed then so too is the pressure at outlet from the turbine and the conditions through the core are determined. Fixing the pressure ratio across the fan is therefore a more direct way of specifying properties inside the engine than, for example, the bypass ratio; the pressure ratio is also the dominant term in the expression for jet velocity and therefore gross and net thrust.

In comparing different designs some parameters need to be held constant. In the present case a uniform technology standard will be assumed, so maximum temperatures and component efficiencies are maintained equal for all the designs. This is the subject of the next section.

16.1 TECHNOLOGY STANDARD

The standard of technology described in this section will be used throughout the treatment of the fighter engine in Chapters 16 – 18. The standard adopted here is believed to be broadly in line with what designers inside major companies are using at the time of writing, but section 16.7 explores the effect of changes in many of the parameters assumed. The most basic parameters defining the technology standard are set out in Table 16.1, but other aspects of the general treatment follow in the remainder of the section.

Table 16.1

Parameters and constraints assumed throughout Chapters 16 –18.

Turbine inlet temperature	$T_{04} \not> 1850$ K
Compressor outlet temperature	$T_{03} \not> 875$ K
LP Compressor *polytropic* efficiency[1]	$\eta_{pc} = 0.85$
HP Compressor *polytropic* efficiency	$\eta_{pc} = 0.90$
HP & LP Turbine *polytropic* efficiency	$\eta_{pt} = 0.875$

The turbine entry temperature is actually the temperature downstream of the HP turbine nozzles and upstream of the rotor, after the cooling air for the nozzles has mixed out. The maximum compressor delivery temperature is appropriate for blades and discs made of titanium alloy; using nickel based alloy a temperature perhaps 100 K higher might be acceptable, but these would be heavier and the maximum rotational speed would be lower for reasons of mechanical stress. The chosen values of both T_{04} and T_{03} are close to the current maxima for this type of engine.

For convenience the *polytropic* efficiencies will be used in this and the following chapter, but the conversion between polytropic efficiency and isentropic (sometimes called adiabatic) efficiency is explained in Chapter 11. Note that the efficiencies are generally lower than those for the civil transport aircraft. These efficiency values are representative for maximum thrust operation; at reduced thrust the Mach numbers inside the engine are lower and the component efficiencies will generally be higher.

It is essential to include the cooling flows in cycle analysis if appropriate magnitudes are to be found for the specific thrust. It will be assumed that 20% of the air which enters the HP compressor is used for turbine cooling. Of this, 8% is used to cool the HP nozzles and its effect is included in the temperature quoted as turbine entry temperature. Another 8% is used to cool the HP rotor; in the simple cycle analysis here this air, at compressor delivery pressure

---

[1]The polytropic efficiency used for the LP compressor is low by modern standards and about 0.90 would be more representative. The effect of this change can be seen in Table 16.3. Conversely the turbine cooling flow specified here is on the low side, consistent with too high a technology standard.

and temperature p_{03} and T_{03}, is taken to mix with the gas leaving the HP rotor where its only effect is to lower the temperature. A further 4% is used to cool the LP turbine and this is mixed out at LP turbine rotor exit.

The turbine inlet temperature can be increased by better cooling technology and by using more cooling air. For the compressor, however, the exit temperature can only be allowed to rise if there are advances in material science, since there is no air at adequate pressure and lower temperature with which to cool the compressor.

A more detailed analysis would include stagnation pressure loss in the combustor, in the bypass duct and in the jet pipe. In each the stagnation pressure falls by of the order of 5% of the local stagnation pressure. There is also a loss in stagnation pressure associated with the temperature rise in the afterburner. For the present purpose all these losses can be neglected in the interest of simplicity without seriously distorting the estimates. Loss of stagnation pressure in the intake is neglected at subsonic flight speeds, but included for supersonic flight by the relation from MIL-E-5007/8 introduced in section 15.2.

The nozzle will be assumed to provide a fully reversible expansion to the ambient pressure, though this is at best an oversimplification even with a convergent–divergent nozzle. Many engines operate with only a convergent final nozzle and some significant thrust loss is inevitable at high pressure ratios; this is more serious for high-speed flight because of the large rise in pressure in the inlet.

When the afterburner is lit the maximum flow of fuel is determined by the ability to consume the oxygen and the temperature approaches the stoichiometric value; here a temperature of 2200 K will be assumed. Some engines in service allow a gradual or staged increase in fuel supplied to the afterburner, giving a variable degree of thrust boost, but it will be presumed here that whenever the afterburner is used this maximum temperature is produced. As noted above, any additional loss in stagnation pressure associated with combustion in the afterburner is neglected.

In treating combat engines the value of specific heat will be obtained by specifying γ and the gas constant R. It is a very satisfactory approximation to take $R = 287$ J/kgK throughout. For unburned air it will be assumed that $\gamma = 1.40$, giving $c_p = 1005$ J/kgK, whilst for products of combustion $\gamma = 1.30$ giving $c_{pe} = 1244$ J/kg/K. (Both these values represent simplifications, with γ being a function of temperature, as shown in Fig.11.1.)

Exercise
16.1 An aircraft flies at the tropopause, 11000 m, at which the ambient temperature is 216.65 K. Find the maximum overall pressure ratio for the engine at flight Mach numbers of 0.9, 1.5 and 2.0 given that the maximum allowable temperature at outlet from the compressor is 875 K.   Assume that the *polytropic* efficiency of the combined compressors is 0.875.   (**Ans:** 45.4; 23.0; 11.9)

Mixing

Whenever mixing takes place there is a loss in stagnation pressure, though there may be a rise in static pressure. The process depends on the mass flow, momentum and energy of each stream. Mixing takes place between the turbine cooling air and the main flow, but this only involves a small fraction of the total air. A much more significant mixing process takes place between the core and the bypass streams downstream of the LP turbine. A detailed calculation could be performed for mixing of core and bypass, though the loss in pressure is only a few per cent and for the present it is assumed that the mixing takes place without loss in stagnation pressure. As discussed in section 15.2 it is assumed that the stagnation pressure of the gas downstream of the LP turbine is equal to the stagnation pressure downstream of the LP compressor, that is

$$p_{05} = p_{013}, \tag{16.1}$$

and this in turn is equal to the stagnation pressure at the nozzle throat p_{08} and the nozzle exit p_{09}. The assumptions for stagnation pressure in the bypass and in the duct leading to the nozzle exit can be justified only by recognising that the losses rise steeply with Mach number and the Mach number of the flow leaving the turbine and in the bypass duct is typically less than 0.3.

The core stream at T_{05} will be assumed to mix with the cooler bypass stream at T_{013} to a uniform temperature T_{06} prior to entering the propelling nozzle.   In 'dry' operation (that is, without the afterburner in use) the jet temperature is given approximately by

$$c_{pe}T_{05} + bpr\, c_p T_{013} = (1+bpr)c_{pm}T_{06}, \tag{16.2}$$

where c_p, c_{pe} and c_{pm} are the specific heats of the air prior to combustion (i.e. in the bypass stream), of the gas leaving the turbine and of the flow mixed out before it enters the propelling nozzle, as in equation 15.4. (This is covered in more detail in equation 16.10.)

When operating with afterburner lit the temperature at the nozzle throat $T_{08} = T_{0ab}$ is effectively fixed by the fuel flow; it will be assumed that the combustion leads to a negligible loss in pressure in the jet pipe and that $\gamma = 1.30$, $c_p = 1244$ J/kgK for the gas. When operating with the afterburner lit the gas temperatures are so high that some of the bypass stream is used to cool the walls of the jet pipe and the walls of the nozzle; this is neglected here and T_{0ab} is taken to be the temperature when the cooling air has mixed out.

16.2 GENERAL ENGINE SPECIFICATION

With a mixed turbofan of the type shown in Fig.15.1 with fixed geometry and a fixed level of technology, two parameters are sufficient to specify the whole engine type. One is the overall pressure ratio p_{03}/p_{02} and the other is the pressure ratio between inlet to the compressor and inlet to the final nozzle, p_{08}/p_{02}. This second pressure ratio is, neglecting pressure losses in the duct and mixing, equal to the LP compressor (fan) ratio p_{013}/p_{02}. Thus specifying p_{03}/p_{02} and p_{013}/p_{02}, together with T_{04}/T_{02}, is sufficient to determine the engine performance, more specifically its specific thrust and specific fuel consumption.

For a given overall pressure ratio p_{03}/p_{02} and temperature ratio T_{04}/T_{02} there is a fixed amount of power to be delivered to the fan. If the chosen fan pressure ratio is raised, the mass flow rate compressed in it must be reduced, in other words the bypass ratio must fall. A small change in pressure ratio can produce a large change in bypass ratio because an increase in pressure ratio not only increases the work per unit mass flow required by the fan but it also increases the back pressure to the LP turbine and therefore reduces its power output.

Overall calculations

With overall pressure ratio and fan pressure ratio selected as input parameters, the calculations of the engine cycle become comparatively straightforward. The stagnation conditions at entry to the compressor, station 2, are determined by the altitude, flight Mach number and losses in the intake (the losses are normally only substantial at supersonic flight speeds). The fan pressure ratio fixes the stagnation pressure at entry to the HP compressor p_{023}. It is assumed here that the fan delivers equal stagnation pressure and stagnation temperature to both the bypass duct and core, $p_{023} = p_{013}$ and $T_{023} = T_{013}$. For a known fan efficiency η, the corresponding temperature and pressure ratio for the core flow in the fan is

$$T_{023}/T_{02} = (p_{023}/p_{02})^{(\gamma-1)/(\eta\gamma)}$$

with a similar expression relating the stagnation temperature and pressure at combustor inlet T_{03} and p_{03}.

The fuel flow to the core combustor $\dot{m}_f$ required to produce a turbine inlet temperature T_{04} is given by the expression found in section 11.6. Not all the air passes through the combustor, since some is used to cool the turbines, but the mass flow rate of fuel should be added; the mass flow of gas leaving the HP turbine nozzle is therefore $\dot{m}_{a4}+\dot{m}_f$ with specific heat c_{pe}. Since the turbine entry temperature is defined as the mixed-out temperature at turbine stator exit, the air used to cool these nozzle guide vanes is included in the balance below,

$$\dot{m}_f LCV \ = \ (\dot{m}_{a4}+\dot{m}_f)c_{pe} \ (T_{04}-298) - \dot{m}_{a4}c_p \, (T_{03} - 298) \qquad (16.3)$$

where $\dot{m}_{a4}$ is the air mass flow at turbine nozzle exit. The combustion is not quite complete by the time the gases leave the combustor and in a more detailed treatment a combustion efficiency would multiply the calorific value. The combustion efficiency is likely to be in excess of 98% over most of the operating regime.

The HP turbine power must equal the HP compressor power, with $\dot{m}_a$ being the total core mass flow of air through the HP compressor, so that

$$\dot{m}_a c_p (T_{03} - T_{023}) = (\dot{m}_{a4} + \dot{m}_f) c_{pe} (T_{04} - T_{045}) \qquad (16.4)$$

and T_{045} can be found. Then, knowing T_{045}/T_{04}, the turbine polytropic efficiency and γ for the combustion products the HP turbine pressure ratio can be calculated using the polytropic relation

$$p_{045}/p_{04} = (T_{045}/T_{04})^{\gamma/(\gamma-1)\eta}, \qquad (16.5)$$

so both T_{045} and p_{045} are known in terms of conditions out of the combustor.

The cooling air to the HP turbine rotor (with mass flow rate equal to $\dot{m}_{a45} - \dot{m}_{a4}$ and compressor delivery temperature T_{03}) is now mixed at a constant pressure to give a mixed temperature $T_{045'}$,

$$(\dot{m}_{a45} + \dot{m}_f) c_{pe} T_{045'} = (\dot{m}_{a45} - \dot{m}_{a4}) c_p T_{03} + (\dot{m}_{a4} + \dot{m}_f) c_{pe} T_{045} . \qquad (16.6)$$

Downstream of the mixing the specific heat capacity and γ are taken to be the values for the burned gas, which is clearly an approximation but one which should be reasonably good since the cooling air only represents a small fraction of the total gas flow.

Across the LP turbine

$$T_{05}/T_{045'} = (p_{05}/p_{045})^{\eta(\gamma-1)/\gamma} = (p_{013}/p_{045})^{\eta(\gamma-1)/\gamma} \underline{\quad} \qquad (16.7)$$

since for the mixed turbofan $p_{05} = p_{013}$, the pressure in the bypass duct downstream of the fan. (Recall that here, for simplicity, we assume that the bypass and core streams from the fan have equal temperature and pressure: $T_{013} = T_{023}$ and $p_{013} = p_{023}$.) Since p_{05} is fixed by the fan pressure ratio, the temperature ratio across the turbine is determined. $T_{045'}$ was given by the mixing equation, (16.5), applied after the HP turbine and so T_{05} is known. There is then a further mixing process downstream of the LP turbine

$$(\dot{m}_{a5} + \dot{m}_f) c_{pe} T_{05'} = (\dot{m}_{a5} - \dot{m}_{a45}) c_p T_{03} + (\dot{m}_{a45} + \dot{m}_f) c_{pe} T_{05} . \qquad (16.8)$$

The power from the LP turbine, passing a mass flow $(\dot{m}_{a45} + \dot{m}_f)$, must equal the power into the fan. The fan passes a mass flow $(1+bpr) \dot{m}_a$ so that the power balance for the LP shaft is

$$(1+bpr)\dot{m}_a c_p (T_{013} - T_{02}) = (\dot{m}_{a45} + \dot{m}_f) c_{pe} (T_{045'} - T_{05}) . \qquad (16.9)$$

When the fan pressure ratio and overall pressure ratio are given as input parameters, application of equations (16.3) to (16.9) allows direct calculation of all temperatures and pressure in the engine. If, however, the bypass ratio were given instead of fan pressure ratio an iterative solution would be necessary.

Downstream of the LP turbine the core and the bypass stream mix out and when the afterburner is not lit the uniform temperature T_{06} is given by

$$(\dot{m}_a + \dot{m}_f)\, c_{pe} T_{05'} + (\dot{m}_a bpr) c_p T_{013} = \{\dot{m}_a(1+bpr) + \dot{m}_f\} c_{pm} T_{06}, \qquad (16.10)$$

where c_{pm} is the specific heat of the mixed flow.

When the afterburner is lit the flow does not mix out to any significant extent before burning. Nevertheless the mixed-out temperature T_{06} without the afterburner can be used in the energy balance to find the fuel flow to the afterburner $\dot{m}_{fab}$ needed to raise the temperature to the required level by the nozzle throat, $T_{08} = T_{0ab}$.

$$\dot{m}_{fab} LCV = \{(\dot{m}_a(1+bpr)+\dot{m}_f)+\dot{m}_{fab}\} c_{pe}(T_{0ab}-298) - \{\dot{m}_a(1+bpr)+\dot{m}_f\} c_{pm}\,(T_{06}-298) \qquad (16.11)$$

The combustion process is not normally complete by the time the gases enter the nozzle and in a more detailed treatment an afterburner combustion efficiency would be introduced, multiplying *LCV*. The value of this efficiency is about 90%.

16.3 THE SELECTION OF OVERALL PRESSURE RATIO

A range of calculations has been run for the three design conditions ($M = 0.9, 1.5$ and 2.0 at the tropopause) using the equations (16.3) to (16.11) and some of these results are shown here in Figs.16.1 and 16.2. It is normally easier and often more appropriate to specify engines according to the fan pressure ratio than the bypass ratio. Nevertheless, because the optimum pressure ratio is so different for the three Mach numbers used for the design points, values of bypass ratio are used instead. The range of bypass ratios selected covers most of the design choices likely: zero (giving the case of a turbojet) 0.5 and 1.0. In all cases the technology level is held at that in section 16.1.

The curves in Fig. 16.1 show specific thrust (units m/s) and specific fuel consumption (sfc, units kg/h/kg) versus overall pressure ratio p_{03}/p_{02} for an engine flying at $M = 0.9$. A curve is drawn for each for the three different bypass ratios. The turbojet has, as expected, a higher thrust and a higher fuel consumption, whereas the highest bypass ratio is the lowest for both specific thrust and sfc. The effect of overall pressure ratio is similar at all three bypass ratios. There is a very slight peak in the thrust for a pressure ratio of about 15, whilst the fuel consumption shows a pronounced decrease with pressure ratio continuing beyond the upper level shown. Using an overall pressure ratio of, say, 30 gives a reduction in specific thrust of about 1.5% relative to the peak, but a corresponding reduction in sfc of about 10%.

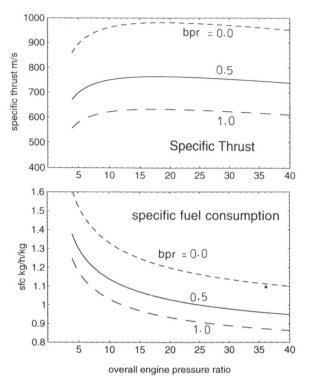

Figure 16.1. Specific thrust and specific fuel consumption (sfc) versus overall
pressure ratio for design point operation at $M = 0.9$ at tropopause.
(Constant technology standard – fan pressure ratio varies.)

Figure 16.2 shows specific thrust and fuel consumption at Mach numbers of 0.9, 1.5 and 2.0
for a fixed bypass ratio of 0.5. The overall pressure ratio for maximum specific thrust decreases
as the Mach number increases, an effect which is understandable because of the increase in
compression upstream of the engine as a result of the forward speed. If the Mach number were
increased to 2.5 it would be found that the pressure ratio for maximum thrust is less than unity,
indicating that a ram-jet would give the highest specific thrust, though still with a high fuel
consumption.

The selection of overall pressure ratio is, for the present purpose, somewhat arbitrary.
An important balance exists between benefits of lower fuel consumption and higher thrust.
There is an additional factor, which is that higher pressure ratio means more turbomachinery
with its attendant weight; the wish to achieve the desired pressure ratio with the least weight
leads to lower component efficiencies than would be possible if the pressure ratios could be
reduced somewhat. As a design decision for the present work pressure ratios will be chosen as
30, 20 and 10 for Mach numbers of 0.9, 1.5 and 2.0 respectively. These pressure ratios are also
acceptable for the compressor delivery temperature, as Exercise 16.1 shows.

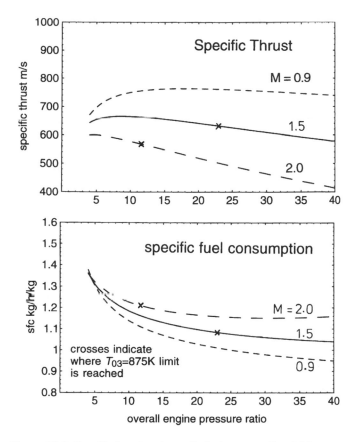

Figure 16.2. Specific thrust and specific fuel consumption (sfc) versus
overall pressure ratio for design point operation, *bpr*=0.5 throughout.
(Constant technology standard – fan pressure ratio varies.)

16.4 THE SELECTION OF FAN PRESSURE RATIO

For the mixed turbofan the selection of the fan pressure ratio fixes the pressure on the downstream side of the turbine; neglecting mixing losses and the pressure loss in the jet pipe, the stagnation pressure entering the propelling nozzle is therefore the stagnation pressure leaving the fan. The jet velocity depends on the pressure ratio across the nozzle and the stagnation temperature entering the nozzle. For a given jet velocity, which implies a given specific thrust, the most efficient engine (i.e. the engine with the lowest specific fuel consumption) is the one with the lowest jet temperature. Raising the bypass ratio lowers the jet temperature so for a given jet velocity the benefit of high bypass ratio is the lower jet temperature. The interconnection of the fan pressure ratio, overall pressure ratio and bypass ratio on specific thrust are shown in Fig. 16.3. With low overall pressure ratios the bypass ratio changes rapidly with fan pressure ratio, and consequently the specific thrust falls most rapidly with increase in overall pressure ratio in this region too.

The selection of fan pressure ratio is complicated by many practical issues. A military fan should be simple and robust but it should also be light. Together these indicate few stages, typically no more than three, with the only variable stators (to cope with off-design operation) being the inlet guide vanes. Together these constraints make it hard to have fan pressure ratios more than about 5.

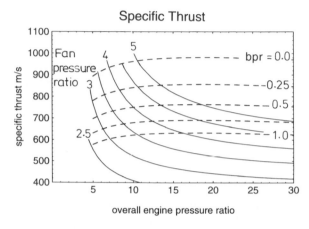

Figure 16.3. Specific thrust versus overall pressure ratio for design point operation at *M* = 0.9 at tropopause. (Constant technology standard – fan pressure ratio varies.)

Specific thrust and specific fuel consumption versus fan pressure ratio are shown in Fig.16.4 for three different overall pressure ratios, all for a flight Mach number of 0.9. For the lowest overall pressure ratio the curve terminates at a fan pressure ratio of about 4.4, since at this condition the bypass ratio has fallen to zero and the engine has become a turbojet. At the overall pressure ratios of 20 and 30 this occurs at higher fan pressure ratios than those shown. Although an increase in overall pressure ratio at a constant fan pressure ratio is beneficial in that it reduces sfc, it also reduces the specific thrust. The reason for this is simple – the jet temperature falls as the overall pressure ratio is raised. In other words the high specific thrust obtained with low overall pressure ratio is obtained at the cost of being less efficient in the conversion of fuel energy into kinetic energy.

While holding overall pressure ratio constant, Fig. 16.4 shows that specific thrust rises sharply with fan pressure ratio, a result to be anticipated from the equation for jet velocity. Because a high jet velocity gives a lower propulsive efficiency, the fuel consumption also rises with fan pressure ratio, as shown in Fig. 16.4. The choice of fan pressure ratio, as for the overall pressure ratio, is a compromise between achieving the highest specific thrust and lowest sfc. For the fan it is less easy because specific thrust rises continuously with fan pressure ratio until the engine becomes a turbojet (whilst the sfc rises continuously) so in this case there is no fan pressure ratio which gives a maximum. The final compromise can only be arrived at when engine weight can be compared with the fuel consumed, the latter depending on mission details.

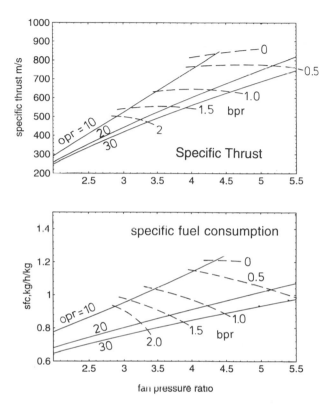

Figure 16.4. Specific thrust and specific fuel consumption (sfc) versus fan pressure ratio for design point operation at $M = 0.9$ at tropopause. Curves show lines of constant overall pressure ratio and bypass ratio. (Constant technology standard.)

Figure 16.5 shows specific thrust and specific fuel consumption versus fan pressure ratio for the three design conditions $M = 0.9$, 1.5 and 2.0 at the tropopause with overall pressure ratios of 30, 20 and 10 respectively adopted earlier. For the two higher Mach numbers the curves terminate when the bypass ratio goes to zero whereas for $M = 0.9$ the bypass ratio is still close to 0.5 at the highest fan pressure ratio of 5.5. The specific thrust varies sharply with fan pressure ratio, increasing more than threefold over the range shown for $M = 0.9$, whereas the fuel consumption increases by about 50% in the same range. At the higher Mach numbers the change in specific thrust is steeper whilst the slope of the sfc is similar. The precise balance of advantage will, as mentioned earlier, depend on the mission, since this determines the weight of fuel used. For the purpose of this design exercise it seems appropriate to select fan pressure ratios of 4.5, 4.0 and 3.0 for flight Mach numbers of 0.9, 1.5 and 2.0 respectively.

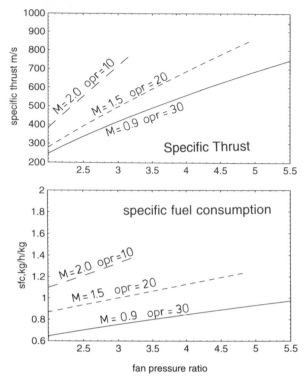

Figure 16.5. Specific thrust and specific fuel consumption (sfc) versus fan pressure ratio
for design point operation at tropopause. Curves show combinations of overall
pressure ratio (opr) and flight Mach number. (Constant technology standard.)

SUMMARY OF SECTIONS 16.1 TO 16.4

It is appropriate to compare designs using a common technology level, giving maximum turbine entry and compressor delivery temperature and component efficiencies. Simple balances for mass flow and power allow conditions in the engine to be found, but although many sources of loss may be neglected in a simple treatment such as this, it is essential to include the effect of the cooling air flow in the power balance if realistic levels of specific thrust are to be obtained. (This is explored below in section 16.7.) Because of the mixing in the jet pipe after the turbine and before the final nozzle, specifying bypass ratio requires an iterative calculation of the engine cycle whereas specifying fan pressure ratio allows a direct calculation.

The design decisions at this stage are the determination of overall pressure ratio and fan pressure ratio. (The bypass ratio is determined when these are given.) As flight Mach number increases the inlet stagnation pressure rises steeply and the overall pressure ratio in the engine required to give maximum specific thrust therefore decreases. The minimum specific fuel consumption occurs at much higher overall pressure ratios than for maximum specific thrust.

The actual design choice must depend on the mission to decide the balance of advantage between high specific thrust (giving low engine weight) and low specific fuel consumption. The overall pressure ratio must be kept below a value which will produce a compressor delivery temperature in excess of that laid down in the technology level, 875 K here.

The fan pressure ratio must also be chosen to give the optimum balance between high specific thrust, calling for the highest fan pressure ratio, and low sfc, calling for low fan pressure ratio. As the flight Mach number increases the allowable range of fan pressure ratios narrows but the rate of increase in specific thrust with pressure ratio is larger than the rate of increase in sfc; this points to operation near the highest possible fan pressure ratio being optimal at high flight speeds; at Mach numbers close to 2.0 the turbojet becomes attractive.

For tropopause conditions, and with the selected technology standard, the design choices adopted for the present exercise are given in Table 16.2.

Table 16.2   Selected overall and fan pressure ratios at design point.

M	Overall pressure ratio p_{03}/p_{02}	Fan pressure ratio p_{023}/p_{02}
0.9	30	4.5
1.5	20	4.0
2.0	10	3.0

Similar arguments lead to the choice of an overall pressure ratio of 30 for take off with a fan pressure ratio of 4.5.

Exercise
16.2 Calculate the fan delivery temperature T_{013} and the compressor delivery temperature T_{03} for the three flight conditions and pressure ratios set out in Table 16.2 (at the tropopause, $M = 0.9, 1.5, 2.0$). Use the efficiencies set out for the technology level in section 16.1.
(**Ans:** $T_{013} = 417.3, 500.5, 564.2$ K; $T_{03} = 762.1, 834.3, 826.8$ K)

16.5 THE SIZE OF THE ENGINE FOR 'MAXIMUM-DRY' OPERATION

In Chapter 15 certain design points were selected for the engine: $M = 0.9, 1.5$ and 2.0 at the tropopause. After selecting the pressure ratio overall and for the fan, sections 16.3 and 16.4, and after having adopted the particular technology standard laid out in section 16.1, the operating cycle of the engine is defined and with it the specific thrust and specific fuel consumption. These are for the engine producing its maximum thrust at the flight condition without the use of the afterburner, the 'Maximum-Dry' condition. The mass flow of air and fuel, and the size of the engine at the design point may then be found.

As pointed out in the introduction to this chapter, the three design points are for distinct 'paper' engines and do *not* correspond to the operating conditions which the same engine would adopt at the different flying conditions – this has to be considered in Chapter 17. For the design point consideration the parameters such as the pressure ratios are taken to be fixed independently, whereas when one engine is operated at different flight speeds off -design the pressure ratios become dependent variables.

Exercises

16.3 Three design cases were introduced in section 16.4. Adopt the standard of technology listed in section 16.1 with the HP turbine entry temperature T_{04} = 1850 K being defined as the mean temperature at HP stator (nozzle) exit after the nozzle cooling air has been mixed. A further 12% of the air compressed will be added downstream. For each of the three design cases:

a) Find the mass flow rate of fuel, per unit mass of air entering the engine core, taking LCV = 43 MJ/kg. Assume γ = 1.30, c_{pe} = 1244 J/kgK for the products of combustion.

(**Ans:** $\dot{m}_f$ = 0.0314, 0.0299, 0.0300)

b) Using an energy balance find the temperature at outlet from the HP turbine, when for each unit of air compressed the mass flow out of the HP turbine nozzle vanes is (0.88+$\dot{m}_f$). Using the efficiency given in the technology level, η_p = 0.875, find the pressure at HP turbine outlet.

(**Ans:** T_{045} = 1544.4, 1553.7, 1617.0 K; p_{045}/p_{04} = 0.409, 0.421, 0.513)

c) The HP rotor cooling air is assumed to mix at constant pressure between the HP rotor and the LP nozzle. If the HP rotor cooling air is 8% of the total compressed, and its temperature is that of the compressor delivery, find the temperature entering the LP turbine $T_{045'}$.

(**Ans:** $T_{045'}$ = 1467.0,1482.6,1540.2 K)

d) The pressure ratio across the LP turbine can be found from the value of p_{045}/p_{04} found in 16.3b and the specified pressure ratio across the fan, p_{023}/p_{02}. Find the temperature T_{05} after the LP turbine. The cooling air to the LP turbine is then mixed with the flow after the turbine (i.e. after T_{05} has been found) to give the mixed out temperature $T_{05'}$.

(**Ans:** T_{05} = 1200.1,1275.5,1381.9 K ; $T_{05'}$ = 1177.4, 1252.1,1354.2 K)

e) Use the LP turbine work to find the bypass ratio by matching it to the work into the fan.

(**Ans:** bpr = 0.997, 0.361, 0.114)

f) The mass flow in the core stream out of the LP turbine is (1+ $\dot{m}_f$) and the corresponding bypass mass flow is equal to bpr. Using a simple mass weighting system find the mean specific heat c_{pm} for the exhaust stream, and thence the mean value of γ. Use c_{pm} to find the temperature entering the propelling nozzle.

(**Ans:** c_{pm} = 1126, 1182, 1222 J/kgK; γ = 1.342, 1.320, 1.308, T_{08} = 844.1, 1086.2, 1289.3 K)

g) Using the results of Exercise 15.3, 16.1 and 16.3d, determine the pressure ratio across the propelling nozzle, p_{08}/p_a.   (**Ans:** 7.61, 14.24, 21.71)

h) Find the jet velocity assuming isentropic expansion in the propelling nozzle. Hence find the specific thrust (based on *net* thrust and including the mass flow rate of fuel) and the specific fuel consumption.

(**Ans:** V_j = 876,1104, 1273 m/s; Spec. thrust = 624.5, 686.5, 717.0 m/s;

sfc = 0.888, 1.127,1.325 kg/h/kg)

16.4   **a)** From Exercise 16.3, which gives the specific thrust at the Maximum Dry condition, determine the mass flow rate of air necessary through each engine to create the thrust for steady level flight (1g acceleration) at the tropopause for Mach numbers of 0.9, 1.5 and 2.0. (The necessary thrusts are given in Exercise 14.4.)   (**Ans:** $\dot{m}_a$ = 12.8, 37.9, 61.9 kg/s)

b) In the inlet duct just ahead of each engine the flow is purely axial and the Mach number is 0.7. (This is equivalent to saying that the LP compressor will be at the same non-dimensional working point at each flight Mach number.) The non-dimensional mass flow $\bar{m}_2 = \dot{m}\sqrt{c_p T_{02}}/(A_2 p_{02})$ is given in terms of γ and Mach number in sections 6.2 – use this to find the area of the duct for each flight Mach number in part (a). Hence, assuming that the diameter of the fan hub is 0.4 times the fan tip diameter, estimate the engine inlet diameter.   **(Ans: fan inlet diameter D = 0.47, 0.58, 0.55 m)**

16.5 Repeat the calculations of Exercise 16.4 for the aircraft making 3g turns at the tropopause at Mach numbers of 0.9 and 1.5.   **(Ans: $\dot{m}$ = 57.3, 57.3 kg/s; fan inlet diameter = 0.985, 0.71m)**

16.6*   a) If the engine is designed for take off at standard sea-level conditions (T_a = 288.15 K, p_a = 101.3kPa) and the same technology level, find the specific thrust and sfc operating 'dry'. Take the fan pressure ratio and the overall pressure ratio to be 4.5 and 30 respectively.

(Ans: 865 m/s, 0.805 kg/h/kg)

b) The static sea-level thrust–weight ratio of the New Fighter is selected to be 0.66 in the 'dry' condition. The mass of the aircraft (18 tonne) and the number of engines (2) is given in section 13.4. Find the mass flow of air required and the inlet diameter of the engine on the same basis as in Exercises 16.4 and 16.5.   **(Ans: 67.4 kg/s, 0.681 m)**

16.6 THE EFFECT OF THE AFTERBURNER

The choice of specific thrust determines the jet velocity and therefore the kinetic energy of the jet. The kinetic energy determines the minimum work which must be supplied to the air for the thrust, but because the static temperature is invariably higher than that of the atmosphere, the fuel consumption will be greater than this hypothetical minimum. While operating in the 'dry' condition the temperature of the exhaust gas can vary and efficient propulsion coincides with comparatively low exhaust temperature; the benefit of high bypass ratio arises because it lowers the temperature of the jet.

When the afterburner is used the exhaust temperature is determined by the amount of fuel burned in the engine and afterburner. In consequence the efficiency of the engine components does not affect the thrust or fuel consumption in the same way, nor does the bypass ratio. With the afterburner in use the specific thrust and specific fuel consumption are uniquely determined by the fan pressure ratio $p_{08}/p_a = p_{023}/p_a$ and exhaust temperature $T_{08} = T_{0ab}$ (neglecting losses in the nozzle itself and incomplete combustion of afterburner fuel).

If c_{pe} and γ_e are the specific heat and ratio of specific heats for the products (taken here to be 1244 J/kgK and 1.30 respectively) then the jet velocity is given by

$$V_{jab} = \sqrt{2c_{pe}T_{0ab}\{1 - (p_a/p_{08})^{(\gamma_e-1)/\gamma_e}\}} \ .$$

With T_{0ab} fixed the only variable is the pressure ratio across the nozzle. It is essential that the propelling nozzle opens when the afterburner is lit and very often this is arranged so that the pressure in the jet pipe remains unaltered; under this condition the engine is unaware of the afterburner. The fan pressure ratio and mass flow therefore do not change between the 'maximum dry' condition (with the afterburner unlit) and the 'combat' condition, with the afterburner lit. Apart from the change in gas properties, the increase in jet velocity with the afterburner is equal to the square root of the temperature ratio T_{0ab}/T_{06}, the ratio of temperature after the afterburner to the mixed temperature ahead of it.

Although the mass flow of air remains unaltered when the afterburner is lit, the mass flow of fuel rises rapidly, and the additional fuel flow $\dot{m}_{fab}$ per unit mass of air through the core $\dot{m}_a$ is then given by equation (16.9). The specific thrust, the net thrust per unit mass flow of air through the engine, is given by

$$\frac{F_N}{(1+ bpr)\dot{m}_a} = \frac{V_{jab}\,\{(1+ bpr)\dot{m}_a + \dot{m}_f + \dot{m}_{fab}\}}{(1+bpr)\dot{m}_a} - V, \qquad (16.12)$$

where $\dot{m}_f$ is the mass flow of fuel in the main combustor and V is the flight speed. Because the net thrust is proportional to the difference between the jet velocity and the flight speed, the proportional increase in net thrust and specific thrust due to the afterburner is much greater than the proportional increase in the jet velocity.

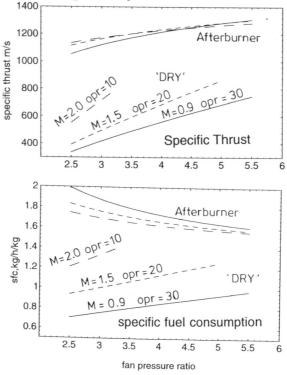

Figure 16.6. Specific thrust and specific fuel consumption (sfc) versus fan pressure ratio for design point operation at tropopause with and without afterburner. Curves show combinations of overall pressure ratio (opr) and flight Mach number. (Constant technology standard.)

Figure 16.6 shows the specific thrust and specific fuel consumption for the engine at the three design points with and without afterburning;the 'dry' results are the same as those shown in Fig. 16.5. Of immediate note is the large increase in specific thrust and in specific fuel consumption when the afterburner is used. Whereas for the 'dry' engine the specific thrust increases quite rapidly with fan pressure ratio, the increase is comparatively slow for the afterburning engine; as a result the boost from the afterburner is noticeably bigger for engines with low fan pressure ratios. This is because the temperature of the exhaust gases is lower for the low pressure ratio fan in the 'dry' condition so the increase in temperature of the exhaust when the afterburner is used is that much greater. With afterburning the specific thrust is almost equal at design for the three flight speeds.

For the 'dry' engine the specific fuel consumption increases when the fan pressure ratio is raised, corresponding to higher jet velocity and specific thrust. With the afterburning engine the opposite is true; the specific fuel consumption falls with fan pressure ratio. The fall may be understood by realising that the stagnation temperature of the jet is fixed with the afterburner lit, whereas the jet velocity, and therefore the thrust, increases as the pressure ratio increases. As a result the thrust for the same fuel input increases with pressure ratio, that is the sfc falls. It may also be noted that for $M = 2.0$ the difference between the sfc 'dry' and with the afterburner is relatively small.

Exercises
16.7 a)For the same three design conditions used in Exercise 16.3 find the specific thrust and specific fuel consumption with the afterburner producing an exit temperature of 2200 K
(**Ans:** Specific thrust = 1250, 1233, 1168 m/s; *sfc* = 1.68, 1.66, 1.69 kg/h/kg)
b*) Find the specific thrust and sfc for the engine of Exercise 16.6 during take off using the afterburner. (**Ans:** specific thrust = 1341 m/s, *sfc* = 2.27 kg/h/kg)
16.8 Determine the mass flow of air and the fan inlet diameter (on the same basis as in Exercise 16.4 and 16.6) when the afterburner is used to create the necessary thrust.
a)For 1g flight at $M = 0.9,1.5$ and 2.0 (**Ans:** $\dot{m}$ = 6.4, 21.0, 37.9 kg/s; D = 0.33, 0.43, 0.43 m)
b) For 3g flight at $M = 0.9$ and 1.5. (**Ans:** $\dot{m}$ = 28.6, 31.8 kg/s; D = 0.70, 0.53 m)
c*) For take off with a thrust/weight ratio equal to unity. (**Ans:** $\dot{m}$ = 65.9 kg/s; D = 0.67 m)

16.9 For the New Fighter Aircraft, Chapter 13, assume that the maximum fuel available during the high-thrust part of combat to be 2 tonne (i.e. 1 tonne per engine). Using the results of Exercises 16.3 and 16.6, with the specified thrust from Exercise 14.4, calculate the maximum time the thrust can be produced with and without afterburning at $M = 2.0$ for 1g flight and at $M = 0.9$ and 1.5 at 3g flight.
(**Ans:** 1g, $M = 2.0$; dry 10.0 minute, afterburning 7.9 minute: 3g , $M = 0.9$; dry 18.5 min, afterburning 9.8 min: 3g, $M = 1.5$; dry 13.2 min, afterburning 9.0 min)

SUMMARY OF SECTIONS 16.5 AND 16.6

The afterburner produces a substantial rise in thrust, approximately in the ratio of the square root of the exhaust temperature with and without the afterburner. The fuel consumption is much higher with the afterburner, but the difference decreases as the Mach number increases and as the pressure ratio across the nozzle increases.

When operating with the afterburner lit, performance is determined entirely by the ratio of fan discharge stagnation pressure to ambient static pressure and the temperature into the nozzle; the efficiencies and losses in the engine become irrelevant, as do the overall pressure ratio and the turbine inlet temperature, provided that the fan pressure ratio can be achieved.

16.7 THE EFFECT OF ALTERATION IN ASSUMED PARAMETERS

It has inevitably been necessary to assume parameter values, including the technology level, to enable calculations to be performed. This immediately begs the question, how sensitive are the conclusions to these assumptions? This section will attempt to answer some of the questions which arise. The calculations will centre on operation 'dry', without the afterburner lit, since this is much more sensitive to the engine variables.

The comparisons will be made for the design case at $M = 0.9$ for which the overall pressure ratio p_{03}/p_{02} is taken to be 30 and the fan pressure ratio p_{023}/p_{02} to be 4.5. With this combination, and with the technology level of section 16.1, the bypass ratio is 0.997 (Exercise 16.3). In the table below one parameter is altered at a time, with the others at their datum values.

Table 16.3   Effect of variation in parameters at design point.

PRESSURE RATIOS HELD CONSTANT	$\Delta(F_N/\dot{m})$ %	$\Delta(sfc)$ %	$\Delta(bpr)$
Reduce turbine inlet temp. from 1850 to 1750 K	1.2	2.0	−0.23
Increase LP compressor efficiency from 0.85 to 0.90	−4.0	−4.5	0.21
Reduce HP compressor efficiency from 0.90 to 0.85	3.4	3.8	−0.17
Reduce HP turbine efficiency from 0.875 to 0.825	2.4	2.7	−0.10
Reduce LP turbine efficiency from 0.875 to 0.825	2.6	2.8	−0.10
Stagnation pressure loss in combustor 5% of p_{03}	2.3	2.5	−0.09
Stagnation pressure loss in bypass 5% of p_{023}	−3.5	−0.9	0.09
Stagnation pressure loss in jet pipe 5% of p_{05}	−1.4	−1.4	0.00
Turbine cooling air increased by 50%	2.5	3.7	−0.25
Dispense with turbine cooling air	−3.5	−5.1	0.48

It can be seen, for example, that reducing the turbine inlet temperature from 1850 K to 1750 K increases specific thrust $F_N/\dot{m}$ by 1.2%, increases sfc by 2.0% and reduces bypass ratio by 0.23. The predicted increase in specific thrust as temperature decreases, and elsewhere in the table when efficiency falls, is counter-intuitive. The reason for this is the stipulation that the pressure ratio of the fan and of the whole engine are constant. A reduction in work output from the core is therefore translated into a reduction in bypass ratio. Taking as an example the 100 K reduction in turbine inlet temperature, the specific thrust has increased to 1.012 times its previous value while at the same time the bypass ratio has been reduced by 0.23 from 0.997 to 0.767. As a result of the drop in bypass ratio the total mass flow through the engine, for the same mass flow through the core, is reduced and because of this the total net thrust has been decreased in the ratio $1.012 \times (1+0.767) \div (1+0.997) = 0.896$, that is a 10.4% drop in thrust.

In all cases in Table 16.3 the changes to specific thrust and fuel consumption are comparatively small and the most significant effect is to the bypass ratio which leads to a reduction in thrust whenever component efficiencies are lowered or losses increased. Of all the components a change in the efficiency of the LP compressor has the largest effect.

Since the magnitudes of the changes in specific thrust, specific fuel consumption and bypass ratio in the table are generally small, when the relatively large size of the changes in efficiency or loss are considered, the trends and general magnitudes found in the chapter may be considered robust. It is noteworthy that the effect of the pressure loss in the combustor, bypass duct or jet pipe are all relatively small, supporting the decision to simplify the treatment by not including arbitrary estimates for these in the body of calculations. The effect of altering turbine cooling air, however, is large; omitting the cooling air would lead to net thrust being over-estimated by 43%, a very substantial error. It is also possible to see from the table that a 'trade' exists between raising turbine inlet temperature and increasing cooling air: from the figures given in Table 16.3 a 100 K reduction in T_{04} gives a 10.4% reduction in thrust, whilst increasing the cooling air by 50% gives a 10.3% reduction in thrust. This is an important balance which designers have to take into account.

Figure 16.7 considers the effect of inlet loss for flight at $M = 1.5$ and 2.0. Curves are shown for both speeds for three cases: when the inlet is assumed reversible, when there is a normal shock and where the loss is given by MIL-E-5007/8. The loss with MIL-E-5007/8 is appropriate for a sophisticated variable inlet decelerating the flow with several shock waves. At a Mach number of 2 the loss in specific thrust with a normal shock is about 6% compared with an isentropic deceleration, and there is a significant increase in sfc, particularly for low fan pressure ratios. An intake conforming to MIL-E-5007/8 introduces a thrust loss of only about 2% compared with the isentropic case for $M = 2.0$. At $M = 1.5$ the loss associated with the normal shock is small, justifying the use of simple fixed inlets for aircraft whose primary mission does not exceed this speed range, for example the F-16.

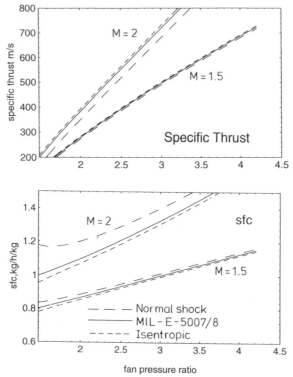

Figure 16.7. Specific thrust and specific fuel consumption (sfc) versus fan pressure
ratio for design point operation at tropopause showing effect of intake loss.
(Apart from intake loss constant technology standard with pressure ratios from Table 16.2.)

SUMMARY CHAPTER 16

Calculations have been carried out for design at three flight Mach numbers at the tropopause
and for take off at sea level. In a design calculation the overall pressure ratio and either the fan
pressure ratio or bypass ratio are selected independently, so the calculations do *not* coincide
with any particular engine designed for one condition operating at some other condition. In
terms of engine non-dimensional operating point the crucial parameter is the ratio of turbine
inlet temperature to compressor inlet temperature, T_{04}/T_{02}.

The calculations were carried out at the same technology level (turbine inlet
temperature, maximum compressor delivery temperature, compressor and turbine efficiencies),
which includes a constant proportion of the core air used for turbine cooling. The values used
are believed to be representative of modern advanced practice. Calculations were also
performed to show that modest changes in any of the assumed values used would not seriously
distort the results, either in the magnitudes of specific thrust and sfc, or in their trends. The
most important additional input, relative to the more elementary calculations carried out for the

civil engine in Chapters 4 and 12, is the mass flow used to cool the turbine; if this is not included the work supplied to the fan is far too large. Losses in stagnation pressure in the combustor, bypass duct and jet pipe, at levels which are plausible, do not significantly alter the magnitudes of specific thrust or specific fuel consumption.

The afterburner can significantly increase the specific thrust, whilst increasing the specific fuel consumption by a rather higher proportion. The increase in both specific thrust and specific fuel consumption with afterburning is proportionally much smaller for an engine designed for $M = 2.0$ than one designed for $M = 0.9$, mainly because the exhaust temperature of the 'dry' engine at $M = 2.0$ is already high. For the engine at $M = 0.9$ at the tropopause the specific thrust is almost exactly doubled with the afterburner; for the engine sized to allow a 3g turn at this Mach number the inlet diameter can be decreased from about 1 m to about 0.7 m with an afterburner. If the engine weight were proportional to linear dimension cubed this would represent an afterburning engine weighing only 35% of the comparable 'dry' engine; to offset this huge reduction is the weight of the afterburner and the long jet pipe and, of course, the additional weight of fuel burned when afterburning is used.

An engine specifically designed for high speed propulsion, $M = 2.0$ for example, will have a very low bypass ratio and such an engine is sometimes colloquially described as a leaky turbojet. (The bypass ratio for the design exercises here is only 0.11.) For $M = 0.9$ the optimum is close to unity at the current technology level. If the technology level were lower the bypass ratio and/or the fan pressure ratio would be lower too.

CHAPTER 17

COMBAT ENGINES
OFF-DESIGN

17.0 INTRODUCTION

The engine for a high-speed aircraft is required to operate over a wide range of conditions and some of these have been discussed in Chapters 13,14 and 15. Of particular importance is the variation in inlet stagnation temperature, which can vary from around 216 K up to nearly 400 K for a Mach-2 aircraft. As a result the ratio of turbine inlet temperature to engine inlet temperature T_{04}/T_{02} alters substantially, even when the engine is producing the maximum thrust it is capable of. In contrast, for the subsonic civil aircraft the value of T_{04}/T_{02} changes comparatively little between take off, climb and cruise, the conditions critical in terms of thrust and fuel consumption, and it is normally only when a civil aircraft is descending to land or is forced to circle an airport that T_{04}/T_{02} is reduced radically.

In Chapter 8 the dynamic scaling and dimensional analysis of engines was considered. There the engine non-dimensional operating point was held constant, for example T_{04}/T_{02} is not constant, so the engine remained 'on-design'. To designate the engine operating condition the value of $N/\sqrt{(c_p T_0)}$ or any of the pressure ratios or non-dimensional mass flows could also be used, but T_{04}/T_{02} has the intuitive advantage since engine thrust is altered by varying fuel flow rate to change T_{04}. In Chapter 12 the more challenging issue of a civil engine operating away from its design condition was addressed, i.e. the case when $T_{04}/T_{02} \neq$ constant, and the subject of this chapter is the similar case for military engines. The principal difference in approach from that of Chapter 12 arises from the practice in the combat engine of mixing the core and bypass flows.

The first part of this chapter develops the theory along the lines of Chapter 12 for the civil transport engine off-design. This is first applied to the engine when it is required to produce maximum thrust: in the terminology of section 15.3 either at combat rating with the afterburner in use, or at maximum dry without the afterburner. Later some consideration will be given to the engine operating at reduced thrust, as it must for most of the flight mission. To simplify discussion two alternative design cases will be considered. In case 1 the design condition is sea-level static, the condition at the start of the take off or on simple test beds, with the inlet temperature at 288 K. In case 2 the design condition is taken to be $M = 0.9$ at the tropopause, one of the important operating conditions of the envelope when the engine inlet temperature is only about 252 K.

17.1   THE SIGNIFICANCE OF OFF-DESIGN OPERATION

For the engine of the combat aircraft the most important variable which alters over the flight envelope is the engine inlet stagnation temperature and Table 15.1 gives some of these temperatures at important operating points. The way that the inlet temperature affects the engine operation and thrust is not always intuitively obvious, and depends on the design of the engine, as will be shown in this chapter. Operation off-design is fixed by the three major constraints on the engine, see Fig.15.8; and Fig. 17.1 shows a corresponding plot of temperatures for the two design cases which are to be considered in this chapter.

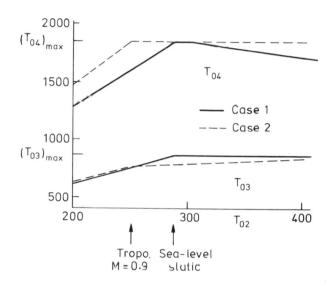

Figure 17.1. Turbine inlet temperature T_{04} and compressor delivery temperature
T_{03} as a function of engine inlet temperature T_{02}.
Design point for case 1 – sea-level static; design point for case 2 – $M = 0.9$ at tropopause.

At low inlet temperature the constraint is on $N/\sqrt{(c_p T_0)}$[1], which is accommodated by adjusting the turbine inlet temperature to keep T_{04}/T_{02} = constant; in this phase the engine operates at the same non-dimensional condition, so all the pressure ratios are also constant. At a higher engine inlet temperature the turbine inlet temperature T_{04} reaches its maximum allowable value at the design condition and further increase in inlet temperature necessitates progressively lower values of T_{04}/T_{02} to keep T_{04} constant, the engine is no longer operating at a constant non-dimensional condition so $N/\sqrt{(c_p T_0)}$ and the pressure ratios in the engine all fall. Further

---

[1] All modern engines will have at least two shafts and there is a value of $N/\sqrt{(c_p T_0)}$ for each shaft, with the T_0 referring to the inlet temperature to the compressor on that shaft. At any non-dimensional operating point there will be a fixed ratio between $N/\sqrt{(c_p T_0)}$ for each shaft; for brevity in the text only a single $N/\sqrt{(c_p T_0)}$ is therefore referred to.

increase in inlet temperature can lead to the compressor delivery temperature T_{03} reaching its upper limit, and if T_{02} is increased still further T_{04} must be lowered to reduce the pressure ratio in the compressors and hence hold T_{03} at its upper limit. The trends and magnitudes in this rematching can be shown by a simple calculation.

17.2   ALTERNATIVE DESIGNS

When the New Fighter Aircraft was introduced in Chapter 13, it was required that the thrust-to-weight ratio at take off should be at least unity when the afterburner was in use and at least 0.66 when the engine was dry. (These are for sea-level, standard atmosphere conditions, 288.15 K and 101.3 kPa). Given a take-off weight of 18 tonne this says that the static sea-level thrust from each engine should be 88.3 kN with the afterburner and 58.3 kN without. In Chapter 14 the thrust requirements in flight were estimated (Exercise 14.4) and these are shown in Table 14.1. These thrust requirements are used to size the two alternative engine designs (Case 1 and Case 2) in this chapter.

Case 1 design

The design point is taken as the sea-level static, the condition commonly used to denote engine size and the thrust requirements for this condition given above. Resting on the experience of Chapter 16 for the design point, it seems appropriate to aim for a fan pressure ratio p_{023}/p_{02} of 4.5 and an overall pressure ratio p_{03}/p_{023} of 30 . Using a simple design calculation, Exercise 16.7, and the technology level set out in section 16.1 the values in Table 17.1 result.

Table 17.1   Case 1: sea-level static design point
fan pr = 4.50, overall pr = 30;T_{02} = T_a = 288.15 K, T_{04} = 1850 K,
T_{04}/T_{02} = 6.42; technology level as in section 16.1

	T_{03}	bpr	V_j	Spec. thrust	sfc
	K		m/s	m/s	kg/h/kg
Dry	872.5	0.471	848	865	0.805
Afterburning	872.5	0.471	1267	1341	1.543

The table shows that the compressor delivery temperature is only 2.5 K below the limit of 875 K (for sea-level operation on a day only slightly warmer than 288 K it would therefore be necessary to reduce the turbine inlet temperature, and therefore T_{04}/T_{02}, below the design values). The engine is at its highest non-dimensional operating point at this condition, in other words T_{04}/T_{02}, $N/\sqrt{(c_pT_0)}$ and all the pressure ratios are at their maximum values. For reductions in inlet temperature, such as would occur at low Mach numbers and increased altitude, the ratio T_{04}/T_{02} is held constant, requiring that turbine inlet temperature is reduced, so the engine is maintained at the same non-dimensional operating point. For example at the tropopause $T_a = 216.65$ K and for an inlet Mach number of 0.9 the inlet temperature T_{02} is only 251.7 K; although this flight condition is not the design condition, the engine is nevertheless at the same non-dimensional condition as design, but with T_{04} reduced to 1616 K. At the tropopause the Mach number must be raised to 1.28 to bring T_{02} back to the sea-level static temperature of 288.15 K, at which point the case 1 engine is at its design point and $T_{04}=T_{04max}$. The temperatures for the compressor delivery T_{03} and turbine entry T_{04} are shown in Fig. 17.1 (similar to Fig.15.8) for both the case 1 and case 2 engines. For the case 1 engine T_{04}/T_{02} must fall if $T_{02}>288.15$ K, because T_{04} has reached its maximum value, and the engine operates below its non-dimensional design point. A further increase in Mach number to 1.35 leads to the compressor delivery temperature reaching its maximum of 875 K, so for still higher Mach numbers T_{04} must be reduced (not just T_{04}/T_{02}) to prevent the compressor deliver temperature becoming any higher. For case 1 the engine is 'throttled back' (i.e. operates at lower non-dimensional conditions) at flight Mach numbers above 1.28 at the tropopause.

Case 2 design

The design is taken here as the engine to give a 3g turn at $M = 0.9$ at the tropopause. From Table 14.1 the thrust requirement per engine for this condition is 35.8 kN. From Chapter 16 for the design point, it is again appropriate to select a fan pressure ratio p_{023}/p_{02} of 4.5 and an overall pressure ratio p_{03}/p_{023} of 30; the case 2 engine is one specified in Exercises 16.3 and 16.5 and the design-point parameters are set out in Table 17.2.

Table 17.2   Case 2: tropopause, $M = 0.9$ design point
fan pr = 4.50, overall pr = 30; $T_a = 216.65$, $T_{02} = 251.7$ K, $T_{04} = 1850$ K,
$T_{04}/T_{02} = 7.35$; technology level as in section 16.1.

	T_{03}	bpr	V_j	Spec. thrust	sfc
	K		m/s	m/s	kg/h/kg
Dry	762.2	0.996	876	625	0.888
Afterburning	762.2	0.996	1430	1250	1.680

The specific thrust and thrust requirement allow the mass flow to be found, so that to produce this thrust dry the mass flow is equal to $(35.8\times10^3)/625 = 57.3$ kg/s and correspondingly, 28.6 kg/s with afterburning. An engine passing only 28.6 kg/s would be small by modern standards (only a little bigger than the Viper, see section 12.2) reflecting the fact that a sustained 3g turn at $M = 0.9$ does not require a high thrust and afterburning would never normally be used to achieve this.

If this is the design point it requires that the non-dimensional rotational speeds and the pressure ratios are at their maximum and cannot be increased. If T_{02} were reduced, which would occur if flight speed were reduced, the turbine inlet temperature would have to be reduced too. If engine inlet temperature were increased the ratio T_{04}/T_{02} must fall, since T_{04} is already at its maximum. At sea-level static conditions, for example, $T_{04}/T_{02} = 1850/288 = 6.42$ (as in case 1) but for case 2 the overall pressure ratio will have been reduced to only 21.4. This will be demonstrated quantitatively below. The temperatures are shown in Fig. 17.1 alongside case 1. The primary limit on performance in case 2 is imposed by turbine entry temperature. Because the engine is 'throttled back' from 251 K (corresponding to $M = 0.9$ at the tropopause) the compressor pressure ratio and temperature rise are so much lower that the compressor delivery temperature never reaches the upper limit.

17.3 A MODEL FOR THE TWO-SHAFT TURBOFAN ENGINE

The approach is so similar to that described in Chapter 12 that much of the explanation can be omitted here. The major difference is the mixing of the core and bypass stream ahead of the final propelling nozzle. In line with the justification in Chapter 15 it will be assumed that the core and bypass streams have equal stagnation pressure where they mix and that this mixing takes place without loss in stagnation pressure.

The mixing of the core and bypass streams downstream of the LP turbine determines that, if the losses in the bypass duct are small enough to neglect, the outlet pressure from the fan and the LP turbine are equal, in other words

$$p_{05} = p_{013}.$$

Furthermore, with small losses in the jet pipe the fan discharge pressure is equal to the stagnation pressure at the nozzle throat, p_{08}. For the design point calculations we could specify the fan pressure ratio and thus the outlet conditions from the turbine. It is then straightforward to find the LP turbine work. Off-design it is less straightforward because the fan pressure ratio is normally not known in advance. It will be realised that a reduction in p_{013} reduces the power input needed to the fan, if the bypass ratio is constant, and simultaneously increases the power output from the LP turbine.   The problem lends itself to a simple numerical iteration.

The efficiencies for the compressor and turbine used in Chapter 16 are retained here, though at lower values of T_{04}/T_{02} and $N\sqrt{(c_p T_0)}$ than at design point the efficiencies might be expected to rise somewhat. The proportion of the core flow used to cool the turbine also remains as it was for the design point: it is assumed that 20% of the core flow does not go through the combustor but 8% is used to cool the HP nozzles and is mixed back in before the stipulation of T_{04} (so this does not need to be addressed specially); a further 8% is used to cool the HP rotor and is mixed in with the HP turbine outlet flow; and 4% is used in the LP turbine and mixed in downstream of the LP rotor.

Choked turbines

Just as in Chapter 12 the HP and LP turbine are assumed to be choked and their polytropic efficiencies are assumed to remain constant. The HP turbine is then constrained so that

$$(T_{045}/T_{04}) = (p_{045}/p_{04})^{\eta(\gamma-1)/\gamma} = \text{constant.} \tag{17.1}$$

The mass flow of air entering the core compressor is $\dot{m}_a$, but some of this is used for turbine cooling so the air mass flow into the HP turbine rotor is $\dot{m}_{a4}$ and the total mass flow into the turbine, including the mass flow rate of fuel, is $\dot{m}_{a4}+\dot{m}_f$. It then follows that the HP turbine power is given by

$$\dot{W}_{HP} = (\dot{m}_{a4}+\dot{m}_f)c_{pe}(T_{04} - T_{045})$$

$$= (\dot{m}_{a4}+\dot{m}_f)c_{pe}T_{04}(1 - T_{045}/T_{04}) .$$

In line with the treatment in Chapter 12 it is convenient to define a coefficient k_{HP} for the HP turbine by

$$T_{04} - T_{045} = k_{HP}T_{04}, \tag{17.2}$$

where the coefficient can found from the turbine performance at the design point.

By equating HP turbine power to the HP compressor power it follows that the HP compressor temperature rise is given by

$$T_{03} - T_{023} = \{(\dot{m}_{a4}+\dot{m}_f)c_{pe}/\dot{m}_a c_p\}k_{HP}T_{04}. \tag{17.3}$$

If turbine inlet temperature T_{04} is given the temperature rise in the HP compressor is fixed; with a known or assumed polytropic efficiency for the HP compressor the pressure ratio depends only on the temperature of the air leaving the LP compressor, T_{023}, which here is assumed equal to the temperature out of the LP compressor in the bypass stream T_{013}.

In the engine with a separate core and bypass nozzle considered in Chapter 12 it was possible to consider the core propelling nozzle to be choked so the LP turbine also operated between choked nozzles and relations similar to (17.1) and (17.2) could be produced. With the mixed turbofan the nozzle is also choked for most conditions of interest, but the balance of flow between the core and bypass is unknown and in this case it is not possible to find an expression for k_{LP}.

Power balance for the LP shaft

The power balance across the LP shaft may be written as

$$\dot{m}_a(1+bpr)c_p(T_{023}-T_{02}) = (\dot{m}_{a45}+\dot{m}_f)c_{pe}(T_{045'} - T_{05}), \tag{17.4}$$

where $\dot{m}_{a45}$ is the mass flow of air entering the LP turbine with temperature $T_{045'}$ after mixing the cooling flow after the HP turbine.

For a fixed LP compressor efficiency,

$$p_{013}/p_{02} = (T_{013}/T_{02})^{\eta\gamma/(\gamma-1)}$$

and for a fixed LP turbine efficiency

$$p_{05}/p_{045} = (T_{05}/T_{045'})^{\gamma/\eta(\gamma-1)}.$$

Using these equations and $p_{05} = p_{013}$ the power balance equation (17.4) may be solved to yield the combination of pressure ratio and bypass ratio appropriate.

Mass flow

There are two mass flows of air to be found, that which goes through the core $\dot{m}_a$ and that which goes through the whole engine, $(1+bpr)\dot{m}_a$. The conditions which determine these are the choking of the HP turbine nozzle and the choking of the final propelling nozzle, respectively. In both cases the mass flow rate of fuel should also be included; when the engine is dry this is approximately 2.5% of the mass flow of core air, whereas when the afterburner is in use it can amount to 10% of the air through the core.

At turbine inlet the non-dimensional mass flow rate is given by

$$\overline{m}_4 \ = \ \frac{\dot{m}\sqrt{c_{pe}T_{04}}}{A_4p_{04}} \ = \ 1.389,$$

based on $\gamma = 1.30$, where the relevant mass flow $\dot{m} = (k\dot{m}_a + \dot{m}_f)$. Here k is the fraction of air compressed in the HP compressor which leaves the HP turbine nozzle guide vanes (88% in the present calculation), the remaining 12% being used to cool the HP rotor and LP turbine. Constancy of $\overline{m}_4$ is what determines the core mass flow of air.

At the throat of the propelling nozzle, with area A_8, a corresponding relation applies

$$\overline{m}_8 \ = \ \frac{\dot{m}\sqrt{c_{pm}T_{08}}}{A_8p_{08}} \ = \ \text{constant},$$

where the relevant mass flow $\dot{m} \ = \ \{\dot{m}_a(1+bpr)+\dot{m}_f\}$. It is the requirement for constant $\overline{m}_8$ which determines the overall mass flow and thence the bypass ratio. The value of the constant depends upon c_{pm} and γ_m. The appropriate value for specific heat c_{pm} needs to be found; if the engine is dry the value of c_{pm} may be estimated by equation (15.5), after which γ_m may be found. If the afterburner is in use $\gamma = 1.30$ and $c_{pe} = 1244$ J/kgK are assumed appropriate. To use equation (15.5) requires the bypass ratio to be known and bypass ratio is something which has to be found. The process is therefore iterative.

Method of solution

For a given engine the design point operation is used to fix k_{HP}, which relates the HP turbine inlet temperature to the drop in temperature across the turbine, as well as $\bar{m}_4$ the non-dimensional mass flow rate into the HP turbine and the non-dimensional mass flow rate into the final nozzle $\bar{m}_8$. The efficiencies and losses are assumed here to remain at the design values. For a particular value of engine inlet temperature T_{02} the turbine inlet temperature T_{04} is selected.

The calculation begins by assuming a value for the fan pressure rise p_{013}/p_{02} and from this the fan temperature rise in the bypass stream and in the core stream T_{023} can be found. Similarly the stagnation pressure downstream of the LP turbine is fixed. The temperature rise in the HP compressor is found from $k_{HP}T_{04}$ and, given T_{023}, the overall pressure ratio p_{03}/p_{02} is known. From the turbine temperature drop $k_{HP}T_{04}$ the HP turbine pressure ratio can be found, and then, since the pressure downstream of the LP turbine is known from the fan pressure ratio, the power per unit mass flow rate from the LP turbine can be found. The LP turbine power fixes the bypass ratio and from this the non-dimensional mass flow in the propelling nozzle $\bar{m}_8$ can be found. The mass flow rate through the core has been determined by the choking of the HP turbine nozzle vanes. If $\bar{m}_8$ is not equal to the value at design the fan pressure rise must be altered and the calculation repeated until there is agreement.

17.4 VARIATION WITH TURBINE INLET TEMPERATURE

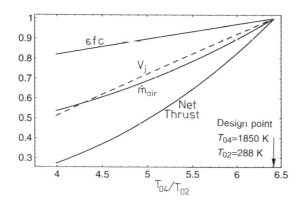

Figure 17.2. Case 1 engine performance as a function of T_{04}/T_{02}, normalised by values at design point (sea-level static, $T_{04} = 1850$ K).

Figure 17.2 shows the variation in various gross parameters of the case 1 engine (design point for sea-level static design) operating on a sea-level static test bed as the turbine inlet temperature is varied between 1850 K (the design point value) and 1150 K. Figure 17.3 shows, for the same case, the variation in fan pressure ratio, overall pressure ratio and bypass ratio. At the lowest value of turbine inlet temperature the fan pressure is such that the final nozzle will be on the point of unchoking and below this pressure ratio it would be necessary to know something more about the nozzle to predict the mass flow. Of particular note is the steep rise in bypass ratio as T_{04}/T_{02} is reduced, an effect seen also for the unmixed engine, Fig.12.9, though the effect is larger in the present case.

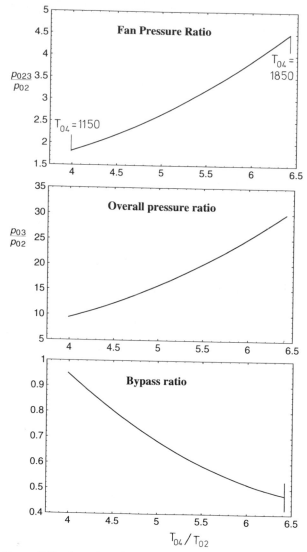

Figure 17.3. Case 1 engine pressure ratios and bypass ratio as a function of T_{04}/T_{02} .

As was discussed in Chapter 12, the major difficulty encountered when engines are operated at reduced pressure ratio is the matching of the compressors. Figure 17.4 shows the working line for the LP and HP compressors in terms of the pressure ratio and non-dimensional mass flow. The non-dimensional mass flow is normalised by the value at the design condition. The two working lines end to the right at the temperature ratio for sea-level static conditions with maximum turbine inlet temperature and this value of $T_{04}/T_{02} = 1850/288 = 6.42$ will also apply to the maximum thrust condition when the inlet temperature is low. As for the unmixed civil turbofan, the HP compressor operating point varies relatively little over the range of T_{04}/T_{02} encountered , whereas the pressure ratio and the non-dimensional mass flow for the LP compressor are reduced to less than half the value at design.

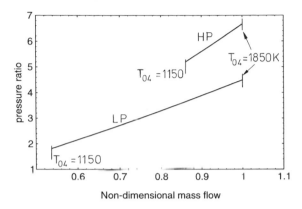

Figure 17.4. Case 1 engine compressor working lines (pressure ratio versus non-dimensional mass flow rate) Extreme turbine inlet temperatures T_{04} shown for engine inlet temperature $T_{02} = 288$ K.

For the case 2 engine (design point for $M = 0.9$ at the tropopause) the corresponding variation in major parameters to is shown in Fig.17.5. The most significant difference, compared with case 1, is the larger extent to which it is throttled back for potentially important operating conditions.

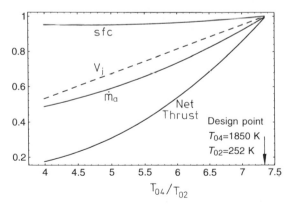

Figure 17.5. Case 2 engine performance as a function of T_{04}/T_{02}, normalised by values at the design point ($M = 0.9$ at the tropopause, $T_{04} = 1850$ K).

One of the things which complicates the treatment is the dependence of some of the parameters such as thrust and sfc on both Mach number and engine operating condition, characterised by the ratio T_{04}/T_{02}. It is possible to simplify and generalise the treatment by recognising that T_{04}/T_{02} (or some other ratio of internal variables characterising the engine operating point) can be used to determine the behaviour of the engine and then simple arguments similar to those used in Chapter 8 can be employed to determine the thrust. This is taken up in the following section.

17.5 DIMENSIONAL REASONING AND DYNAMIC SCALING

The crucial step is to recall that so long as the final nozzle is choked, which covers all important operating conditions, the engine is unaware of the external static pressure. The inputs to the engine are then the inlet stagnation pressure and temperature, p_{02} and T_{02} and the fuel flow $\dot{m}_f$. The temperature ratio T_{04}/T_{02}, which is varied by changing the fuel flow, may be used as a surrogate for the non-dimensional group based on fuel flow. It then follows that *all* the pressure ratios and temperature ratios inside the engine are determined by T_{04}/T_{02}. So too are the non-dimensional rotational speeds $N\sqrt{(c_pT_0)}$ and the non-dimensional mass flow into the engine $\dot{m}\sqrt{c_pT_{02}}/(A_2p_{02})$.

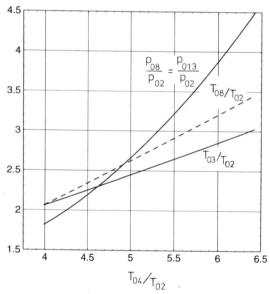

Figure 17.6. Case 1 engine pressure ratio and temperature ratios as a function of T_{04}/T_{02}

It has already become clear that one of the most important and commonly encountered constraints on the operation of the engine is the compressor delivery temperature, T_{03}. From what has been written above it should be clear that T_{03}/T_{02} is a unique function of T_{04}/T_{02} for a given engine; so too is, for example T_{08}/T_{02}, where T_{08} is the stagnation temperature of the gas entering the final propelling nozzle. In the case of the mixed turbofan a numerical iteration is needed to obtain all the pressure ratios and temperature ratios, as described in section 17.2.

Figure 17.6 shows non-dimensional groups for the case 1 engine which were obtained after iterative calculations and which facilitate hand calculations of thrust in the exercises of this chapter.   The fan pressure ratio, the ratio of compressor delivery temperature to compressor inlet temperature (a simple function of overall pressure ratio if the efficiency is constant) and the exhaust stagnation temperature ratio T_{08}/T_{02} are shown as a function of T_{04}/T_{02}.

The procedure for using the curves in Figs. 17.6 to find the engine operating point and thrust are as follows:

(i) Determine the inlet stagnation temperature and pressure T_{02} and p_{02} from the ambient static conditions using the expression in terms of Mach number and allowing for inlet loss, if any;

(ii) Determine the maximum value of T_{03}/T_{02} which can be allowed, assuming here that $T_{03} = 875$ K. This effectively finds the maximum overall pressure ratio and using Fig. 17.6 this provides one upper bound on the value of T_{04}/T_{02};

(iii) For the given maximum allowable T_{04} determine T_{04}/T_{02}. If this ratio is in excess of the value at the design point it means that the limit is not on T_{04} itself but on $N\sqrt{(c_p T_0)}$. T_{04} must be reduced to give the design value of T_{04}/T_{02};

(iv) Using the lower of the values for T_{04}/T_{02} from ii) and iii) find fan pressure ratio, p_{013}/p_{02} on Fig. 17.6. The pressure ratio across the propelling nozzle is given by

$$p_{08}/p_a = (p_{08}/p_{02})(p_{02}/p_a) = (p_{013}/p_{02})(p_{02}/p_a);$$

(v) The jet velocity can be obtained, on the assumption that the expansion is reversible, from

$$V_j = \sqrt{2c_{pm}T_{08}\{1- (p_a/p_{08})^{(\gamma-1)/\gamma}\}} .$$

The appropriate values of c_{pm} and γ_m for the exhaust gas must be used. (If a simple convergent nozzle were used there is some loss of thrust and the expression of gross thrust in terms of exit velocity and exit static pressure must be used ($\dot{m}V_9 + p_9A_9$), as discussed in section 8.2.) When the afterburner is in use $T_{08} = T_{0ab}$ and suitable values of c_p and γ should be used;

(vi) The overall air mass flow can be found from the ratio of $\bar{m}_8$, whilst the core mass flow can be obtained from $\bar{m}_4$;

(vii) Net thrust follows from

$$F_N = \{\dot{m}_f + \dot{m}_a (1+bpr)V_j - \dot{m}_a(1+bpr)V\},$$

and the sfc may be obtained from the knowledge of calorific value and the temperature difference $T_{04} - T_{03}$.

It will be seen that steps (ii) and (iii) are involved in determining what is the maximum operating condition; step (iv) uses the curves of Fig. 17.6 to find the fan pressure ratio and thereby avoid the iteration.   After determining p_{013}/p_{02} a procedure can be used which is identical to that used in Exercises 16.4 –16.7 at the design point.

Exercises

17.1 The case 1 engine introduced in section 17.1 was calculated in Exercise 16.6 as a design based on sea-level static conditions. Use the design point calculation of Exercise 16.6 to find k_{HP}, $\bar{m}_4$ and $\bar{m}_8$ for the engine.   (**Ans:** $k_{HP} = 0.190$, $\bar{m}_4 A_4 = 0.455 \ 10^{-3}$,$\bar{m}_8 A_8 = 3.55 \ 10^{-3}$)

17.2 The case 1 engine is to operate at the tropopause for $M = 0.9$, 1.5 and 2.0. Variations in engine parameters during operation at Maximum Dry conditions are shown in Fig. 17.6: use this to find the maximum allowable turbine inlet temperature at the tropopause for each Mach number. (At $M = 0.9$ the engine is restricted by $N/\sqrt{(c_p T_0)}$ and so T_{04}/T_{02} must be at the design value; for $M = 1.5$ and 2.0 it is the compressor delivery temperature which restricts operation.)

(**Ans:** $T_{04max} = 1616, 1828.5, 1732.5$ K)

Use Fig. 17.6 to determine the pressure ratio p_{013}/p_{02} across the LP compressor and the corresponding compressor exit temperature T_{013}. Use the value of T_{03} to find the overall compression ratio. The efficiencies given in the technology level in Table 16.1 should be assumed to be still valid.

(**Ans:** $p_{013}/p_{02} = 4.5, 3.64, 2.18$; $T_{023} = 417, 485, 507$ K;
$p_{03}/p_{013} = 6.66, 6.41, 5.58$; $T_{03} = 762, 875, 875$ K)

17.3 At each of the three flight Mach numbers for the Maximum Dry condition:
   a) Find the fuel mass flow rate per unit mass flow of air entering the HP compressor.
(**Ans:** $\dot{m}_f = 0.0250, 0.0283, 0.0257$)

   b)Use the maximum turbine inlet temperatures and corresponding pressures found in Exercise 17.2 to determine the ratio μ of the core mass flow to that at design point. Use $\bar{m}_4$ from Exercise 17.1

(**Ans:** $\mu = 0.407, 0.626, 0.684$)

   c) Find the temperature drop across the HP turbine and thence the exit temperature T_{045}. Use the ratio T_{045}/T_{04} to find the pressure at LP turbine entry. .

(**Ans:** $T_{045} = 1310, 1482, 1404$ K; $p_{045}/p_a = 17.7, 29.3, 31.1$)

By mixing 8% of the flow at compressor delivery temperature, find the temperature into the LP turbine, $T_{045'}$.

(**Ans:** $T_{045'} = 1253, 1419, 1347$ K))

   d) Knowing the pressure ratio across the LP turbine, find the temperature at exit T_{05}.

(**Ans:** $T_{05} = 1057, 1203, 1175$ K)

By mixing 4% of the core flow at compressor delivery temperature, find the mixed temperature out of the core $T_{05'}$.

(**Ans:** $T_{05'} = 1039, 1184, 1157$ K)

   e) From the temperature drop in the LP turbine and, knowing the temperature rise in the LP compressor, find the bypass ratio.

(**Ans:** $bpr = 0.449, 0.546, 0.803$)

   f) Knowing the bypass ratio find the value of c_{pm} and γ_m after the core and bypass flows mix.

(**Ans:** $c_{pm} = 1171, 1161, 1139$ J/kgK; $\gamma = 1.324, 1.328, 1.337$)

   g) Find the stagnation temperature into the propelling nozzle T_{08} after the core and bypass streams mix.

(**Ans:** $T_{08} = 877, 974, 905$ K)

   h) Knowing jet pipe stagnation pressure and temperature find the jet velocity and the specific thrust.

(**Ans:** $V_j = 897, 1030, 1016$ m/s; spec. thrust = 636, 606, 441 m/s)

   i) Using fuel mass flow and the specific thrust find the specific fuel consumption sfc.

(**Ans:** $sfc = 0.958, 1.067, 1.142$ kg/h/kg)

   j) The ratio of total mass flow entering the engine compared to that at the design point is given by $mr = \mu(1+bpr)/(1+bpr_{des})$ Find the value of mr.   (**Ans:** mass flow ratio = 0.401 0.658,0.839)

   k) Determine the ratio of net thrust to that at design point, that is
$$F_N/F_{Ndesign} = mr \times (spec\ thrust)/(spec\ thrust\ at\ design)$$

(**Ans:** 0.295, 0.462, 0.428)

Notes: 1) It might seem that the choking condition for the final nozzle is not needed, but this only appears to be the case because the iterative part is avoided by deriving the fan pressure ratio from Fig. 17.6.

2) The bypass ratio for the case 1 engine at $M = 0.9$ at the tropopause found in section (j) is not equal to the value at design 0.471. This is surprising since the engine is expected to be at the same non-dimensional operating point at design and at this flight condition. The explanation for this is that the fuel flow per unit mass flow of air needed at the $M = 0.9$ tropopause case is about 20% smaller than at sea-level static since the rise in temperature in the combustor is smaller. This gives a smaller mass flow through the turbine which is sufficient to alter the bypass ratio by this amount.

17.4 For the case 1 engine calculate performance at the Combat condition, with the afterburner raising the final stagnation temperature to 2200 K for flight at the tropopause for $M = 0.9$, 1.5 and 2.0. Assume that the engine operates at the same conditions as in the Maximum Dry condition. (The solution can therefore begin from 17.3h.) At the three Mach numbers find the jet velocity, specific thrust, sfc and net thrust (as a fraction of the thrust with the afterburner at sea-level static conditions, Exercise 16.7b).
  (Ans: $V_j = $ 1431, 1563, 1600 m/s; spec.thrust = 1250, 1210, 1098 m/s; $sfc = $ 1.66, 1.67, 1.78 kg/h/kg;
        ratio of net thrust to that at design with the afterburner in use 0.374, 0.594, 0.687)

17.6 CASE 1 AND 2 ENGINES AT MAXIMUM DRY AND COMBAT RATING

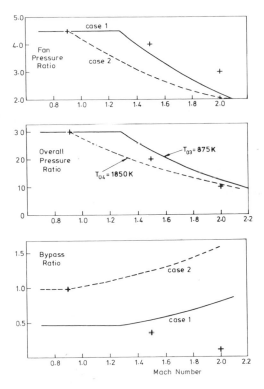

Figure 17.7. Case 1 and Case 2 engines at Maximum Dry condition as a function of Mach number for flight at tropopause. Comparison of p_{023}/p_{02}, p_{03}/p_{02} and *bpr*. Crosses show *design point* engines from Chapter 16, Exercises 16.3 and 16.7.

It may be recalled that the case 1 engine took sea-level static conditions, for T_a=288 K to define the design point. For case 2 the design point was taken to be M=0.9 at the tropopause. Figures 17.7 and 17.8 show comparative results for the operation of the case 1 and case 2 engines at the tropopause for a range of Mach numbers. The discontinuities in the curves correspond to the different constraints coming into play. Also shown by bold crosses are the results for the three design-point studies of Chapter 16 at M = 0.9, 1.5 and 2.0 at the tropopause. (In Figs.17.7 and 17.8 M = 0.9, of course, coincides with the design point for the case 2 engine.)

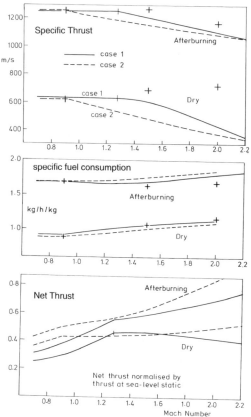

Figure 17.8. Case 1 and Case 2 engines at Maximum Dry and Combat condition as a function of Mach number for flight at tropopause. Comparison of specific thrust, net thrust and *sfc*. Crosses show *design point* engines from Chapter 16, Exercises 16.3 and 16.7.

In Fig. 17.7 the pressure ratio across the fan and the overall pressure ratio are shown. For case 1 it follows that at M = 0.9 the engine is at its non-dimensional design point because the inlet temperature T_{02} is below the design value for the sea-level test bed and the turbine inlet temperature can be reduced to 1616 K (compared with 1850 K on the test bed) to hold T_{04}/T_{02} at the design value. At their design conditions the pressure ratios are the same for both engines, so the pressure ratios also coincide at M = 0.9. The pressure ratios for the case 2 are, however, below those at design point for a much greater part of the operating range. The change in pressure ratio, particularly for case 1, approximately follows the values selected for the design- point calculations of Chapter 16.

At its design point at $M = 0.9$ at the tropopause the core of the case 2 engine is producing more work per unit mass of air compressed because the turbine inlet temperature is much higher. Since, however, the fan pressure ratio is equal for both case 1 and case 2 at this condition, this extra work appears as a higher bypass ratio; in other words for the same mass flow of air through the core more air is passing through the engine. For both engines the bypass ratio increases as the engine is 'throttled back' and for the case 2 engine, for which the bypass ratio is always higher, a value of 1.57 is reached at $M = 2.0$. In contrast to the way an engine behaves off-design the design-point calculation indicated that the optimum engine would have *lower* bypass ratio at high speeds, going as low as 0.11 at $M = 2.0$. In this respect the off-design engine behaves very differently from what would be chosen at design.

Specific thrust, net thrust and specific fuel consumption are shown in Fig. 17.8. As expected, the specific thrust for the case 2 engine is lower than the case 1 engine except at Mach numbers below 1.285. (This Mach number was shown in Exercise 15.6 to be sufficient to raise the inlet stagnation temperature at the tropopause to 288 K, the sea-level static temperature.) The lower specific thrust for case 2 can be related to the lower fan pressure ratio (it is entirely due to this effect when the afterburner is in use) and the lower jet pipe temperature associated with the higher bypass ratio. The largest difference is between the specific thrust for engines off-design and the design point engines. The explanation for this lies with the big difference in bypass ratio and pressure ratio for engine in high Mach number flight when they are on-design and off-design. Whereas on-design engines at high M have low bypass ratio and relatively high fan pressure ratio, the trend is in the opposite direction for engines off-design. The difference between case 1 and case 2 is less straightforward for the ratio of net thrust to the net thrust while static at sea level; the higher mass flow for the case 2 engine leads to a higher thrust despite the lower specific thrust.

The specific fuel consumption curves given in Fig. 17.8 show remarkably little difference between cases 1 and 2; the difference is even quite small between cases 1 and 2 off-design and values calculated for engines at design points at Mach numbers of 0.9, 1.5 and 2.0 at the tropopause. (Case 2, of course, coincides with the design at $M = 0.9$.) This is because pressure ratio plays such a large part in determining the sfc and these were not generally very different. The design-point engine with afterburning does, however, show a significantly lower sfc for the highest Mach number.

Case 1 and case 2 set fairly extreme examples of different design assumptions. As the results in Figs.17.7 and 17.8 show, the trends for the different engines are not in any sense obvious or intuitive. Whether case 1 or case 2 is better depends of the parts of the mission which are most critical; it is generally desirable to have the design point close to the condition which is critical. Perhaps the most striking feature of the results shown in Figs.17.7 and 17.8 is the relatively small difference between the performance of two engines when the difference in their design point is so pronounced. This brings an important general conclusion; if the differences between the engines is relatively small the almost endless search for an optimum engine may be fruitless. The differences between different missions, and the lack of knowledge about what form any mission would actually take when the aircraft and engine are being designed over many years with a service life of 20 or more years, far outweigh the small differences between engines.

Exercises

17.5 In Exercise 16.8 the mass flow of air needed for the case 1 engine with the afterburner to provide a thrust–weight ratio of unity at take off at sea level on a standard day was calculated to be 65.9 kg/s. Assuming that the aircraft weight for subsequent manoeuvres is 15 tonne, values of drag were calculated in Exercise 14.4 and are given in Table 14.1. Use these with the off-design thrust ratios in Exercises 17.3k and 17.4 to calculate the specific excess power (SEP) for the aircraft. If SEP <0 find the amount by which the mass flow must be increased to achieve the sustained flight condition. Do this for the engine at Maximum Dry and at Combat (i.e. with the afterburner) for the following:

      a) 11 k m altitude and $M = 0.9$, 1.5 and 2.0 in straight and level (i.e. 1 g) steady flight;

      b) 11 km at $M = 0.9$ and 1.5 for *sustained* 3g turns;

          (**Ans:** a) dry: $SEP = 33.2$, 5.4 m/s, increase $\dot{m}_a$ by 78%; a/b: $SEP = 90.2$, 158.8, 131.5 m/s;

           b) dry: increase $\dot{m}_a$ by 108%, 46%;   a/b: increase $\dot{m}_a$ by 16%, $SEP = 78.8$ m/s)

Note: The results show that 'supercruise' (supersonic flight without the afterburner) would be possible for the case 1 engine at $M = 1.5$ at the tropopause, but 'supercruise' at $M = 2.0$ would require a much bigger engine. For 3g sustained turns at the tropopause – this is impossible with the case 1 engine at $M = 0.9$ even with the afterburner, and possible only with the afterburner at $M = 1.5$.

17.6 For $M = 0.9$ at sea level a calculation for the case 1 engine shows that the ratio of net thrust to that at sea-level static conditions is 0.798 when dry and 0.979 with the afterburner. Exercise 14.4b gave the drag at this speed and altitude for 5g and 9g turns. Determine the specific excess power for both turn rates dry and with afterburner - if the SEP is negative determine the increase in engine size which would be necessary.

      (**Ans:** dry; 5g turn, $SEP = 27.8$ m/s, 9g turn, increase $\dot{m}_a$ by 80%:

          a/b; 5g turn, $SEP = 194$ m/s, 9g turn $SEP = 12.5$ m/s)

Note: Close to sea level the case 1 engine allows 5g sustained turns with the engine dry, but 9 g sustained turns are possible only with the afterburner.

17.7 OPERATION AT REDUCED THRUST

In section 14.3 some attention was directed to the operation of the aircraft when the drag is much less than the maximum and the amount of engine thrust needed is very small. In Exercise 14.6 the altitude for loiter (to maximise the time spent flying in a vicinity) at a Mach number of 0.6 was found to be about 5.6 km, at which condition the thrust from each engine was only required to be 7.1 kN. In Exercise 14.7 the altitude for maximum range at $M = 0.8$ was found to be about 5.7 km at which condition the thrust should be 8.1 kN. Since minimising fuel consumption is the primary goal for both these flight conditions the engine must obviously operate dry.

To determine the condition at which the engine should operate to produce the required thrust, more specifically the temperature at entry to the turbine, it is necessary carry out an iterative analysis of the type described earlier in section 17.4. It is also possible to use the ideas of section 17.5 to generalise a small number of results and to use a graphical solution to remove the need to carry out an iteration.

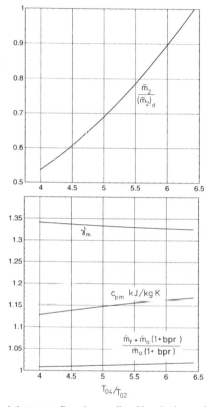

Figure 17.9. Case 1 engine. Inlet mass flow (normalised by design value), exhaust mass flow for dry operation normalised by inlet mass flow, average c_p and average γ as functions of T_{04}/T_{02}.

For operation with a choked final nozzle which expands the exhaust reversibly the nozzle divergence must alter with pressure ratio across the nozzle p_{08}/p_a. The pressure ratio is determined by the engine operating condition and by flight Mach number, the latter determining the pressure in the intake. The change in the nozzle area ratio represents a change in the geometry of the entire engine and to obtain simple non-dimensional relations one must stop at the nozzle throat. The primary relations will then be the pressure ratio $p_{08}/p_{02} = p_{013}/p_{02}$ and the temperature ratio T_{08}/T_{02} in terms of the familiar ratio of turbine inlet temperature to compressor inlet temperature T_{04}/T_{02}; these are shown in Fig.17.6. Knowing

p_{02}/p_a from the Mach number it is possible to calculate the jet velocity; but to facilitate this it is convenient to have the mean specific heat and the mean specific heat ratio plotted versus T_{04}/T_{02}. To obtain the specific thrust the additional mass flow attributable to fuel is needed and the ratio

$$\frac{\dot{m}_f + \dot{m}_a(1+bpr)}{\dot{m}_a\,(1+bpr)}$$

can also be plotted versus T_{04}/T_{02}. Specific thrust alone is not sufficient to determine net thrust since the total mass flow rate is needed, but non-dimensional mass flow $\bar{m}_2$ can also be given as a curve versus T_{04}/T_{02}. All of these curves are shown in Fig. 17.9.

Exercise
17.7 It was calculated in Exercise 14.6 that the New Fighter Aircraft would need 7.1 kN thrust from each engine when flying at $M = 0.6$ at an altitude of 5.6 km. (At this altitude take the ambient pressure and temperature to be 50.5 kPa and 251.75 K.) Show that the case 1 engine will produce this thrust with a turbine inlet temperature of 1075 K. (This may be achieved using the curves in Figs. 17.6 and 17.9.) If the bypass ratio is 0.958 find the specific fuel consumption.   **(Ans: 0.87 kg/h/kg)**

SUMMARY CHAPTER 17

In a manner similar to that described in more detail in Chapter 12 it is possible to calculate fairly easily how a given engine will behave as either the inlet conditions are altered or the fuel flow is altered. So far as the inlet conditions are concerned the most important variable is the stagnation temperature, which is a function of both altitude and Mach number. (It is also dependent on the climatic conditions, but throughout the book we have assumed only standard conditions.) At Mach numbers substantially greater than unity the inlet temperature rises rapidly. Under these conditions the turbine inlet temperature must generally be decreased to prevent the compressor discharge temperature exceeding the value which the material can stand; in addition the engine non-dimensional performance is determined by the temperature ratio T_{04}/T_{02} and as T_{02} rises it is inevitable that the ratio falls. In consequence all the pressure ratios, non-dimensional mass flows $\bar{m}$ and non-dimensional rotational speeds also fall. At high flight speeds the engine is therefore inevitably operating at a condition equivalent to 'throttled back' at sea level.

For exactly equivalent reasons an engine operating on a stationary sea-level test bed (or taking off) when the ambient temperature is above the design value will be in a 'throttled-back' condition. Throttling back has a much larger effect on the LP shaft than the HP; the LP compressor has to accommodate large excursions in non-dimensional mass flow as its operating point changes, but the range for the HP may only be on the order of a quarter as large. The speed changes for the LP shaft are similarly much larger than for the HP shaft.

The calculations of the engine performance off-design are here based on two major simplifications. It is assumed that both the HP and LP turbines are choked, as is the final propelling nozzle. It is also assumed that the turbine polytropic efficiencies are constant. To calculate the operating point of an engine off-design requires a simple iterative calculation. To calculate the net thrust at an off-design condition requires more than the specific thrust, since it is necessary to know the mass flow through the engine; the mass flow falls with altitude in proportion to the ambient pressure, but rises steeply as the speed increases due to the rapid rise in inlet stagnation pressure. The bypass ratio also rises as the engine is 'throttled back'.

The designer can choose where to base the design. Very commonly this is at sea-level static conditions with a selected representative ambient temperature. If a much lower inlet temperature is selected for the design point, such as that for $M = 0.9$ at the tropopause, the net thrust is higher and the sfc is lower at the design point than an engine designed at sea-level static. Away from the design point the benefits arc less clear, with the engine operating more 'throttled back' over most of the range. The optimum design will emerge from a full analysis of many possible engines for off-design operation over the full mission (in fact, over a large number of candidate missions) to see how many of the intended missions are possible and the extent of the compromises necessary.

CHAPTER 18

TURBOMACHINERY
FOR COMBAT ENGINES

18.0 INTRODUCTION

This chapter will look only briefly at the design of the compressors and turbines for combat engines, following on from Chapter 9 for the civil transport engine. The flow Mach numbers inside the compressors tend to be higher than for the subsonic transport and this makes the treatment of each blade row rather special; the design rules must take account of the presence of strong shock waves. A very important design consideration is how the compressor will behave at off-design conditions since combat engines are off-design for so much of their operation. The problem arises because of the large density ratio between inlet and outlet of the compressor, and the reduction in this ratio when $N/\sqrt{(c_p T_0)}$ is decreased. The turbine stages do not suffer from this off-design problem. The turbines are required to produce large work output in relation to the blade speed, that is $\Delta h_0/U^2$ must be high, but at off-design conditions for the engine the turbine condition is essentially unaltered from the condition at design. This, it may be recalled from Chapters 12 and 17, is because the turbines and the propelling nozzle are effectively choked, so the turbines are forced to operate at the same non-dimensional condition.

In this chapter the consideration will be based on the case 1 engine of Chapter 17, with fan pressure ratio 4.5 and HP compressor ratio 6.66 at design point, sea-level static. This was calculated as Exercise 16.6 with the technology standard given in section 16.1, in particular the polytropic efficiency for the fan is 0.85, for the HP compressor 0.90 and for the turbine 0.875.

18.1 THE COMPRESSORS

It can be seen from Fig.15.1 that the LP compressor or fan has three stages and the HP compressor 5 stages and these numbers will be adopted here.

In Exercise 16.6 the mass flow of air into the engine to give the necessary thrust at take off was found to be 67.4 kg/s and with a Mach number in the duct upstream of the fan equal to 0.70 the fan tip diameter was fixed at 0.681 m. (It was also assumed that the hub diameter was equal to 0.40 times the casing diameter.)

Exercises

18.1* If the fan tip diameter at inlet is equal to 0.681 m and the fan tip speed is 500 m/s find the rotational speed of the LP shaft. If the inlet flow is axial with Mach number 0.70 and the inlet stagnation temperature on a sea-level test bed is 288 K, find the relative velocity and relative Mach number at the tip of the first stage fan.   (**Ans:** 14022 rpm, 549 m/s, $M = 1.69$)

18.2* The pressure ratio across the fan is 4.5 . Assume that the stagnation temperature rise in each of the three stages are equal and that the stage efficiencies are the same and equal to 0.85. Find the stagnation temperature rise in each stage and the stagnation pressure rises in each of the stages.
(**Ans:** 63.2 K; pr = 1.804, 1.636, 1.525)

18.3 At design the Mach number of the flow at inlet to the fan is 0.7. Through the LP compressor the Mach number is progressively reduced so that at outlet it is 0.3. (The flow is axial at both inlet and outlet.) Use the expression given in section 6.2 to find $\bar{m}_2$ and then $\bar{m}_{23}$ and hence find the cross-sectional area of the flow at outlet assuming that the flow is uniform. If the hub diameter is increased by 20% from the front of the first stage to the rear of the last fan stage, find the hub and casing diameter at fan outlet.   (**Ans:** $\bar{m}_2 = 1.171$; $\bar{m}_{23} = 0.629$, $A_{out} = 0.163$ m^2, $D_{hub} = 0.327$ m, $D_{casing} = 0.561$ m)

18.4 In Exercise 17.2j the mass flow into the engine at $M = 2.0$ at the tropopause is 0.839 times that at sea-level static conditions. Find the value of $\bar{m}_2$ at this condition and verify that the flow Mach number into the fan is about 0.340. From Exercise 17.2 the pressure ratio across the fan at this condition is 2.18; use this to find value of $\bar{m}_{23}$ assuming the flow to be uniform and the fan efficiency to be unchanged. Verify that the Mach number at exit from the fan is about 0.346.

(**Ans:** $\bar{m}_2 = 0.703$; $\bar{m}_{23} = 0.717$)

For simplicity in the above exercises the flow is treated as uniform at fan exit. This is an oversimplification and the error increases at off-design conditions. It will be noted from Fig.17.7 that the bypass ratio increases from 0.47 at design to about 0.80 at $M = 2.0$ which leads to a radial shift of the streamlines passing through the fan so as to pass a smaller fraction of the total flow through the core.

The HP compressor varies its operating point relatively little over the entire operating range of the aircraft, as shown in Fig.17.4 and it is only for starting that any special measures to ensure satisfactory performance are needed. Typically bleed valves are used for this condition.

Exercises

18.5* From Fig.15.1 it is possible to estimate that the tip diameter of the first stage of the HP compressor is equal to 0.65 times the corresponding diameter of the fan. Assume that the flow into the HP compressor is uniform and axial with a Mach number at design of 0.5. If the relative Mach number onto the first rotor tip is 1.2 find the rotational speed of the HP shaft.
(**Ans:** $D_{tip} = 0.443$ m; $N_{HP} = 20109$ rpm)

18.6 The pressure ratio across the HP compressor at design is 6.66. Assume that the stagnation temperature rises in each of the five stages are equal and that the stage efficiencies are the same and equal to 0.90. Find the stagnation temperature rise in each stage and the stagnation pressure rises in the first and last stages.   (**Ans:** 78.9 K; stage pr = 1.619, 1.348)

18.7 At HP compressor inlet $M = 0.5$ and at outlet $M = 0.3$; the values of $\bar{m}$ are 0.956 and 0.629 respectively. Find the inlet and outlet area necessary when the overall mass flow into the engine is 67.4 kg/s and the bypass ratio is 0.471. With the inlet tip diameter given from Exercise 18.5 find the inlet hub diameter. Assuming a constant casing diameter find the hub diameter at the HP compressor inlet and outlet and then the blade height h.

(**Ans:** area = 0.0728, 0.0224 m^2; D_{hub} = 0.322, 0.4095 m; h = 60, 17 mm)

18.8 From Exercise 18.7 the mean radius at inlet and outlet is determined. Find the mean blade speed U_m for the first and last stages and thence the value of the loading parameter $\Delta h_0/U_m^2$ for these stages.
(**Ans:** U_m = 403, 448 m/s; $\Delta h_0/U_m^2$ = 0.489, 0.394)

18.2 THE TURBINES

The turbines of the combat engine bear a close resemblance to those of a civil engine. The non-dimensional loading $\Delta h_0/U_m^2$ and flow deflection will normally be somewhat higher for the combat engine, with correspondingly lower efficiencies. The higher cooling flow generally leads to somewhat higher loss and lower efficiency.

The high speeds and temperatures produce very high stresses in the turbine. Whereas the compressor can be made of lightweight titanium alloys, the turbine can only be made of more dense nickel alloys. The stress in the turbine is crucial in fixing the diameter of the turbine rotors. From Fig.15.1 it can be deduced that the *mean* diameters of the HP and LP turbines are approximately 0.675 and 0.725 times the fan inlet tip diameter respectively. With these estimates it is possible to deduce many of the aspects of the turbines.

Exercises

18.9* **a)** Find the blade speed at mean radius U_m for the HP and LP turbines. Using the temperature drops given in Exercise 16.6 find the values for $\Delta h_0/U_m^2$ for the HP and LP turbines . If the axial velocity is equal to 0.5 times the mean blade speed for each turbine use Fig.9.3 to estimate the stage efficiencies of each turbine.

(**Ans:** U_m = 484.3, 362.5; $\Delta h_0/U_m^2$ = 1.86, 2.16; η_{HP}= 0.895, η_{LP} = 0.893)

b) On the assumption that the *absolute* velocity at outlet from each turbine is purely axial, use the Euler work equation (see section 9.2) to calculate the *absolute* swirl velocity into each rotor at the mean radius.
(**Ans:** V_θ = 901, 783 m/s)

c) Assuming that the axial velocity in the turbines is equal to 0.5 times the mean blade speed find the resultant velocity leaving the HP and LP nozzle guide vanes and thence the *static* temperature. Use this to calculate the local speed of sound and hence find the corresponding Mach numbers. What are the flow deflections in the nozzle guide vanes?

(**Ans:** V = 933, 803 m/s; T = 1500, 1176 K; M = 1.247, 1.212; 75.0°, 77.0°)

18.10 For the mean diameter draw the velocity triangles at entry to and exit from the HP and LP rotor. Assume that the axial velocity is equal on both sides of each rotor. Hence find the deflection in each rotor at the mean diameter
(**Ans:** 123°, 130°)

18.11* **a)** Find the area of the throat for the HP and LP turbine stators. The mass flow through the HP turbine throat is equal to 0.88 times the mass flow of air into the HP compressor plus the mass flow of fuel; for the LP turbine the mass flow is 0.96 times the compressor flow plus the fuel flow. The mass of fuel is given in Exercise 16.6 to be 2.9% of the air into the HP compressor. (**Ans:** 0.0150, 0.0406 m^2)

b) As an approximation assume that the flow direction out of the nozzle guide vanes is the same as that at the throat. Hence using the mean diameters for the HP and LP turbines find the blade height at the throat.   (**Ans:** 40.1, 116 mm)

SUMMARY CHAPTER 18

This has been a brief chapter, resting as it does very firmly on Chapter 9. Use has been made of the drawing of an engine in Fig.15.1 to assess the relative diameters of the compressors and turbines; with these diameters fixed the ease or difficulty of designing the stages is more or less determined.

The diameters of the turbomachinery components largely determine the overall size and shape of the engine and if suitable information were accessible its weight could be estimated. Naturally the treatment here in Chapters 15–18 has skirted over many of the aerodynamic difficulties and all the mechanical ones; in fact it is to be expected that every aspect will be as aggressively designed as possible to optimise performance and minimise weight. It is a consequence of this aggressive push to optimise that the cost of developing combat engines is so high.

Part 4

A Return to the

Civil Transport Engine

CHAPTER 19 · A RETURN TO THE CIVIL TRANSPORT ENGINE

19.0 INTRODUCTION

When the engine for a new civil[1] transport, the New Large Aircraft, was considered in Chapters 1 to 10 many assumptions were introduced to make the treatment as simple as possible. In the treatment of the engine for a New Fighter Aircraft in Chapters 13–18 the level of complexity was increased. This increase in complexity included allowing the properties of the gas to be different for burned and unburned air; the effect of the mass flow of fuel added to the burned air passing through the turbine was included; the effect of the cooling air supplied to the turbines was allowed for; and the effect of the pressure losses in the combustor, the bypass duct and the jet pipe were demonstrated. It is appropriate to recalculate the performance of an engine for the civil aircraft with some of these effects included.

Figure 19.1. An unmixed engine on the Airbus 330 and a mixed engine on the Airbus 340-300

Another difference between the treatment for the civil engine in Chapters 1–10 and the treatment for the combat aircraft was the mixing of the core and bypass streams upstream of the final propelling nozzle in the combat engine. Some engines on subsonic transport aircraft also have mixed streams; Fig. 19.1 shows photographs of an unmixed and a mixed engine on the wing of two contemporary aircraft, the Airbus-330 and Airbus-340-300. By a simple treatment it is possible to demonstrate the advantages which the mixed configuration brings. The chapter begins by showing calculations for some behaviour of the mixed bypass engine and then comparing the working line for the fan of both mixed and unmixed engines.

---

[1]Civil transport is the British usage; the equivalent US usage is commercial transport.

The mixed engine is then used as a vehicle for examining the effect of the component performance (efficiency for the compressors and turbines, pressure loss for the combustor) on the engine performance. This is first done when the efficiency change is accommodated in the design (so the calculations are of the on-design type with unaltered pressure ratios), then when efficiency alters during operation (so the calculations are of the off-design type with the pressure ratios in the engine altering). The mixed engine facilitates this because some of the freedom is removed; once the fan pressure ratio is selected the bypass ratio is determined for a given turbine inlet temperature and overall pressure ratio.

19.1 THE MIXED-FLOW HIGH BYPASS ENGINE

The approach for calculating the variables inside the mixed civil engine and the overall performance is exactly that developed to cope with the mixed combat engine. In line with the notation of Chapters 1 to 10 the condition downstream of the fan in the bypass duct is denoted 13, whilst the condition at outlet from the LP booster stage, in line with Chapter 12, is denoted 23. Downstream of the fan the stagnation pressure p_{013} is equal to the stagnation pressure at LP turbine outlet p_{05}, neglecting losses in the bypass duct. With the mixed engine specification of the pressure downstream of the LP turbine fixes not only the power output from the LP turbine, but the pressure ratio across the fan. For a given choice of turbine entry temperature and overall pressure ratio the LP power is fixed and, since the fan pressure ratio is also specified, the bypass ratio is determined. (This is in contrast with the unmixed engine for which the bypass ratio and fan pressure ratio can be selected independently, at least over some range.)

For the present calculation a two-shaft engine with core booster compressor stages on the LP shaft is considered. The polytropic efficiency for all the compressors (the fan in the bypass, the fan and booster stages for the core and the HP compressor) have been taken to be 0.90; likewise the polytropic efficiencies of the HP and LP turbines are taken to be 0.90. At the design point the overall pressure ratio is $p_{03}/p_{02} = 40$ and the pressure ratio for the core flow through the fan and booster stages is 2.5. It is assumed that when operating off-design the temperature rise of core flow through the fan and booster is proportional to the temperature rise of the bypass flow through the fan. Cooling flow to the HP turbine is included here, just as it was for the combat engine, with 5% of the air passing through the HP compressor being assumed to be used to cool the rotor. The cooling air to the HP nozzle guide vanes is assumed to be fully mixed with the main flow prior to the nozzle exit and its effect is therefore included in the definition of the turbine inlet temperature. As in Chapters 16 and 17 the rotor cooling air, at compressor delivery temperature, is mixed with the gas which leaves the HP turbine rotor; during mixing it is assumed that the stagnation pressure remains constant whilst the stagnation temperature decreases. It is also assumed that there is loss in stagnation pressure in the combustor so that at turbine entry $p_{04} = 0.95p_{03}$, but pressure loss in the bypass duct or jet pipe is neglected here. For the unburned air it is assumed that $\gamma = 1.40$ (giving $c_p = 1005$ J/kgK) while for the mixture of air and burned fuel $\gamma = 1.30$ (giving $c_p = 1244$ J/kgK) is used.

Design-point calculations

Figure 19.2 shows the result of design-point calculations for mixed flow engines at cruise (chosen here to be $M = 0.85$ flight at 31000 ft), with the calculations performed for two values of turbine inlet temperature, $T_{04} = 1450$ K and 1550 K. The calculations for specific thrust (net thrust per unit mass flow of air through the engine) were also performed at both temperatures but the curves for the two values of T_{04} are indistinguishable, reinforcing the point made in the context of combat engines that it is the fan pressure ratio which overwhelmingly determines the specific thrust. Raising fan pressure ratio increases specific thrust and leads to a corresponding increase in the fuel consumption. (The specific fuel consumption and specific thrust shown in this figure are for the bare engine and ignore drag around the engine nacelle.)

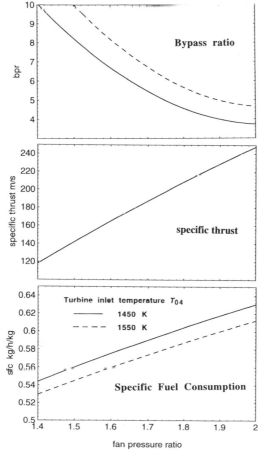

Figure 19.2. Mixed turbofan performance parameters as function of fan pressure ratio.
Cruise at $M = 0.85$ at 31000 ft. Conditions as for datum on Table 19.2

As Fig. 19.2 shows, the bypass ratio is determined by the fan pressure ratio for a given turbine inlet temperature: with higher turbine inlet temperature (actually T_{04}/T_{02}) the power available from the turbine is larger and the bypass ratio is greater for the same pressure ratio. Increasing the bypass ratio in this way does not affect the propulsive efficiency, because the specific thrust is unaltered, but it does lead to a reduction in the temperature of the exhaust. Consequently an increase in turbine inlet temperature gives a reduction in specific fuel consumption when pressure ratios are held constant.

Off-design calculations for the fan

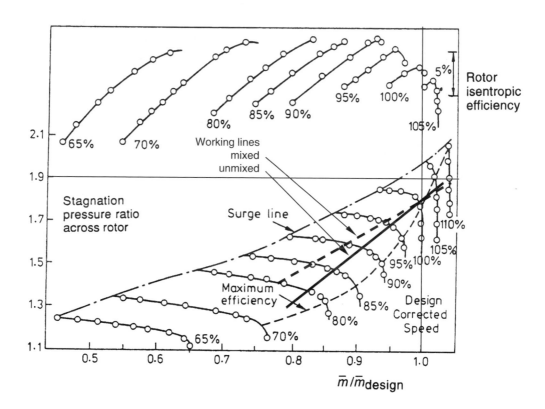

Figure 19.3. Working lines superimposed on the performance map for a modern
fan (Fig.11.4). $M = 0.85$ at 31000 ft. At design point fan pressure ratio 1.81,
overall pressure ratio 40, component polytropic efficiencies 90%

For this section attention is directed at the fan and the design point corresponds to top of climb at $M = 0.85$ at 31000 ft. At this condition a fan pressure ratio of 1.81 at the 100% corrected mass flow condition has been selected since, for the mixed engine used above, this pressure ratio can be achieved with a bypass ratio of 6.00 for a turbine inlet temperature of 1572 K,

$T_{04}/T_{02} = 6.058$. Figure 19.3 shows the measured performance characteristics of a fan (earlier shown as Fig.11.4) with two fan working lines superimposed, one working line is for the mixed engine, the other for an equivalent engine having a separate core and bypass nozzle. For both the mixed and unmixed engines the fan pressure ratio was equal at the design point $p_{013}/p_{02} = 1.81$, with the core fan and booster pressure ratio equal to 2.5 and overall pressure ratio 40; in both cases the engines had the same component efficiencies, cooling air flow rate and pressure loss giving a bypass ratio of 6.00 at the design point.

The working lines were calculated by the approach used in section 12.4 for the unmixed engine and in section 17.3 for the mixed engine. The turbine inlet temperature in each case was allowed to vary from 1155 K to 1622 K ($4.45 > T_{04}/T_{02} > 6.25$). It is immediately apparent that for the mixed engine the working line is much steeper, so that the pressure ratio falls more rapidly as turbine inlet temperature decreases, but the mass flow through the fan decreases proportionally less. This steeper working line of the fan is a primary reason for the use of the mixer. It will be seen from the curves for fan efficiency that by having a steeper working line the fan is able to operate nearer to its maximum efficiency over a wider range. The maximum fan pressure ratio occurs during the climb, but of far greater concern to the overall fuel consumption are conditions when the engine is throttled back somewhat, such as at cruise, because the engine operates for much longer at this condition. The benefit conferred by the working line being further to the right for the mixed engine augments the small increase in thrust arising directly from the mixing of core and bypass flow, see exercise 19.2. A further benefit is that there is a smaller drop in rotational speed required to reduce the thrust with the steeper working line; this higher rotational speed, relative to that for the unmixed engine, means that at the reduced thrust condition for cruise the LP turbine will operate at a lower value of $\Delta h_0/U^2$ which is likely to give a further increase in efficiency.

Exercises

19.1 Use the working lines in Fig. 19.2 to estimate the increase in fan efficiency achieved by using a mixed nozzle when the fan corrected speed is reduced to 85%.   (**Ans:** 1.5%)

19.2 Compare the gross thrust from two engines, each having a bypass ratio of 6, one mixed, the other unmixed. In each case the stagnation pressure at exit from the bypass and core are equal to 1.65 times the inlet pressure p_{02} and $p_{02} = 1.60 p_a$, where p_a is the ambient pressure. The stagnation temperature of the core stream at exit from the LP turbine is 737K, while for the bypass stream it is 305K. Take $\gamma = 1.40$ for the bypass and 1.30 for the core streams; for the mixed stream take $\gamma = 1.38$. Calculate the gross thrust per unit mass flow through the core (ignoring the mass flow of fuel) for an unmixed engine (when the two streams expand separately to ambient pressure) and for a mixed engine (when they first mix without loss in pressure to a uniform pressure and temperature before expanding to ambient pressure). Assume isentropic expansion in each case.

(**Ans:** Unmixed $F_G = 2919$N/kg/s; Mixed $F_G = 3014$N/kg/s)

If the flight speed is 256 m/s, find the proportional change in net thrust F_N.   (**Ans:** 8%)

Note: The increase in net thrust calculated in exercise 19.2 looks very significant, but in reality the benefits are much smaller and the 3% increase in gross thrust calculated here is an overestimate for several reasons. If a more realistic value of γ is used for the core stream ($\gamma = 1.345$) the increase in F_G is reduced to about 2%. This figure must be reduced to take account of the stagnation pressure reduction associated with the heat exchange between hot and cold gas and the stagnation pressure loss associated with viscous fluid mechanic effects. Furthermore, the mixing of core and bypass streams is incomplete, so not all of the potential benefit will be achieved, and the gain in F_G is only about 1%, which translates into an increase in *net* thrust of about 2.5%. But even this is not achieved, because the longer cowl around the outside of the engine increases the drag around the outside of the engine nacelle, and the residual increase in F_N is less than 1%.

19.2 THE EFFECT OF PARAMETER CHANGES AT THE DESIGN STAGE

In this section the impact of changes in component performance will be assessed when these are incorporated at the design stage. The pressure ratios are fixed, and so too is the ratio of turbine inlet temperature to engine inlet temperature T_{04}/T_{02}, so a change in the performance of one component has to be compensated by an alteration in some other component. This means that if, for example, the compressor efficiency falls, the power input to it must be increased to maintain the same overall pressure ratio. The net power from the LP turbine will be reduced by the increase in power to the compressor; since the pressure ratio of the fan is being held constant the bypass ratio must therefore be decreased. The effect is *not* the same as an alteration in any quantity after the engine has been constructed, such as might occur by wear, damage or incorrect manufacture – in that case the pressure ratios will not remain constant, and this forms the subject of section 19.4 below.

It is convenient to carry out the comparison using the mixed engine because the constraint that the stagnation pressure at outlet from the LP turbine equals that downstream of the fan removes a degree of freedom; once the fan pressure ratio is specified the bypass ratio is also determined. The procedure adopted is to have a datum engine and then to vary individual parameters in turn. The engine is similar to that introduced in section 19.1, assumed to be flying at $M = 0.85$ at 31000 ft, ($T_{02} = 259.5$ K). The overall pressure ratio is kept constant at $p_{03}/p_{02} = 40$, the fan pressure ratio held constant at $p_{013}/p_{02} = 1.65$ and the pressure rise of the core flow on the LP shaft is $p_{023}/p_{02} = 2.5$. In the unperturbed datum engine the component polytropic efficiencies are taken to be 90%, with 5% of the air compressed in the core used to cool the HP turbine rotor and a pressure loss of 5% assumed to occur in the combustor. These pressure ratios and a turbine inlet temperature of 1450 K ($T_{04}/T_{02} = 5.588$) give a bypass ratio of 5.99 with the datum efficiencies. With this combination the jet velocity is 432 m/s, the specific thrust $F_N/\dot{m}_a = 176$ m/s, the net thrust per unit mass of core flow is 1234 m/s and the specific fuel consumption $sfc = 0.583$ kg/h/kg.

Reducing the fan efficiency by 1% to 89%, for example, has the effect of reducing the bypass ratio from 5.99 to 5.92. The jet temperature is therefore raised slightly and, because the pressure ratio p_{013}/p_{02} is held constant, the jet velocity rises from 431.8 m/s to 432.4 m/s with a corresponding small *increase* in specific thrust. In terms of thrust from a given size of engine core, however, the drop in bypass ratio more than compensates for the rise in jet velocity, so when fan efficiency falls by 1% the thrust decreases by about 0.6%. The thrust per unit core flow is a better indicator or measure of the engine thrust from a given size or weight of engine, so it is this which is given in Table 19.1. The specific fuel consumption is of primary significance and the effect of changes on this are also tabulated. The bypass ratio, also shown, helps explain some of the changes observed.

Table 19.1 Design point variations.
Cruise at $M = 0.85$ at 31000 ft.
Fan pressure ratio $p_{013}/p_{02} = 1.65$, overall pressure ratio $p_{03}/p_{02} = 40$.
Where not perturbed: $T_{04} = 1450$ K, component polytropic efficiencies 90%,
cooling air flow rate to HP rotor 5% of core compressor air, $p_{04}/p_{03} = 0.95$
$\gamma = 1.40$ for unburned air, $\gamma = 1.30$ for combustion products

PRESSURE RATIOS HELD CONSTANT	bpr	Thrust per unit core flow m/s	$\Delta(sfc)$ %
Datum engine	5.99	1234	0
Reduce fan bypass η_p to 0.89	5.92	1226	0.63
Reduce HP compressor η_p to 0.89	5.80	1209	1.13
Reduce core fan & booster compressor η_p to 0.89	5.93	1226	0.36
Reduce HP turbine η_p to 0.89	5.92	1226	0.69
Reduce LP turbine η_p to 0.89	5.92	1226	0.66
Reduce cooling air to HP turbine to 2.5%	6.36	1287 −	1.61
Reduce pressure loss in combustor to zero	6.21	1258 −	1.90
Increase turbine inlet temp. by 25 K to 1475 K	6.33	1288 −	0.78

So far as the treatment in this book is concerned, the most encouraging conclusion from this table is that the results are not so sensitive to the input parameters that the overall conclusions from trends would be compromised by the imprecise value of the efficiencies or losses used. The alterations in the input variables and the changes in the outputs are sufficiently small that the variation of one with another may be assumed linear. This means that for such small changes the effect may be assumed linearly proportional to the input and the effects produced by several inputs may be added.

For this engine the turbomachinery component with the largest impact on the fuel consumption for a change at the design point with pressure ratios held constant is the HP compressor; a loss of 1% in compressor efficiency translates into 1.1% increase in sfc and a 2% loss in thrust for the same size core flow. The changes associated with the fan and the HP and

LP turbines are similar to one another. Raising the turbine inlet temperature has a marked beneficial effect on the fuel consumption and thrust, largely because of the increase in the bypass ratio. If during the design and development stage it is found that a component is low in its efficiency it is possible to correct for this to some extent by raising the turbine inlet temperature. Treating the effects as linear, Table 19.1 shows that the loss in thrust consequent upon a reduction of HP compressor efficiency of 1% can be restored by increasing the turbine inlet temperature by $25 \times (25 \div 54) = 11.6 \, \text{K}$. There would nevertheless be a net increase in the specific fuel consumption, approximately equal to $1.13 - 0.78 \times (11.6 \div 25) = 0.77\%$.

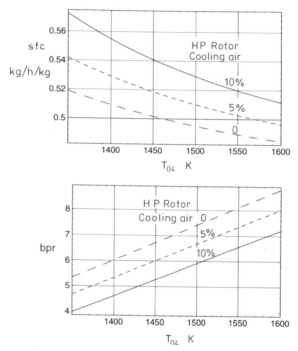

Figure 19.4. Variation in *sfc* and *bpr* with turbine inlet temperature varying the cooling air flow rate. Conditions as for datum engine in Table 19.2

The table also shows how important are the cooling flow and the pressure loss in the combustor. In the specification of turbine operating conditions there is a trade between the benefits of higher temperature and the detrimental effects of increased cooling air flow rate. The effect of cooling air flow rate is illustrated in Fig. 19.4, where in one case sfc is shown versus T_{04} and in the other case bypass ratio versus T_{04}. Curves are shown for no HP rotor cooling flow and for rotor cooling flow equal to 5% and 10% of HP compressor mass flow. (The total cooling air flow rate would probably be about double this because of the need to cool the HP turbine nozzle, but this is allowed for in the stipulation of turbine inlet temperature at nozzle guide vane exit.) In terms of bypass ratio it can be seen that the benefits of a 50 K rise in turbine inlet temperature are approximately undone by a 5% increase in cooling air flow rate to the HP turbine rotor. For the sfc the effect is about twice as large, approximately a 100 K increase in T_{04} being required to compensate for a 5% increase in rotor cooling air flow rate. Alternatively, at constant turbine inlet temperature the sfc can be said to increase by about 4% for a 5% increase in the rotor cooling flow.

The cooling flows were ignored in Chapters 1 to 13 and this helped to compensate for the underestimate of power from the turbine when the value of $\gamma = 1.40$ was used in the turbine. This is illustrated in the Table 19.2, showing the separate effects of cooling air flow rate and γ on the bypass ratio, thrust per unit mass flow through the core and sfc.

Table 19.2   Effect of gas properties and turbine cooling flow
At design point for the datum engine of Table 19.1
($\gamma = 1.40$ for unburned gas in all cases)

PRESSURE RATIOS HELD CONSTANT	bpr	Thrust per unit core flow m/s	$\Delta(sfc)$ %
Datum: $\gamma = 1.30$, cooling air to HP rotor 5%	5.99	1234	0.0
$\gamma = 1.30$, no cooling air to HP rotor	6.72	1340	–3.0
$\gamma = 1.40$, no cooling air to HP rotor	5.30	1027	–13.2
$\gamma = 1.40$, cooling air to HP rotor 5%	4.68	944	–10.0

It is apparent from this table that the choice of the value of γ has a considerable effect on the magnitude of thrust and sfc. It is also clear that neglect of the cooling flow leads to substantial overestimate of thrust and underestimate of sfc. In Chapters 1 to 10 the emphasis was on keeping the calculations simple, in particular neglecting the cooling air flows. It was because of the need to include cooling flow, if accurate results were to be obtained, that the use of $\gamma = 1.40$ for the combustion products seemed adequate.

19.3 THE EFFECT OF PARAMETER CHANGES TO AN IN-SERVICE ENGINE

This, the off-design case, considers what happens when an existing engine experiences changes during operation. Pressure ratios do *not* now stay constant when parameters vary. The results are presented in Table 19.3. The most obvious is changes in turbine inlet temperature, but deterioration in component efficiency (and flow rate) is expected during the operating life of an engine, due to dirt, foreign object damage, wear and to increased clearances. The comparison is performed for the same datum engine as that used in section 19.3 flying at $M = 0.85$ at 31000 ft. In this case the change in net thrust is significant.

The effect of the alteration in parameters here is significantly different from the variation on-design shown in Table 19.1, where all the pressure ratios inside the engine were kept constant as the efficiencies where changed. As in Table 19.1 the alterations in input parameters are sufficiently small that the consequent changes may be assumed to be proportional to the imposed perturbations and effects of perturbations to more than one component may therefore be added. The large loss in thrust associated with a drop in efficiency of the booster stages and, more particularly, a drop in efficiency of the HP compressor, are because of the large effect that the reduction in overall pressure ratio has on the mass flow through the core. The core mass flow is determined by the choked HP turbine nozzle guide vanes, so that $\dot{m}\sqrt{c_p T_{04}}/A p_{04} = 1.389$, and the core mass flow therefore falls in proportion to the drop in compressor delivery pressure. The reduction in core mass flow shows up in the increase in bypass ratio when the booster or core efficiencies fall.

Table 19.3.   Off-design variations
Datum engine as in Table 19.1
Where not perturbed: T_{04} = 1450 K, component polytropic efficiencies 90%

	fan pr	bpr	Overall pr	$\Delta(F_N)$ %	$\Delta(sfc)$ %
Datum	1.650	5.99	40.00	0	0
Reduce Fan η_p to 0.89	1.646	5.96	40.05	−0.43	0.53
Reduce HP compressor η_p to 0.89	1.622	6.17	38.51	−4.01	0.51
Reduce core fan & booster comp. η_p to 0.89	1.641	6.05	39.51	−1.33	0.16
Reduce HP turbine η_p to 0.89	1.644	5.98	39.94	−0.65	0.55
Reduce LP turbine η_p to 0.89	1.645	5.98	39.95	−0.62	0.53
Increase turbine inlet temp. 25 K to 1475 K	1.688	5.88	41.71	5.72	0.00

The operational procedure to cope with fall in component efficiency is to increase the turbine inlet temperature to maintain thrust, and one sign that an engine is due for overhaul is when the turbine inlet (or exit) temperature exceeds a set value for a given thrust or given fan pressure ratio. Concentrating just on the HP compressor, for example, Table 19.3 shows that a 1% drop in its efficiency would be approximately compensated for in terms of thrust by a rise in T_{04} of about 55×(4.01÷5.72)= 17.5 K. Similarly, a 1% loss in HP turbine efficiency would be corrected for thrust by a rise in T_{04} of 25×(0.65÷5.72)= 2.84 K. Although the increase in turbine entry temperature can recover the thrust, the specific fuel consumption would increase by approximately 0.5% relative to the datum engine for a loss in efficiency of 1% in the fan, HP compressor or either of the turbines. If each component fell by 1% the linear character of the changes means that the combined reduction in thrust would be the sum of the individual reductions $0.43+4.01+1.33+0.65+0.65 = 7.07\%$. Then the corresponding rise in turbine inlet temperature needed to compensate for this would be approximately $25\times(7.07\div5.72) = 30.9$ K, with a rise in sfc of about 2.3%.

19.4 HIGH-SPEED CIVIL TRANSPORT

From the Wright brothers' first flight to the time of the Boeing 707 entering service there has been a continual increase in flight speeds. For civil transport the speeds have not increased from the time of the Boeing 707 (with the notable, but uneconomic, exception of Concorde) and in some cases have fallen. Many of the big twins (e.g. Airbus A300 and Boeing 767) cruise near $M=0.80$ rather than $M=0.85$ for the older big jets. There has been long-standing interest in a supersonic transport which is bigger than Concorde (Concorde can only carry about 100 passengers), with longer range (Concorde has a maximum range of 3500 nm, just adequate for Paris to Washington DC) and possibly with a higher cruise speed (Concorde cruises at $M=2.0$). (For $M=2.0$ sonic boom prevents flights over populated regions and above $M=2.2$ the skin has to be cooled or else titanium has to be used in place of aluminium.) There are many difficulties, including airport noise, sonic boom, fuel consumption and the cost of the highly complicated aircraft. More recently concern for the environment puts another hurdle in the way of a new supersonic transport.

In the spring of 2001 Boeing announced that they were preparing designs for a medium sized aircraft (about 200 seats) with a range of 9000 nm and a cruise Mach number of 0.98, dubbed the 'sonic cruiser'. Not everyone in the aviation industry, however, believed that these targets were realisable. The favoured configuration appeared like a slender delta with forward (canard) wings. The aircraft was aimed at the business traveller who is prepared to pay more for speed. The increase in Mach number from 0.85 to 0.95 could take about one hour off a journey from London to Los Angeles. The engine companies (Rolls-Royce, General Electric and Pratt & Whitney) prepared engine proposals for propelling this aircraft.

The first matter to consider is whether the choice of Mach number was sensible - would it alter significantly as the design refines? At first sight a Mach number of 0.98 was an odd choice, since aircraft drag rises very steeply from about $M=0.85$. Beyond $M=1.0$ the drag levels off and may even fall – at $M = 1.2$ perhaps three hours could be taken off the journey from London to Los Angeles. At these speeds sonic boom would not be an issue and the stagnation temperature would not be high enough to affect the strength of aluminium alloys.

The engineering problems that the sonic cruiser presented were large and consideration was given to employing a range of new technologies to improve engine performance (in particular, reduced fuel consumption and engine weight), improved aircraft aerodynamics and reduced airframe weight (largely by extensive use of composites).

With aircraft of conventional shape and Mach number it is possible to estimate the drag at cruise with considerable precision. This means that the engine thrust (and engine size) can be optimised with some confidence in advance. For a new shape of aircraft the drag will not be known with such confidence. Worse still, it is difficult to measure drag accurately in a wind tunnel at speeds close to sonic; the best that can be achieved is to use a small model in a very large wind tunnel.

The initial responses from some airlines to the sonic cruiser proposal were enthusiastic, but over time there was a gradual cooling. This was partly because of the financial pressure that most airlines were under in 2002, but it was also because the economic benefits to the airlines turned out smaller than might at first be imagined. It turns out that it is hard to take advantage of the reduced journey time in many long flights because of constraints with airports at each end. As a result, Boeing cancelled the sonic cruiser project at the end of 2002, while announcing a new aircraft initially called the 7E7. This plane is intended to replace the Boeing 767 with a cruise mach number of 0.85. The 7E7 will exploit the advanced technology that had been directed towards the sonic cruiser, but will offer the benefits in lower operating costs and lower environmental impact.

19.5 DESIGN FOR LONG RANGE

The first part of this book addressed the New Large Aircraft designed to fly with a full payload for 8000 nautical miles. This was a bigger aircraft than those currently in service with a range at the very upper bound of what is possible for smaller aircraft currently in service. It has been implicitly assumed by people in the aircraft industry when considering new aircraft that longer range is desirable, since it opens up the possibility of non-stop flight between more remote pairs of cities.

Over the last few years the concern over the environmental impact of aircraft has been growing. There is still uncertainty and controversy about the impact of present and future aircraft on climate and on the ozone layer, but few people are likely to believe that dumping oxides of nitrogen or additional quantities of water vapour or carbon dioxide in the upper troposphere or lower stratosphere is desirable. The oxides of nitrogen can, to some extent, be reduced by improved designs of combustor, but water and carbon dioxide are unavoidable results of burning hydrocarbon fuels. The burning of hydrogen as a fuel would certainly remove CO_2, but there are such practical and logistic problems in the use of liquid hydrogen that few believe it to be realistic for many years to come, if ever. In any case burning hydrogen would introduce water vapour into the upper atmosphere and this is believed to be a cause of climate change. Consequently the only way to reduce emissions of CO_2 and water is to reduce the amount of fuel burned per passenger-mile travelled.

This issue has been considered in a recent paper by Green(2002), presented at the *Greener by Design* conference held in London in July 2001. The scope for reducing CO_2 and water by improving further the efficiency of new engines is, by common consent, limited. Likewise the scope for improving the lift-drag ratio of aircraft is limited. What has emerged in this paper is the very considerable scope for reducing the fuel burned per passenger-mile by reducing the range of the aircraft. This is illustrated by considering an aircraft designed for a full-payload range of 15000 km (8095 nm). The fuel burned per passenger is then compared to the fuel burned if the same payload were carried the same distance in 3 steps in an aircraft designed for a maximum range of 6000 km. The reduction in fuel burn, per passenger mile, is of the order of 40%, far larger than the projected benefits of improved engine efficiency or improved aircraft lift-drag ratio. The explanation is simple: for a long range flight much of the weight at take off and for the early part of cruise is fuel and by reducing the range the same payload can be carried in a smaller, lighter aircraft. There are obvious benefits in time saving and reduced landing fees by flying very long legs, but what this study has revealed is that there is a potential for reducing fuel burned on long journeys if the incentives are sufficient.

SUMMARY CHAPTER 19

When the civil engine was considered in the early chapters there were many simplifications introduced which can be removed with the level of complication introduced for treating the combat engine. It is also possible to consider the mixed high-bypass engine, such as are being fitted to many current aircraft.

There is an improvement in sfc if the core and bypass are mixed before the final nozzle. Some of the advantage comes from the mixing itself, whereas some comes from the difference in the working line of the fan; with a mixed configuration the fan working line stays closer to the locus of the maximum efficiency, making the cruise condition more favourable.

Calculations were performed for a mixed engine at a constant fan pressure ratio and overall pressure ratio (on-design calculations) to assess the significance of the various parameters, such as component efficiencies, cooling flow and combustor pressure loss. These calculations showed that at constant pressure ratios and T_{04}/T_{02}, variations in component performance have a marked effect on bypass ratio. The alteration in bypass ratio has a marked effect on the thrust per unit mass flow through the core and on specific fuel consumption. The HP compressor has the largest effect on both thrust per unit mass flow through the core and sfc; for a plausible design of engine the sfc rises by about 1.1% for each per cent reduction in compressor efficiency. Halving the cooling air flow rate would, if it were possible, lead to a reduction in sfc of about 1.6%; there would also be a substantial increase in bypass ratio leading to a 4% greater thrust from the same size of core.

For simple calculations, in which the gases are treated as perfect with constant properties, it is better to take $\gamma = 1.30$ for the products of combustion. If $\gamma = 1.40$ is used for the products of combustion (as well as for the unburned air) the thrust and sfc are underestimated.

If $\gamma = 1.40$ is used for combustion products the error in thrust per unit mass flow rate through the core is less if the cooling flow is neglected.

For a fixed configuration of engine, with the constant T_{04}/T_{02}, the pressure ratios are altered as the component efficiencies are reduced; such changes in efficiency might be as a result of in-service deterioration. Calculations for the same datum engine showed that the largest alteration was produced by the change in HP compressor efficiency – in this case a 1% decrease in the HP compressor efficiency led to a reduction in thrust of 4.0%; this loss in thrust could be recovered with a 17.5 K increase in turbine inlet temperature but with an increase in sfc. The large loss in thrust is because of the reduction in mass flow through the core consequent on the fall in the pressure ratio as the compressor efficiency is reduced. The specific fuel consumption was increased by about 0.53% for a 1% reduction in fan, HP compressor, HP turbine or LP turbine efficiency.

There are a number of issues to be addressed in designing engines for an aircraft designed to fly at a higher Mach number than current subsonic aircraft. It is not clear what these alterations will be, but satisfactory fuel consumption at cruise and acceptable noise at take off are the key issues.

There is potential for large fuel savings if aircraft are designed for relatively short legs and long journeys are accomplished in several steps; these savings could far outweigh the benefits likely to be achieved by improvements in engine sfc.

CHAPTER 20   TO CONCLUDE

20.0 INTRODUCTION

This chapter returns to some general issues related to both civil and military engines. These are topics which can more satisfactorily be addressed with the background of earlier chapters.

20.1 CHOICE OF NUMBER OF ENGINES FOR CIVIL AIRCRAFT

The design for the New Large Aircraft assumed that there would be four engines and indeed the Airbus A380 does have four engines. When the first jet aircraft were beginning to fly across the Atlantic ocean four engines were necessary to cope with the possibility of loss in thrust in one engine at take off; to have had to accommodate loss of one engine with fewer than four engines would have involved carrying a lot of unnecessary engine weight (and available thrust) during cruise. Even when the Boeing 747 entered service around 1970, four engines were considered optimum for a transatlantic flight; the large twin-engine aircraft introduced by Airbus, the A300, seemed suited only to shorter routes. By the 1990s the most widely used aircraft on the transatlantic route were twin-engine aircraft, such as the Boeing 767 or Airbus 300, providing non-stop flights from, for example, London to Los Angeles (4700 nautical miles). The more recent twin, the Boeing 777, is used mainly for flights under 5000 nautical miles but the newer long-range version will have a range of 8810 nautical miles. It is reasonable to ask what has changed to make it possible and second what has made it attractive to use twin-engine aircraft on such long routes?

The second question is answered more easily, and it centres on cost. Engines are expensive to buy and to maintain, so reducing the number of engines offers an immediate advantage. Implicit in this is the observed trend that the cost of a given quantity of installed thrust drops as the size of the engine increases. Only when the flight becomes very long, substantially more than 4700 nautical miles, does the logic which was originally used to justify four engines lead to this configuration being more economic; it seems that for the longest ranges the four-engine aircraft will remain the first choice. The Airbus A340 and the A380 are both four-engined aircraft specifically aimed at very long routes.

The other question is how is it possible for a twin-engine aircraft to be practical. One part of the answer is the phenomenal reliability achieved by the engines operating under the strict certification and operational rules governing ETOPS (Extended-range Twin Operations).

When an aircraft and engine is suitably certificated it can fly anywhere a suitable airport can be reached with one engine inoperative within 180 minutes, which covers almost anywhere in the world. The occurrence of engine failure is very rare: statistics show that up to 1995 there had been 35 million ETOPS flights and only 39 in-flight engine shut downs. (It is known, however, that in substantially more cases one engine has had to be throttled back to idling conditions, but not shut down completely, and therefore does not show in these statistics.)

There is another part to the answer which affects both economy and practicality. It was shown in Chapter 8 that a twin-engine aircraft with the engine sized to produce just enough thrust to climb adequately at 31000 ft would not have enough engine thrust to cope with an engine-out at take off. The net thrust available at take off was shown in section 8.4 to be approximately 268 kN whereas the required thrust for a one-engine take off was about 406 kN. (This was making the assumption that the engine would be at the same non-dimensional operating point at take off with only one engine as it was designed for at cruise. In fact the engine would produce some more thrust than this.) It was also shown in Chapter 8 that for cruise at a fixed Mach number the net engine thrust is proportional to ambient static pressure. It is common for long-range twin-engine aircraft to begin cruise at substantially higher altitude than four-engine aircraft; commonly 37000 or 39000 feet, with the wing area selected to put optimum ML/D (Mach number times lift–drag ratio) at this altitude. (In other words the aircraft has larger wings than an aircraft designed to cruise at 31000 ft.) At 39000 ft the ambient pressure is 19.7 kPa, whereas at 31000 ft it is 28.7 kPa; scaling the result from section 8.4, the available thrust at sea level from an engine able to propel an aircraft of the same weight at 39000 ft, at the same engine non-dimensional operating point, is therefore $268 \times 28.7 \div 19.7 = 390$ kN. This is not far short of the value of 406 kN shown to be needed for a one-engine take off and this shortfall in thrust could easily be made up by increasing the temperature ratio to a value that is allowed for take off. It should also be noted from Fig. 8.2 that operating engines 'throttled back' at thrusts below their maximum is quite good so far as sfc is concerned, though it does imply carrying unnecessary weight.

The selection of wing area can be realised as fixing the required thrust of the engine at cruise. As the aircraft is developed during its life to allow heavier payloads it is unusual to alter the wings significantly. To maintain optimum ML/D with the heavier payload the aircraft cruises in denser air, at lower altitude. For the same Mach number the lift coefficient is inversely proportional to ambient pressure, so if aircraft weight is increased the ambient pressure during cruise must be higher in proportion. For constant lift–drag ratio the necessary thrust must also rise in proportion to the weight, but it has already been noted that at constant Mach number the net thrust from an engine is proportional to ambient pressure. In other words the engine produces additional thrust in the same proportion that the wings produce additional lift and the engine non-dimensional operating condition remains constant as the aircraft weight is increased. Although the cruise non-dimensional thrust requirement does not change as the aircraft is developed to have a higher take off weight, the thrust actually required at take off must increase in proportion to this weight. This provides some justification for the normal practice of quoting engine thrust at the sea-level static condition rather than that at cruise.

Given the cost advantage and the practicality of long-range twin operation, why should four-engines still be envisaged for the very largest aircraft? One reason is that with four engines mounted under the wing the weight of the outboard engines reduces the bending moment in the wing relative to that with only engines and allows more fuel to be carried, enabling longer range. However the major reason for four engines is size. The preferred position for the engines is on pylons beneath the wings, and the preferred position for the wings is below the passenger compartment. Twin engines large enough to propel the New Large Aircraft or the A380 would be so large that the aircraft would need to be higher off the ground. This would require the landing gear to be higher, which would increase weight, but it would also require alterations to the passenger handling facilities at airports. These are important drawbacks. There is a further drawback to a very large engine, which is that it might find application only on the New Large Aircraft; it is unattractive to develop a new engine for a single application and in most cases a manufacturer hopes to spread the cost over applications to several different aircraft, a topic picked up below.

20.2 SOME REAL ISSUES OF ENGINE DESIGN AND DEVELOPMENT

The cost of engine development is a strong disincentive to doing it very often. Whenever possible the engine from one aircraft application is adapted to another aircraft and even adapted for electrical power generation or other industrial use. Looking back, the General Electric CF6, designed for the DC10 was installed on the Boeing 747 and 767 and many Airbus designs, most recently the A-330. The Rolls Royce RB211 was designed for the Lockheed L1011 and has found application on Boeing 747, 757 and 767 aircraft; as a more major derivative it has evolved into the Trent to power the Boeing 777 and the Airbus A330, A340 and A380. In the cases of both the CF6 and the RB211 the engines have been modified considerably, but much of the original design architecture has remained. Going back further the CF6 core is itself a derivative of a core for a military transport aircraft. Perhaps more remarkably, the smaller CFM56 powering later versions of the Boeing 737 uses a core developed from an engine which was originally designed to propel the supersonic B-1 bomber.

It would therefore be disingenuous to leave the impression that most engines have followed a logical design path from a specification in terms of the optimum new engine. The optimum provides a reference for comparison, but there is great pressure to adapt what already exists. Although the aircraft operator may be forced to use more fuel because of this, there are advantages to the operator because the use of existing developed technology reduces risk and increases reliability.

For the civil engine there is another aspect which confuses the logical development. The time taken to develop a new engine has, until recently, been longer than the time taken to design, test and produce a new civil aircraft, so the engine development needed to begin sooner. The engine manufacturer is a supplier to the airframe manufacturer and is dependent on the airframe maker to give the engine specifications; the specifications of the aircraft, and therefore the size of engine needed, may be altering right up to the time when the aircraft is complete. There is a tendency for aircraft to become heavier as the design progresses and problems are encountered; it is not unusual for engines to become heavier too, leading to the requirement for still more thrust. For the maker of engines this requires judgement and experience to reduce the risks that are entailed: the engines may be designed for a higher thrust than that originally asked for on the expectation that thrust requirements will rise. If this conservatism is carried too far, however, the engines are too big and too heavy and as a result are likely to be uncompetitive.

For military aircraft there are similarities and differences. One similarity is the huge cost, so whenever possible a developed engine will be used in another application, perhaps in a wholly different aircraft. On other occasions only part of the engine, such as the core, may be used with a new LP spool. When a new engine is to be designed and built the situation differs from the civil market. First the government is likely to be involved in stipulating the mission requirements, in making the decision on engine parameters and type and in deciding who shall build the engine. There are now marked differences between US practice and that in most of Europe. In the USA the trend now is to run competitions between engines, normally produced by General Electric and Pratt & Whitney. In Europe the trend is to have a single engine designed and built by an international consortium; the exception here is France which has until now tended to develop its own fighter aircraft and engines. The huge costs of the aircraft and their engines leads to extensive studies to optimise the combination for a range of duties, many of them conflicting in their requirements, and this takes a long time, typically more than a decade. The situation is complicated by the changing geo-political situation, so that the threat foreseen as dominant at the start of the process may have receded by the end, and by the changes in funds available. As Chapters 16 and 17 showed, a wide range of engine types could meet many of the goals and the search for the best possible may lead to a substantial increase in cost. Certainly the requirement for an aircraft to perform widely different roles necessarily leads to compromises which will make it much less effective in each of the roles than an aircraft more tailored to that mission.

20.3 OVERVIEW

This short book has addressed a lot of different topics and has therefore needed to avoid too much detail. As a result specialists inside companies may find many areas where the book stops short of the level of detail they require. The emphasis has been towards understanding what is the design practice of aircraft engines and why the design choices are made the way they are. So far as possible, empiricism has been avoided, but some is essential: one has to take realistic values of turbine inlet temperature and component efficiencies, for example. Where empiricism has been used it has often been possible to look back at how sensitive the calculated results are to these inputs and in all cases the trends are correctly predicted and the magnitudes are sensible.

Another choice has been made to keep the book manageable: it addresses only the aerodynamic and thermodynamic issues involved. It should hardly need saying that the mechanical and materials issues are every bit as important as aerodynamics and thermodynamics, but to have introduced these with a satisfactory level of detail would have made the book substantially longer. In any case it is the goals set by the thermodynamic and aerodynamic considerations which tend to drive the overall specification of the engine (such as pressure ratio and bypass ratio) and the mechanical and materials issues come in as constraints; this book has therefore chosen to concentrate on the aerothermal aspect.

As the materials get stronger and their tolerance of high temperature increases the specific fuel consumption of subsonic civil engines will decrease whilst the specific thrust of military engines will increase. At present the limit of operation of an engine is as likely to be provided by the compressor delivery temperature as by the turbine inlet temperature. Whereas the turbine inlet temperature is open to increase by increasing the amount of cooling air, the compressor delivery temperature alone defines the conditions which the material must stand. If overall pressure ratios are to increase there must be major materials changes or substantial increases in compressor efficiency.

For the military engine, progress in producing better materials will, to a large extent, determine the progress towards higher thrust–weight ratio. Not only does a stronger material make direct weight savings possible, it also makes higher blade speeds feasible. With higher blade speeds it is possible to reduce the number of stages for a given pressure ratio, further reducing the overall weight.

Environmental issues will very likely come to play an increasing part in the choice of cycle for both the civil and military engine and the publication of the IPCC report on the effect of aviation in 1999 has given added weight to concern for emissions at high altitude. This report showed that not only is CO_2 a greenhouse gas, but NOx can be of similar importance and the effect of the contrails induced by the water vapour in the engine exhaust is another potentially important source. The release of combustion products up around the tropopause, it appears, is several times more serious than the burning of a similar amount of fuel at sea level. The creation of NOx (oxides of nitrogen) is enhanced when the temperature of combustion increases, as it does when the compressor delivery temperature is raised. It may be that reducing the emissions, and not material limitations, is going to be the ultimate restriction on compressor delivery temperature and therefore on the engine cycle. It seems highly probable that new civil aircraft which lead to seriously greater levels of emissions at high altitude will not be allowed on environmental grounds, even though the rules for cruise emissions have not yet been formulated. It then seems probable that the onus will be on the aircraft makers to prove that no greater harm is being done by a new supersonic transport or a 'sonic cruiser' than is produced by current large aircraft.

Noise is likely to favour lower fan pressure ratios, and higher bypass ratios, and already the specification of the engines for the Airbus A380 to meet noise rules produces a higher fuel burn than would be the case if noise could be ignored. Further in the future one could imagine aircraft geometries being adopted which are not optimal for aerodynamic performance but which minimise noise; for example, putting engines over the wings and using the wings to shield the ground. It is highly likely that airport noise rules are going to constrain the design of subsonic aircraft in ways not yet fully understood and no scheme for a supersonic transport with acceptable noise has yet been found which is capable of bearing serious examination.

Appendix NOISE AND ITS REGULATION

History and regulation

As jet air transport increased in the 1960's the annoyance to people living and working around major airports was becoming intense. Regulations affecting international air transport are governed by the International Civil Aviation Organisation (ICAO), but this body was moving so slowly that in 1969 the US Federal Aviation Agency (FAA) made proposals for maximum permitted noise levels. After extensive discussions in the USA these were formally approved as Federal Aviation Regulation (FAR) Part 36 in 1971, retroactive with effect from 1969, but only for new aircraft. Shortly afterwards the ICAO Committee on Aircraft Noise published similar recommendations, to be known as Annex 16, a formal addendum to the 1944 Chicago Convention on Civil Aviation; each member state had then to accept the rules in Annex 16 and write them into their legal framework. The underlying principle for the noise certification of aircraft under FAR Part 36 and Annex 16 are similar and has remained unchanged ever since, with the levels under the US and ICAO rules subsequently becoming virtually identical.

The certification for noise relies on measurements at three positions, two for take off (referred to as lateral and flyover) and one for landing (referred to as approach). The levels are expressed in decibels (EPNdB) using effective perceived noise level (EPNL), described in outline below. The layout for testing is shown in Fig. A1.

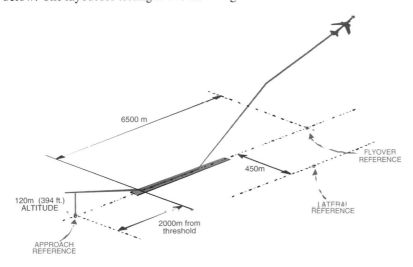

Figure A1   The three noise certification reference positions.

The noise at the lateral position is the highest noise measured along a line parallel to the runway whilst the aircraft is departing at full power and the maximum usually occurs when the aircraft has climbed to about 1000 feet. Flyover noise is measured directly under the flight path after take off and at an altitude where it is normal to cut-back the power to reduce the noise whilst still maintaining a safe rate of climb. The approach noise is also measured directly under the flight path as the aircraft prepares to land, with the glide slope carefully controlled. The flights are for the maximum allowed weight of the aircraft and correspond to standard day temperatures (which will generally require corrections to be made to the measurements since tests are rarely carried out at precisely the standard conditions). Needless to say, aircraft do not always operate as specified for the tests, but the tests do at least provide a standard way of comparing aircraft and thereby regulating airport operations.

Around 1977 noise levels for certification were lowered and these were known in the USA as FAR Part 36 Stage 3 and elsewhere as Chapter 3 of Annex 16. The levels of noise to qualify for certification increase with gross take off weight up to 400 tonne for lateral noise, 385 tonne for flyover and 280 tonne for approach; at greater weights the levels are constant. The allowable flyover noise is highest for four-engine planes and lowest for twins to make allowance for the slower climb rate with four engines. Figure A2 shows the noise levels permitted by ICAO for Chapter 3 of Annex 16 for the three conditions used. Shown on the figures are indications of the measured certification levels for a selection of modern aircraft, revealing very clearly that the recent aircraft types are well below the certification levels, notwithstanding the regulations failing to increase the allowed levels beyond maximum take-off weight of 400 tonne. The ICAO Chapter 3 levels are therefore a challenge only for older designs of aircraft and engine.

It may be noted that the absolute levels of the certification levels shown in Fig.A2 are not equal for the different conditions. This is to reflect the different distances between the aircraft and measurement point for each condition

In January 2001 Chapter 4 of Annex 16 was agreed by the Committee of Aviation Environment Protection, CAEP, part of ICAO. This was ratified by ICAO in the autumn of 2001, and is expected to be incorporated into law by all its members. Chapter 4 requires reductions a cumulative margin of 10 EPNdB from the levels in Chapter 3. (Cumulative means the numerical sum of the margins relative to Chapter 3 for the three noise measuring conditions: lateral, flyover and approach.) In addition at no condition must the level exceed that for Chapter 3 and there must be a cumulative margin of at least 2 EPNdB from Chapter 3 for any two conditions. At first Chapter 4 will apply only to new aircraft designs, i.e. those which are not already in production, but the industry is assuming that eventually the rules will be extended to apply to all aircraft.

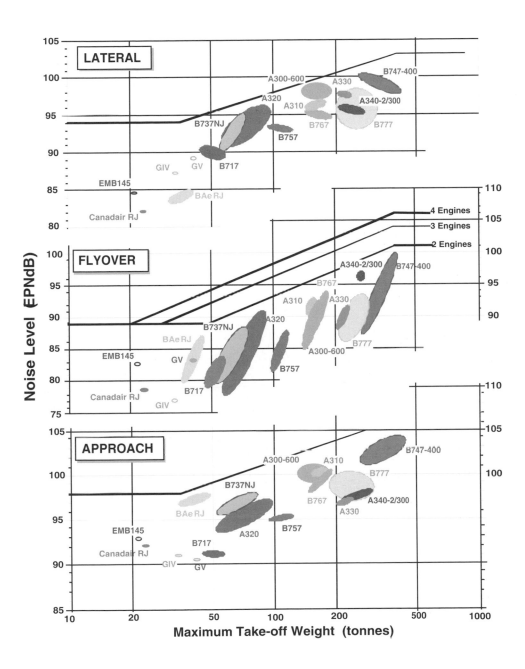

Figure A.2. Certification noise levels for FAR Part 36 and ICAO Annex 16, Chapter 3

The measurement of noise

Noise is annoying sound and consists of pressure fluctuations propagating through the air as acoustic waves. The sound pressure level (SPL) relates the level of the pressure amplitude to that which the human ear can just detect at its most sensitive frequency, $p_{ref} = 20 \times 10^{-6}$ Pa. Because the ear is able to accommodate a wide range of amplitudes it is customary to use a logarithmic scale so the sound pressure level for a signal of amplitude p is written

$$SPL = 20 \times \log_{10}(p/p_{ref})$$

and expressed in units of decibel (dB). A change of 3 dB is readily noticed whereas a change of 1 dB is normally imperceptible. Well away from the source of noise in an open environment the sound pressure decreases with the square of distance from the source and so for a doubling of the distance the SPL can be shown to fall by about 6 dB.

The human ear is most sensitive to frequencies in the middle range (peaking at about 3 kHz) and to get a measure of annoyance it is necessary to correct the SPL to allow for this. The simplest approach is merely to adjust the levels in line with the sensitivity of the ear, and a commonly used measure based on this gives dBA, which can be read directly from a dial on an instrument. A more complex method, allowing for noise amplitude as well as frequency and requiring processing by a computer, can be used to give Perceived Noise Level (PNL), measured in PNdB. PNL forms the basis of the aircraft noise certification measurements, but it also turns out the humans are more annoyed by noise with a tonal content than noise of a broadband nature, and the measured noise can be corrected for this too to give PNLT. Finally the annoyance is affected by the duration over which the noise is present and allowance is made for this. The result is that a set of defined procedures are made to convert the instantaneous measurements of sound pressure level into a single number, the Effective Perceived Noise Level, EPNL, which is measured in units of EPNdB. The regulations adopted by the FAA, by ICAO, and by many airports, are couched in terms of EPNdB.

There is still quite widespread use of dBA, however, and this is the basis of the day-night level, LDN, which integrates the dBA and arbitrarily adds 10 dB at night to allow for the greater annoyance produced by aircraft noise during the night. LDN is strongly favoured by the FAA in the USA.

Local noise regulation

Some residents living near busy airports have felt that the international agreements are moving too slowly to address their concerns and as a result local regulations have sprung up. The most important of these affecting the design of large aircraft, and currently the most onerous for large aircraft, are those for the London airports. Similar schemes have been adopted by some other

European airports These are based on a system of quotas for arrival and departure. At departure the procedure for calculating the quota count (QC) is to take the arithmetic average of the certification level in EPNdB for lateral and flyover conditions. For calculating QC for arrival 9.0 EPNdB is subtracted from the certification EPNL at approach to allow for the difference in measurement distance. The level of EPNL at the bottom of each QC band is as follows:

	QC0.5	QC1	QC2	QC4	QC8	QC16	
Minimum EPNL	-	90	93	96	99	102	(EPNdB)

The exact rules of operation of the quota-count system have been varied over the years, but there are some general features which can be illustrated by the proposals becoming effective in 2002 and which serve to demonstrate how important noise has become to the airline operator. The total quota of night flights (i.e. product of the number of flights and the appropriate QC for the type of aircraft) for Heathrow for a whole year has been set at 9750. No scheduled departures for QC4 will be allowed between 11:30 pm and 6:00 am, whilst for QC8 and QC16 this prohibition is from 11:00 pm to 7:00 am. Delayed QC4 departures are allowed after 11:30 pm but QC8 departures will not be permitted between 11:30 pm and 6:00 am, even if the flight departure has been delayed. At Heathrow airport there is also a monitoring system, measured in dBA, to ensure that actual levels of noise are not exceeded, regardless of aircraft weight, with fines imposed in cases of infringement.

For the Airbus A380 operation in London was considered essential for many of the customers and similarly stringent local rules may be introduced at other airports. Airbus and the engine manufacturers therefore agreed to target QC2 at departure, and QC1 at arrival, a cumulative margin from Chapter 3 of at least 20 EPNdb, much greater than the 10 EPNdB margin required for ICAO Chapter 4

Noise generation

The understanding of the generation of noise in a quantitative way is one of the most challenging tasks in fluid mechanics. Since noise is merely propagating pressure fluctuations, any non-steady flow or any moving object, such as a fan blade, is capable of being a noise source. The problem is greatly increased for aircraft engines by the common tendency for the blade speeds and the flow velocities inside the engine to be close to sonic Even in a qualitative sense there is not full agreement on what the noise sources are, nor what is their relative importance; this is most clearly the case with broadband noise from the fan and from the LP turbine. The greatest broadband noise source is jet noise, attributable to the turbulent mixing which necessarily takes place. It was shown many years ago that jet noise is proportional to the eighth power of jet velocity; i.e. the acoustic power more than doubles (increases by 3dB) for a 10% increase in jet velocity. This needs to be modified to allow for forward motion, high Mach numbers, the effects of separate core and bypass streams (each with different velocity) and the temperature of the core stream, but the dependence on a high power of velocity remains.

Figure A3 shows the rough breakdown of the noise sources at the three conditions used for certification for a large engine. It will be seen that with a moderately high bypass ratio (and therefore relatively low jet velocity) the jet noise is still the largest source at the lateral position and it is comparable with the fan (inlet and aft radiated fan noise) at cut-back conditions during flyover. (It may be added that the cut-back manoeuvre, which is to reduce the engine thrust when a safe altitude has been reached, is precisely intended to reduce jet noise.) The fan noise from the rear (i.e. exhaust) is more important than from the intake at departure for this engine, a feature which does not concur with the subjective impression obtained standing near the runway during takeoff. At approach the jet noise is negligible, but the fan is now the most important source from the engine. (Because the fan is rotating much more slowly during approach than take-off the *cause* of the fan noise will be different at the different conditions.) A more modern engine would have rather lower jet noise relative to fan noise than the one in Figure A3, but with fan noise also lower in absolute terms.

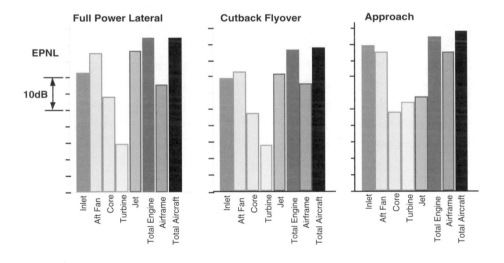

Figure A3. An approximate breakdown of major noise sources at the three certification conditions from NASA Contractor Report 198298, June 1996. (Bypass ratio ≈ 4)

An important feature shown in Fig. A3 for the approach condition is the high level of noise from the airframe. In the case shown it is almost comparable to the noise from the engines, but with more modern engines, with higher bypass ratio and lower jet velocity, such as for the Airbus A380, the airframe noise is likely to dominate at approach. The high level of airframe noise is not altogether surprising when the configuration of the aircraft during the approach is considered. For stability reasons the flaps are lowered sufficiently far to give high drag as a result of large-scale separations, with the turbulence and noise that this entails. The high lift coefficient at approach, because of the low forward speed, gives strong trailing vortices which are turbulent and noisy. Lastly the undercarriage and the wheel wells are serious noise sources. Although the noise from the undercarriage and wells does seem amenable to reduction, it is not obvious how to reduce the other two sources which are the result of fundamental features of aircraft operation.

Noise reduction

Quite remarkable reductions have been achieved in aircraft noise over the last 30 or so years. A reduction at each condition of about 20 EPNdB has been achieved, notwithstanding the increase in aircraft size and weight which has occurred. Most of this has been achieved by the selection of the engine cycle to reduce jet noise, careful choice of the numbers of blades in the fan and LP turbine, and the use of acoustic treatment on solid walls of the inlet and outlet ducts. Whereas the design of blade shape to give good aerodynamic performance and maximum efficiency is well advanced; the similar selection of blade shape to minimise noise is currently in its infancy. The more sweeping steps that can and are taken to lower noise are listed below with some limitations and difficulties noted.

Lower the jet velocity, which for a given core means choosing a higher bypass ratio and lower pressure ratio across the nozzle. This has the effect of increasing the propulsive efficiency and the change from low to modest values of bypass ratio led to a significant reduction in specific fuel consumption. For a fixed level of engine thrust, lowering the jet velocity requires a larger mass flow through the engine. A value of jet velocity is reached where the increase in engine nacelle drag (and degradation of wing performance) and increased weight leads to a greater total fuel burn; for the latest engines proposed for the A380 this limit has been crossed, so the minimising of noise is leading to about 1% higher fuel burn. Furthermore installation effects (can the engine fit under the wing? or will the engine be transportable in a 747 freighter?) come to limit the extent to which jet noise can be reduced.

Lower the speed of the rotating components, notably the fan. Lower blade speeds are generally, though not always, associated with lower noise. With the move towards lower jet velocity, the pressure ratio required across the fan is reduced and lower fan speeds can be tolerated. The problems are now associated with the LP turbine which, as fan rotational speed is reduced, experiences higher non-dimensional stage loading. One solution is to add additional LP turbine stages, but these are heavy and expensive. There can also be a limit on the torque carrying capacity of the LP shaft, the diameter of which is fixed by the bore diameter of the compressor and turbine discs.

Avoidance of flow distortion into the fan. This is desirable, but some aspects are outside the engine designers control. At take off the intake is subject to natural cross winds and at both take off and approach the flow into the intake is at incidence, which tends to cause flow non-uniformity. There has to be some structure or pylon arrangement downstream of the fan and this imposes a static pressure non-uniformity on the fan. Steps are taken to minimise all these effects, but some distortion seems inevitable at certain flight conditions.

Choose the numbers of rotor blades and stator blades to avoid interference patterns which rotate much faster than the blades and which produce strong unattenuated tones. This is inclined to make the turbomachinery designs less than optimum, either for cost or efficiency, but is already widely practised and there is little scope for further introduction. In practice this means having more than twice as many stator blades as rotor blades for fans and about 1.5 times as many stators as rotors in the LP turbine. The avoidance of resonance due to forced vibration also imposes constraints on blade number.

Large axial gap between rotors and stators to weaken the interaction between the moving and stationary blades. This is most important for the bypass stream of the fan, but is also important for the core stream of the fan and in the LP turbine. Large axial gaps are also desirable for the reduction of forced vibration. However axial gaps come at a cost in terms of engine length, stiffness and weight, so there are limits on how much can be done to weaken the rotor-stator interaction.

Acoustic liners in the intake, in the bypass duct and sometimes in the nozzle of the core. They are normally a perforated metal layer under which is a honeycomb structure and then a solid backing. The porosity of the perforated plate and the depth of the honeycomb are tuned for the application so as to maximise the impact on EPNL. For large civil engines liners are universally used in the intake, the bypass duct and sometimes in the duct downstream of the LP turbine. They substantially increase the cost and can be a cause of increased maintenance. Liners of improved performance, such as those which are in two layers (i.e. a perforated plate over a honeycomb over another perforated plate and honeycomb of different depth) are already in service. The amount of coverage by liners is also being increased. The scope for further noise benefits from the liners are now generally limited.

BIBLIOGRAPHY

ANDERSON J D. *Introduction to Flight*. McGraw-Hill Book Co. Inc. 3rd Edition, 1989.
    A useful and highly readable treatment of the aerodynamics of flight.

BAHR D W & DODDS W J. *Design of Modern Turbine Combustors*. Academic Press Inc., San Diego CA. 1990.
    An excellent book on the field of combustor technology.

COHEN H, ROGERS G F C & SARAVANAMUTTOO H I H. *Gas Turbine Theory*. Longman, 5th Edition, 2001.
    A wide ranging book addressing many issues of relevance recently updated. A good basic introduction to turbomachinery aerodynamics and detailed treatment of off-design matching.

CUMPSTY N A. *Compressor Aerodynamics*. Longman, 1989.
    Directed at the compressor, it also gives much of the basic theory of turbomachinery at a relatively high level. There is a solid treatment of dynamic scaling and dimensional analysis.

DIXON S L. *Fluid Mechanics, Thermodynamics of Turbomachinery*. Butterworth-Heinemann. 3rd Edition 1995.
    A useful introductory text to turbomachinery.

GREEN J E. *Greener by Design - the technology challenge*. The Aeronautical Journal, Vol. 106, pp57-113, 2002.
    A stimulating and original recent look at aircraft and the potential to reduce emissions.

GUNSTON B. *The Development of Jet and Turbine Aero Engines*. Patrick Stephens Ltd, Sparkford, England, 2nd Edition, 1997.
    Lots of interesting history and useful pictures of what the hardware looks like.

HÜNECKE K. *JET Engines: Fundamentals of Theory, Design and Operation*. Airlife Publishing Ltd, Shrewsbury, England, 1997.
    Many good pictures and a particularly good treatment of intakes, nozzles and afterburners.

HILL P G & PETERSON C R. *Mechanics and Thermodynamics of Propulsion*. Addison-Wesley, 2nd Edition, 1992.
    A book which covers a very wide range of topics, including many beyond air-breathing jet propulsion. A good physical treatment of gas turbine engines, including the non-dimensional behaviour. Very good introductory treatment of turbomachinery and combustion.

INTERGOVERNMENTAL PANEL ON CLIMATE CHANGE. *Aviation and the Global Atmosphere.* Special Report of IPCC Working Groups I & III. 1999. (Cambridge University Press)
 A detailed, and very cautious, assessment of the current effect of aviation on the atmosphere by experts from around the world.

ISO *Standard Atmosphere ISO 2533* International Standards Organisation, Geneva 1975.
 The details of the International Standard Atmosphere are given here. The significance of this standard is that it is widely used by engine makers, aircraft makers and aircraft operators.

THE JET ENGINE. Rolls Royce, 4th Edition, 1986.
 A highly unusual book with exceptional illustrations of many aspects of engine and some highly relevant practical information.

KERREBROCK J L. *Aircraft Engines and Gas Turbines.* MIT Press, 2nd Edition, 1992.
 A wide ranging treatment touching on many subjects. It includes one of the few clear accounts, at an elementary level, of mechanical aspects of engine design. A good treatment of emission control.

LEFEBVRE A H. *Gas Turbine Combustion* . Taylor & Francis, Philadephia, 2nd Edition, 1998.
 A much more detailed coverage of combustion.

MATTINGLY J D. *Elements of Gas Turbine Propulsion.* McGraw-Hill Book Co. Inc, 1996.
 A long book (960 pages) with a detailed coverage of many aspects.

MATTINGLY J D, HEISER W H & DALEY D H. *Aircraft Engine Design.* AIAA Education Series, 1987.
 More obviously a design oriented book than most, this is directed mainly towards the military engine. A useful reference for information, but rather heavily algebraic in places.

MUNSON B R, YOUNG D F & OKIISHI T H. *Fundamentals of Fluid Mechanics.* John Wiley and Sons Inc. 2nd Edition, 1994.
 There are many fine text books on engineering fluid mechanics – this is nice modern one.

SHAPIRO A H. *Compressible Fluid Flow,* Vol 1. John Wiley and Sons Inc. 1953.
 A classic text, written shortly after the topic became important and was put on a sound footing.

SHAW R L. *Fighter Combat.* United States Naval Institute, Annapolis, Maryland. 1988.
 An interesting book from a different point of view, with good treatment in the appendix of military thrust requirements.

SMITH M J T. *Aircraft Noise*. Cambridge University Press, 1989.

Gives a good practical background to a very complicated subject, including subjective and regulatory aspects.

VAN WYLEN G J & SONNTAG R E. *Fundamentals of Classical Thermodynamics*. John Wiley and Sons Inc. 3rd Edition, 1985.

This is one of the many fine text books on engineering thermodynamics.

WALSH P B & FLETCHER P. *Gas Turbine Performance*. Blackwell Sciences Ltd, Oxford. 1998.

This recent book contains a mine of information on gas turbines, including jet engines.

WHITFORD R. *Design for Air Combat*. Jane's Information Group Ltd., 1989.

A useful reference on aspects of military aircraft, engines, intakes and nozzles.

REFERENCES

BANES R, McINTYRE R W &SIMS J.A. *Properties of air and combustion products with kerosine and hydrogen fuels.* AGARD 1967.

BUCKNER J K AND WEBB J B. *Selected results from the YF-16 wind tunnel test program.* AIAA Paper 74–619, 1974.

DENNING R M AND MITCHELL N A. *Trends in military aircraft propulsion.* Proceedings of Institution of Mechanical Engineers, Vol 203. Part G: Journal of Aerospace Engineering. 1989.

GARWOOD K R, ROUND P AND HODGES G S. *Advanced Combat Engines – Tailoring the Thrust to the Critical Flight Regimes.* AGARD Conference: Advanced Aero-Engine Concepts and Controls, AGARD CP-572, PAPER NUMBER 5. 1995.

KUNASAKA H A, MARTINEZ M M AND WEIR D S. *Definition of 1999 Technology Aircraft Noise Levels and Methodology for Assessing Airplane Noise Impact of Compared Noise Reductions Concepts.* NASA Contractor Report 198298, 1996.

The Northrop F-20 Tigershark, L'Aeronautique et L'Astronautique No 102, 1983–5

SMITH S F. *A simple correlation of turbine efficiency.* Journal of Royal Aeronautical Society, Vol.69, 1965.

Software

Kurzke, J. *Gasturb* . Available from www.gasturb.de
    This convenient and inexpensive package runs on a PC. It allows realistic on-design and off-design calculations to be run simply for a wide range of engines, including both engines for power production and jet engines.

INDEX

NEW LARGE CIVIL AIRCRAFT   Design sheet for the engine

This table gives quantities which are likely to be needed repeatedly.   Numbers in parenthesis indicate the exercise in which the data is evaluated.

Initial Cruise altitude = 31000 ft,   Cruise Mach number = 0.85

Ambient temperature = 226.7 K, ambient pressure = 28.7 kPa   (Standard Atmosphere)

Speed of sound at cruise altitude $a =$ m/s; cruise speed of aircraft $V =$ m/s   (1.2)

Aircraft mass at start of cruise = 635.6 10^3 kg, aircraft lift/drag ratio at cruise = 20.

Thrust per engine for level flight at 31000ft (assume 4 engines) F_N =   kN.   (2.3)

Entering engine: Stagn. temp. and press. $T_{02} = T_{01}$ =   K, $p_{02} = p_{01}$ =   kPa   (6.2)

Turbine inlet stagnation temperature at cruise, T_{04} = 1450 K

Engine pressure ratio at cruise for flow through core, p_{03}/p_{02} = 40

Gas turbine cycle efficiency η_{cy} =   (4.3e)

Entering core (after fan): stagn. temp. and press. T_{023} =   K, p_{023} =   kPa   (5.1)

Leaving core compressor: stagn. temp. and press. T_{03} =   K, p_{03} =   kPa

Leaving HP turbine: stagn. temp. and press. T_{045} =   K, p_{045} =   kPa   (5.1)

Leaving LP turbine: stagn. temp. and press. T_{05} =   K, p_{05} =   kPa   (7.1)

Velocity of core jet (taken equal to vel. of bypass jet)V_j =   m/s   (7.1)

Bypass ratio bpr =   (7.1)

Bypass leaving fan: stagn. temp. and press, T_{013} =   K, p_{013} =   kPa   (7.1)

sfc of bare engine   =   kg/hkg   (7.1)

sfc of installed engine   =   kg/hkg   (7.2)

Mass flow through the engine $\dot{m}_{air}$ =   kg/s

Diameter of fan   =   m   (9.2)

Rotational speed of LP shaft   = 3180 rev/min

Mean diameter of core compressor   =   m   (9.2)

Core compressor blade height   at inlet =   m; at outlet =   m   (9.2)

Rotational speed of HP shaft   =   radian/s   (9.3)

Core compressor mean blade speed   =   m/s   (9.3)

Number of core compressor stages   =   (9.4)

HP turbine mean diameter   =   m   (9.5)

HP turbine mean blade height   =   mm   (9.7

LP turbine mean diameter   =   m   (9.8)

Number of LP turbine stages   =   (9.8)

NEW FIGHTER AIRCRAFT Design sheet for the engine

This table gives quantities which are likely to be needed repeatedly.   Numbers in parenthesis indicate the exercise in which the data is evaluated.

Aircraft mass at take off = 18 10^3kg, aircraft mass at tropopause = 15 10^3kg.

Tropopause 11 km, $p_a=$ 22.7 kPa, $T_a=$ 216.65 K $\rho_a=$ 0.365 kg/m^3

Flight Mach number M	0.9	1.5	2.0	
Flight speed V m/s				(14.1b)
Inlet stagn. temp. T_{02} K				(15.3)
Stagn. pressure relative to aircraft kPa				(15.3)
Stagn. pressure into engine p_{02} kPa				(15.3)
Min. net engine thrust F_N for 1g turn kN				(14.4a)
3g			–	(14.4c)

Sea level $p_a=$ 101.3 kPa, $T_a=$ 288.15 K $\rho_a=$ 1.223 kg/m^3

Flight Mach number M	0.9	1.5	2.0	
Min. net engine thrust F_N for 5g turn kN			–	(14.4b)
9g			–	(14.4b)

Design point for Take off, $p_a=$ 101.3 kPa, $T_a=$ 288.15 K,

HP turbine stator exit temp $T_{04}=$1850 K; fan pr$=p_{013}/p_{02}=$4.5; overall pr$=p_{03}/p_{02}=$30,

Fan delivery temperature $T_{023} =$ K	(16.6)
Core compressor delivery temperature $T_{03} =$ K	

HP turbine exit pressure $p_{045}=$ kPa; temperature $T_{045}=$ K

Mixed out temperature at LP turbine inlet $T_{045'}=$ K; $c_{pm} =$ J/kgK; $\gamma_m=$

LP turbine exit pressure $p_{05}=$ kPa; temperature $T_{05}=$ K

Mixed out temperature at LP turbine exit $T_{05'}=$ K; $c_{pm} =$ J/kgK; $\gamma_m=$

Mixed out temperature at nozzle inlet (dry) $T_{09}=$ K; $c_{pm} =$ J/kgK; $\gamma_m=$

Mixed out temperature at nozzle inlet (a/b) $T_{09}=$2200   K; $c_{pm} =$1244 J/kgK; $\gamma_m=$1.30

	dry	a/b, $T_{09}=$2200 K
Net thrust F_N kN	(16.6)	(16.8c)
Specific thrust m/s	(16.6)	(16.7b)
bypass ratio		
sfc kg/h kg	(16.6)	(16.7b)
$\dot{m}$ air kg/s	(16.6)	(16.8c)
fan inlet dia. m	(16.6)	(16.8c)

For dry design

Fan first stage $N =$ rpm; tip speed $U_t=$ m/s, tip Mach number	(18.1)
Fan stage temp rise = K; first-stage pressure ratios	(18.2)
HP compressor $N =$ rpm; first-stage tip diameter $D =$ m	(18.5)
HP turbine mean speed $U_m=$ m/s; $\Delta h_0/U_m^2=$	(18.9)
LP turbine mean speed $U_m=$ m/s; $\Delta h_0/U_m^2=$	(18.9)
HP and LP turbine blade heights at throat mm; mm	(18.11)